Studies in Systems, Decision and Control

Volume 106

Series editor

Janusz Kacprzyk, Polish Academy of Sciences, Warsaw, Poland
e-mail: kacprzyk@ibspan.waw.pl

About this Series

The series "Studies in Systems, Decision and Control" (SSDC) covers both new developments and advances, as well as the state of the art, in the various areas of broadly perceived systems, decision making and control- quickly, up to date and with a high quality. The intent is to cover the theory, applications, and perspectives on the state of the art and future developments relevant to systems, decision making, control, complex processes and related areas, as embedded in the fields of engineering, computer science, physics, economics, social and life sciences, as well as the paradigms and methodologies behind them. The series contains monographs, textbooks, lecture notes and edited volumes in systems, decision making and control spanning the areas of Cyber-Physical Systems, Autonomous Systems, Sensor Networks, Control Systems, Energy Systems, Automotive Systems, Biological Systems, Vehicular Networking and Connected Vehicles, Aerospace Systems, Automation, Manufacturing, Smart Grids, Nonlinear Systems, Power Systems, Robotics, Social Systems, Economic Systems and other. Of particular value to both the contributors and the readership are the short publication timeframe and the world-wide distribution and exposure which enable both a wide and rapid dissemination of research output.

More information about this series at http://www.springer.com/series/13304

Aleksander Nawrat · Damian Bereska
Karol Jędrasiak
Editors

Advanced Technologies in Practical Applications for National Security

Editors
Aleksander Nawrat
Institute of Automatic Control
Silesian University of Technology
Gliwice
Poland

Karol Jędrasiak
Institute of Automatic Control
Silesian University of Technology
Gliwice
Poland

Damian Bereska
Institute of Automatic Control
Silesian University of Technology
Gliwice
Poland

ISSN 2198-4182 ISSN 2198-4190 (electronic)
Studies in Systems, Decision and Control
ISBN 978-3-319-87848-5 ISBN 978-3-319-64674-9 (eBook)
https://doi.org/10.1007/978-3-319-64674-9

Printed on acid-free paper

This Springer imprint is published by Springer Nature
The registered company is Springer International Publishing AG
The registered company address is: Gewerbestrasse 11, 6330 Cham, Switzerland

Knowing is not enough; we must apply.
Willing is not enough; we must do.

Johan Wolfgang von Goethe

Preface

Recent terrorist attacks in France have shown that the Global War on Terrorism started after the September 11, 2001 attacks is far from being over. The security of US and EU citizens has been threatened by terrorists using multitude of ways including attacks by armed individuals, vehicles used as weapons or victim-operated Improvised Explosive Devices (IED). The reader of this book will be presented with advanced technologies used in practice to enable early recognition and tracking of various threats for national security.

Undeniably fast advances in development of sophisticated sensory devices, significant increase of computing power available to embedded designs and development of airborne and ground unmanned vehicles give almost unlimited possibilities to fight various types of pathologies affecting our societies.

The book shall address several innovative solutions and algorithms for tracking of moving objects in visual light as well as thermographic video streams and distinguishing objects form its surroundings under difficult field conditions. Visual and thermographic tracking and monitoring is often carried out by unmanned vehicles which equipped with intelligent algorithms become autonomous and automatically report about the place and conditions of detection of a potential security incident in complex multi-agent environments. Unmanned vehicles require control algorithms which must be carefully designed and properly tuned.

Present-day national security greatly benefits from available techniques of modeling and simulation used for training of security special forces as well as in analysis of past and possible security threats and incidents. The practical solutions may range from various types of calibration of DLP projectors through analysis of RF propagation under urban conditions to analysis of dynamics and kinematics of human arm.

Finally, the design of innovative control, tracking and monitoring algorithms most often require prior knowledge of dynamical characteristics of equipment being used in many different national security tasks carried out in extremely difficult field conditions. From thermal characteristics of IMU modules and static tests of IED interrogation arm towards reducing the impact of IED on patrol vehicles used in war areas, one equipped in latest technology and algorithms becomes a strong opponent and hopefully the winner in the Global War on Terrorism.

Gliwice, Poland
November 2016

Aleksander Nawrat
Damian Bereska
Karol Jędrasiak

Acknowledgements

This work has been supported by National Centre for Research and Development as a project ID: DOB-BIO6/11/90/2014, Virtual Simulator of Protective Measures of Government Protection Bureau.

Aleksander Nawrat
Damian Bereska
Karol Jędrasiak

Contents

Part I
Practical Applications of Object Tracking Algorithms

Present-day national security greatly benefits from available techniques of object tracking and recognition. One of the approaches of object tracking is based on the calculation of an indicator which describes the color features of the object, for instance, the human skin. The ratios between the red and green components as well as the ratio between red and blue components are the indicators which defined these features. Such approach is particularly useful in the cases when the object and terrain colors were significantly different.

One of the most challenging tasks in object tracking field is to design flexible and fast algorithms for thermal imaging cameras. The chapter presents algorithms of detection and object tracking designed for thermographic video streams. Described tracking methods were implemented in real life systems and experimentally verified. Tracking allows effective determination of the trajectory of moving objects in thermographic images.

Many research groups work on design and implementation of control algorithms allowing computer controlled A.I to play decision game. In the chapter a solution allowing a group of small mobile robots playing decision game is presented. The whole process of extracting vision information from input images is discussed in detail. The method for calculating objects orientation based on the special shape of color markers on top of each robot, is presented. Experimental results obtained with use of the algorithm presented are also provided.

As well as in today's and tomorrow's army, devices for contactless monitoring of soldier's parameters will be one of the most important equipment. The chapter presents usefulness of myGaze to analyze eye movement during optokinetic stimulation. There is a lot of possible applications of gaze tracking like human-computer interface or concentration analysis.

Military operations in order to complete successfully require a vast amount of information and high quality algorithms capable of using preliminary knowledge of a target object. Tracking of such object may be conducted with the method pattern vector modification. The object which has the same visual images and different thermal images may be an example of the abovementioned case. The pattern vector and current feature vector for an image of a given type were used to compute the distance between the object pattern vector and feature vector calculated for a given location of the aperture. Visual and thermal pattern vectors are used to calculate the distance.

This chapter includes a number of important challenges in the fields mentioned above. At the same time valuable suggestions and conclusions from authors are presented and discussed in detail.

Accurate Tracking of Fast Objects with a Weak Video Input Signal

Robert Bieda and Krzysztof Jaskot

1 Introduction

Over 80% of all perceptual information being received by the human's brain comes from his eyes. So it is quite natural that engineers try to add vision to robot systems in order to improve their capabilities. The vision algorithm and its implementation described in this work are part of research and development of multi-agent systems in complex, dynamic environments [1]. A good example of such multi-agent system is a soccer match of two teams of three robots on small playing field. Although, at first sight, the RoboSoccer tournament seems to be nothing more than building and playing with toys, practical attempts to participate in it reveal hundreds of fascinating and challenging problems from the domain of mechanics, electronics, automatic control and artificial intelligence that have to be solved and their solutions implemented [2, 3]. One of these problems is the task of building proper vision system. Vision system could be also used for planning collision free path of other types of robots e.g. UAV's [4, 5].

From the control algorithm's point of view, the vision system works as the feedback loop, providing measurements of values being inputs to the controller. Because the vision processing is done on the on the host computer and not on board of a robot, it is possible to develop quite sophisticated and very precise vision algorithm, as the computational power of contemporary PCs is thousands as big as of micro-controllers used on-board the robots.

In our system there will be a central control engine which analyses situation on the playing field, and decides what action should be taken by robots. This engine

R. Bieda · K. Jaskot (✉)
Institute of Automatic Control, Silesian University of Technology,
Akademicka 16, 44-100 Gliwice, Poland
e-mail: krzysztof.jaskot@polsl.pl

R. Bieda
e-mail: robert.bieda@polsl.pl

A. Nawrat et al. (eds.), *Advanced Technologies in Practical Applications for National Security*, Studies in Systems, Decision and Control 106,
https://doi.org/10.1007/978-3-319-64674-9_1

Fig. 1 A typical image captured by camera

obviously needs data about positions and directions of robots and a ball on the playing field. That data can be obtained from a camera installed above the playing field (Fig. 1). However, camera gives us only a bitmap—an array of colors assigned to each pixel which is not convenient representation of needed data. As a matter of fact extraction of positions and directions of objects from such bitmap is one of the most challenging tasks in the whole system. In addition analysis of the image should be done very quickly (hundreds of milliseconds is far too long). In this work the application accomplishing this task for needs of our system is presented in details.

The project started a few years ago and since this time two applications dealing with this difficult task was created. Although some goals was achieved in that work by [6, 7], after many tests it was still not good enough to be used in the final system.

The general idea was as follows. At the beginning (after some simple filters), image is compared to the background (learned earlier) and some distance in RGB space for each pixel is computed. Then, such image of distances is thresholded. In this way we obtain a mask which theoretically is a set of points belonging to some objects on the screen (robots, ball). In the next step too dark pixels are removed from the mask and in effect only colorful marks on the top of robots ("T" shaped) remain in the mask. In most cases (but not all) for each robot and for ball we have one blob. Then, these blobs are assigned to robots by comparison of their colors histograms to some models. Position is determined basing on the position of blob, and it's direction is obtained by mean of shape of blob. This is typical approach in the domain of computer vision, however there are some efforts to get rid of the learning phase and to build an algorithm capable of autonomous online learning [8].

At first glance presented algorithm seems to be correct, robust and fast. In fact, it has many drawbacks. The critical ones are listed below.

- **Angle determination**. Error of determined angle of rotation of robot is often above 15°. In addition even when nothing changes in the scene there are big variations i.e. recognized angle changes significantly in time (Fig. 2a, b). So the random error is quite big much too high for our purposes. Moreover—sometimes the recognized direction is opposite to the true one (Fig. 2c).

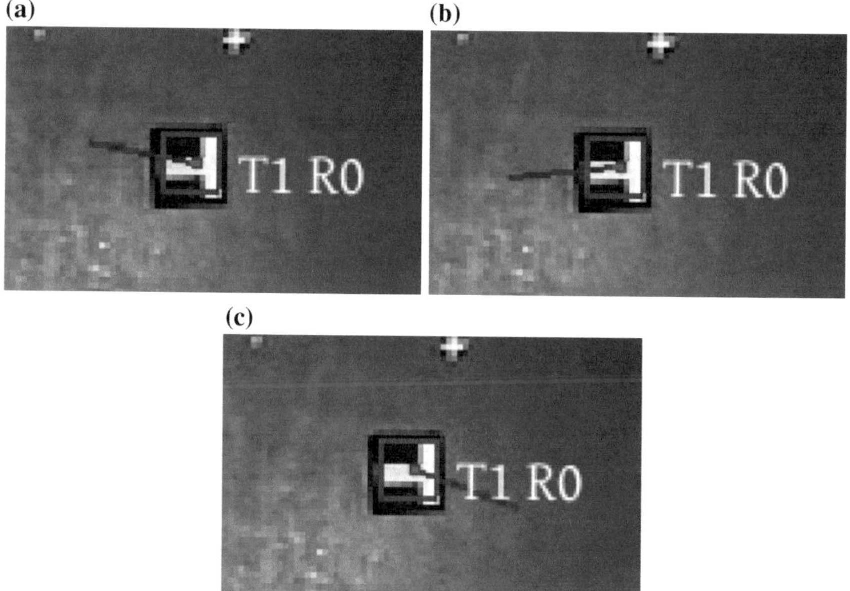

Fig. 2 Angle variations for static object

- **Failure when objects in contact**. Often when objects (robots, ball) are too close to each other they visually merge into one blob (on the final motion mask). In such situation algorithm cannot recognize them properly so position and direction of these objects are lost.

If listed above problems were not present, the system would be usable, although still not very good. The other ones (though not critical) are:

- **Limitation of background color**. In the algorithm there is an assumption that shadows are darker than marks on the top of objects. If the playing field was light (e.g. white) that might not be true and shadows would disturb the shape of blobs.
- **Fixed type of marks**. The most important part of algorithm depends on assumption that marks on robots have specific shape. They have to be in shape of letter "T". However this is not assured that the other team, which can potentially play a match with our team, would use similar marks.
- **Algorithm is not stable**. If it happen that classification of blobs (assignment of blobs to robots) fails for a longer time (e.g. in case of temporary occlusion), then histogram models are updated with wrong data and finally they doesn't match to true model. If such case occur, the robot is lost (by the vision system) until a manual action is taken (restoration of histograms).
- **Long learning procedure**. Before each use (after application is started) the long learning procedure must be done. It contain calibration of a few values (thresholds, radial correction factor, etc.) and background learning for which all objects must

be removed from playing field. This drawback could be easily fixed by modifying application in way allowing to save last settings. Then learning procedure would be needed only when conditions are changed.

Although list of drawbacks is long, this system could obtain some results in short time, which is really a challenge in image recognition science. The frame rate was 30 fps (limited by the frame grabber) and it was taking approximately 50% of CPU time on the machine. Probably the accuracy would be better if quality and resolution of camera were better it was tested on industrial camera with resolution 320×240 pixels.

2 The Main Algorithm

This section describes in details all parts of algorithm used to finding positions and direction of robots. Figure 3 shows the general block diagram.

Fig. 3 Block diagram of algorithm

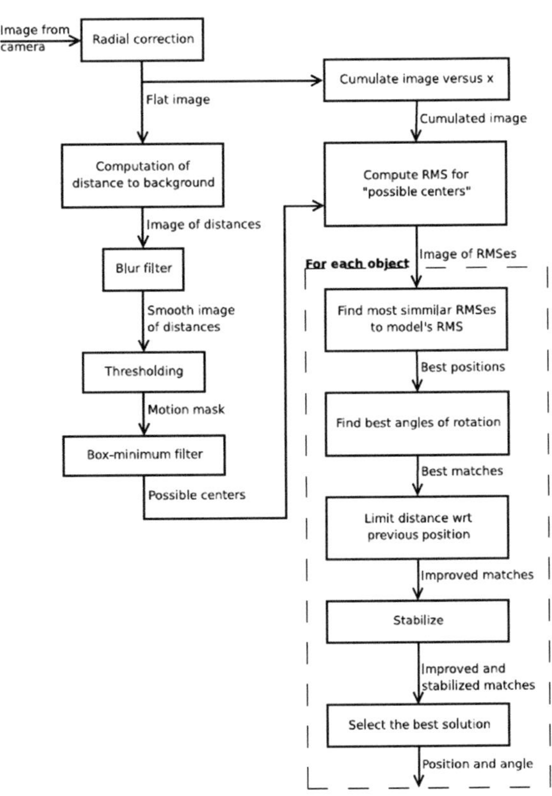

2.1 Correction of Radial Distortion

The first step, which in our case is not very important, is the first one in whole processing chain. The method we used is not ideal one but quite simple. It uses inverse mapping for finding colors of pixels in the output image. Each point on the output (corrected) image is mapped to a point in the input one (distorted) [9–11].

Usually coordinates of the mapped point are not integer numbers so to obtain the color of pixel in the output image we need to do some interpolation [12, 13]. There are two interpolation which can be used. First the nearest neighbourhood interpolation (1) uses color of the nearest pixel. The second one—bilinear interpolation (2) makes use of colors of four nearest pixels. Let's assume the real coordinates are (x, y). Let's define $\Delta x = x - \lfloor x \rfloor$ and $\Delta y = \lfloor y \rfloor$ and denote a color of the pixel with integer coordinates (x_i, y_i) as $I(x_i, y_i)$. Then interpolated color for point (x, y) is [14]:

- **For nearest neighbourhood interpolation**:

$$I'(x, y) = I(round(x), round(y)) \tag{1}$$

- **For bilinear interpolation**:

$$\begin{aligned} I'(x, y) = I(\lfloor x \rfloor, \lfloor y \rfloor) \cdot (1 - \Delta x)(1 - \Delta y) + \\ I(\lfloor x \rfloor + 1, \lfloor y \rfloor) \cdot \Delta x(1 - \Delta y) + \\ I(\lfloor x \rfloor, \lfloor y \rfloor + 1) \cdot (1 - \Delta x)\Delta y) + \\ I(\lfloor x \rfloor + 1, \lfloor y \rfloor + 1) \cdot \Delta x \Delta y \end{aligned} \tag{2}$$

Results achieved by NN interpolation are not satisfactory (Fig. 4) and may lead to big errors of angle, while the bilinear one (Fig. 5) is very slow. The algorithm of radial correction can be described as follows. Let $I_d(x, y)$ denote interpolated color at point (x, y) of input (distorted) image and $I_c(x, y)$ denote color of pixel at (x, y) of the corrected image. Then:

$$I_c(x, y) = I'_d(x', y') \tag{3}$$

where:

$$x' = (x - x_c)(1 - \gamma) + x_c \tag{4}$$

$$y' = (y - y_c)(1 - \gamma) + y_c \tag{5}$$

$$\gamma = \lambda \frac{(x - x_c)^2 + (y - y_c)^2}{x_c y_c} \tag{6}$$

for each integer pair of coordinates (x, y) (in range given by image dimensions). The parameter is to be chosen by the user. Figures 4, 5 and 6 show the result of algorithm. This part is not found as very important, but it might be on other cameras or with other lens.

Fig. 4 Result of correction of radial distortion, before correction

Fig. 5 Result of correction of radial distortion, correction with NN interpolation

Fig. 6 Result of correction of radial distortion, correction with bilinear interpolation

2.2 Motion Mask

In this part the mask of pixels which belong to some mobile objects is obtained [15–19]. Then all blobs are reduced (using box-minimum filter) in order to obtain mask of possible central points of objects.

2.3 Computation of Deviation from Background

We assume there is a model of background and the input image. The image of distances (deviations) is computed as follows:

$$I_d(x, y) = d(I(x, y), B(x, y)) \tag{7}$$

where I_d is an image of distances, I is the input image and B is the background model. The function d is a Tchebyshev metric defined as:

$$d(x, y) = \max_{i \in \{r,g,b\}} | x_i - y_i | \tag{8}$$

Result of this step is illustrated in the Fig. 7.

2.4 Blur

To achieve smooth motion mask a blur filter is applied before [20]. In this case the rectangular blur is used because of implementation issues. Rectangular blur is a filter which assigns to color of pixel in the output image the average color on rectangle with center at this point and given (as parameters) width and height. In more formal way:

$$I_b(x, y) = \frac{\sum_{\delta y=-h}^{h} \sum_{\delta x=-w}^{w} I(x + \delta x, y + \delta y)}{(2w + 1)(2h + 1)} \tag{9}$$

where I_b is the blurred image and, I is the input image.

If this algorithm was implemented directly from the formula (9), tthe complexity is too high. There is commonly known algorithm to do that in much faster way.

Fig. 7 Image of distances from background (the darker color the distance larger)

Fig. 8 Image of distances blurred with the radius $r = 2$

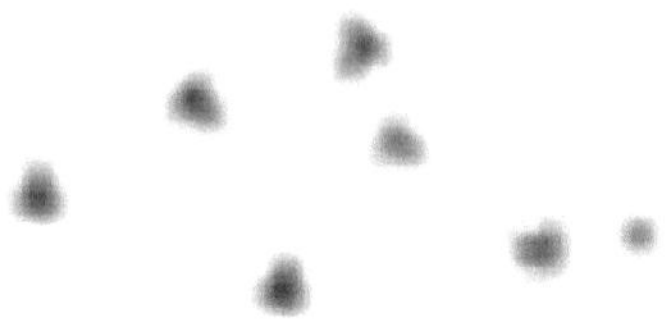

The idea is to transform the image at the beginning according to the formula (10) (accumulation):

$$I'(x, y) = \sum_{\delta i=0}^{x} I(\delta i, y) \tag{10}$$

which can be done in linear time with respect to number of pixels (summing colors of pixels in row from the left to the right for each row separately). Then each sum of $2w + 1$ consecutive pixels colors can be computed in constant time by simple subtraction of two numbers in the transformed image:

$$\begin{aligned} I''(x, y) &= \sum_{\delta x=-w}^{w} I(x + \delta x, y) \\ &= I'(x + w, y) - I'(x - w - 1, y) \end{aligned} \tag{11}$$

Then, analogously, the same rule can be used vertically to image I'' which is in fact image of sum of horizontal segments (with center at given point). Then we obtain image of sums of pixels in proper rectangles. After that we divide each pixel's color by size of the rectangles ($(2w + 1)(2h + 1)$) obtaining finally the blurred image I_b. The result is shown in the Fig. 8.

2.5 Thresholding

The next step is very simple one. The output image is the binary mask i.e. for each pixel only two values are possible (0 or 1). Precisely speaking we have some threshold level τ. If the value of pixel in the input image I is greater or equal to τ then value of the same pixel in the output (thresholded) image I_t is 1. In other cases it's 0. It can be described in the formal way:

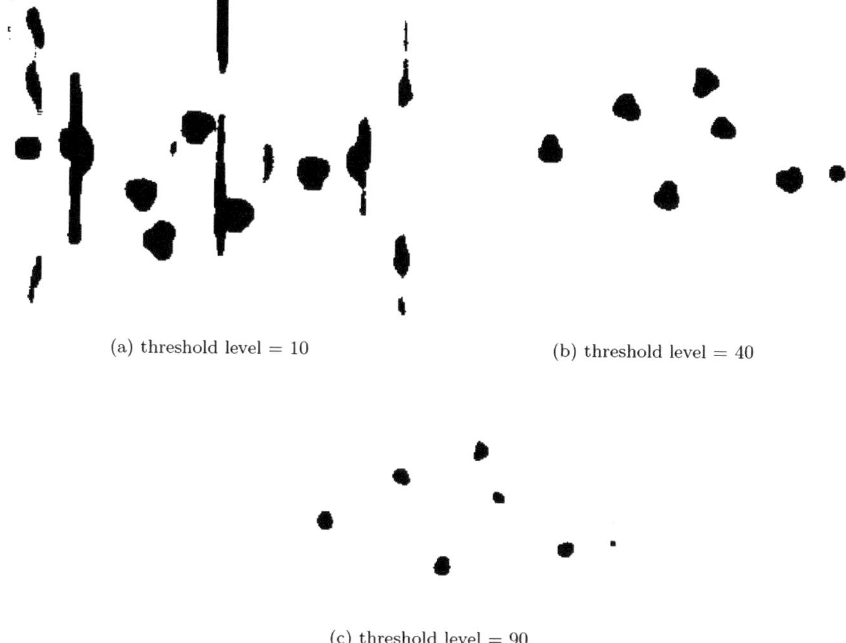

(a) threshold level = 10

(b) threshold level = 40

(c) threshold level = 90

Fig. 9 Motion mask

$$I_t(x,y) = \begin{cases} 1 \text{ for } I(x,y) \geq \tau \\ 0 \text{ for } I(x,y) < \tau \end{cases} \tag{12}$$

This value (τ) should be set in such way that all object that should be visible on is as possibly regular and smooth shapes. Its not a big problem if some small areas which are not objects are also in this mask it can only make analysis a bit slower (if areas are large enough). Figure 9 shows the result of this step.

2.6 *Box-Minimum Filter*

In effect of previous step (12) we have a motion mask which theoretically contains pixels belonging to some objects. These pixels form blobs. As we want only pixels which could be center of some object (robot) we can ignore points which are too close to edges of those blobs. It should be stressed that it's still only optimization and the aim is to reduce a set of points where analysis will be done. It's not a problem if there are left some points which are not centers of object but it's a serious problem if true centers are not contained in the final mask.

To remove pixels which are too close to edges we use box-minimum filter which simply assigns to each pixel of output image the least value of pixels in a box with center at this point.

$$I_m(x, y) = \min_{\delta y=\langle -h,h\rangle} \min_{\delta x=\langle -w,w\rangle} I(x + \delta x, y + \delta y) \tag{13}$$

The implementation is based on idea of finding first minimum in horizontal direction and then in vertical. Like in case of blur, also in this case $w = h = r$. Note, that this variable has different values for robots and for ball.

2.7 *Round Mean Sequences*

The idea of round mean sequences (RMS) was based on computing average color of all pixels on successive distances from the center. Speaking more precisely we sum colors (each of RGB channel separately) of pixels lying on the circle with radius of k pixels. Then we divide this sum by number of pixels taken into account and that's our k'th "round mean". If the model has radius of n we can compute n different "round means". This sequence should be a good statistics describing an appearance of the model independently of its rotation. We cannot expect that this sequence will be identical for model and for captured image. We needed to find some metric which allows to define a distance between two sequences of round means. The lower distance, the more similar they are. Many experiments were done with various metrics. Whereas that this metric should be more sensitive to change of color chromaticity than it's brightness. Finally obtained metric can be defined as follows:

$$\begin{aligned} d(x, y) = \frac{1}{3} \sum_{i=1}^{n} i \cdot \Bigg(& \max_{c=\{r,g,b\}} \mid x_{ic} - y_{ic} \mid + \\ 2 \cdot \Bigg(& \max_{c=\{r,g,b\}} \mid x_{ic} - y_{ic} \mid - \min_{c=\{r,g,b\}} \mid x_{ic} - y_{ic} \mid \Bigg)\Bigg) \end{aligned} \tag{14}$$

where: x, y—round mean sequences, x_i—i'th element of sequence, x_{ic}—given (by c) color channel of i'th element, n—number of elements in sequences x and y.

In order to not to loose any pixel which belongs to disc of radius r (the radius of model) the k-th element of RMS will be computed basing on pixels which belongs to disc of radius k and don't belong to those of radius $k - 1$. The trivial algorithm can just sum colors of all pixels belonging to a disc of successive radices. If we compute at the beginning a accumulated image we can then compute a sum of any continuous horizontal segment in constant time:

$$I'(x, y) = \sum_{\delta i=0}^{x} I(\delta i, y) \tag{15}$$

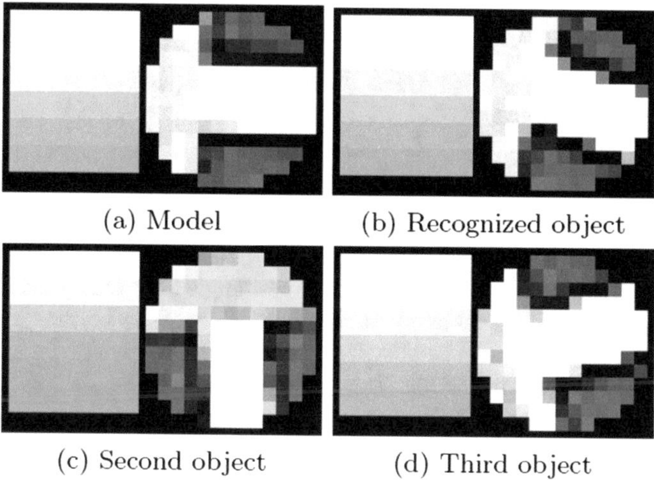

(a) Model (b) Recognized object

(c) Second object (d) Third object

Fig. 10 Round mean sequences

$$\sum_{\delta x=x_1}^{x_2} I(x+\delta x, y) = I'(x+x_2, y) - I'(x+x_1-1, y) \tag{16}$$

Figure 10 presents some RMS'es with images for which they was computed. Such RMS'es are computed once for each model and for each point present in the final mask of possible centers. Then for each model a distance to each of these points is computed according special metric for RMS'es (14). Some number of best-fitting points (with least distances) is taken into account in the further fitting (usually 5–10). RMS of the model is computed one time (after object is captured) so it doesn't affects the complexity of main algorithm.

2.8 *Searching for Best Fitting*

Then for each of points selected in previous step angle fitting is done. In each step of this procedure some directions of model are checked i.e. a distance between rotated model and part of captured image is computed. Because we use a distance between two images in angle fitting we need some metric for this purpose. The best results were obtained for squared euclidean metric with normalization (17). In our case we substitute a formula for distance between colors into formula for distance between images we obtain a simpler form, much convenient to compute because square roots are not present:

$$
\begin{aligned}
D(I_1, I_2) &= \sqrt{\sum_{p \in D} \sqrt{\sum_{i \in \{r,g,b\}} (x_i - y_i - \eta_i)^2}^{\,2}}^{\,2} \\
&= \sum_{p \in D} \sum_{i \in \{r,g,b\}} (x_i - y_i - \eta_i)^2
\end{aligned}
\tag{17}
$$

First of all all multiplicities of 30° are checked. Then the interval is limited to 60° with center at the best fitting angle (with the lowest distance) and each 10° are checked. Then in the range of 10° from the best angle all possible rotations (with assumed resolution, 1° in our case) are checked.

Size of the steps in the successive phases of narrowing down of the interval was well-chosen for certain resolution. The optimal values can be also other for other types of marks (on robots). Assuming that these values and resolution are constant the speed depends only on radius of model and number of considered points.

2.9 Modifications and Final Selection

In such a way we obtain a best angle and distance related to it for each considered position. We have therefore some value which tells us how good was fitting at each considered point. Let's call this value as scores. The easiest way is to select the one with the lowest number of scores and return it's position and angle for which fitting was the best. However in order to improve results some modifications are done before this selection.

2.10 Stabilization

In order to reduce in-time variations of position and angle of static object some value can be subtracted from scores of position where object was found last time. In result this position is selected as the best one even if it has a bit more scores than the other ones. The number of scores subtracted can be defined as percentage of original value.

2.11 Distance Limit

There is also a possibility to define the maximal distance which object can travel between two successive captures of frames. The scores for points which are farther from previous position the number of scores is doubled, effectively reducing their chances to be selected as best fitting ones.

2.12 *Background Model Extraction*

Background is updated using the method of moving average [21, 22]. Pixels which are closer than $2r$ from center of any recognized object (where r is radius of model of this object) are not taken into account. Let's denote set of all other points by A. Background after frame i will be denoted as B^i and the current frame is denoted by I (18). Then update procedure can be written as follows:

$$B^{i+1}(x, y) = B^i(x, y) \cdot (1 - \rho) + I(x, y) \cdot \rho \, for(x, y) \in A \tag{18}$$

where ρ is a factor having influence on the rate of updating process.

3 Results

The application was tested on a PC working under Ubuntu Linux. The computer was equipped with Pentium 4 3.00 GHz Hyper Threading processor, 512 MB DDR RAM and and GeForce 5200 FX graphics card. Input images were acquired with Samsung SCC331 CCD camera equipped with Ernitec lens ($f = 612$ mm, $\phi = 35.5$ mm, 1:1.4). Images were then digitized by simple BT878-based frame-grabber card. The lighting conditions were medium several fluorescent lamps on the ceiling.

3.1 *Tracking Reliability*

This experiment shows how many times and for how long time the vision system looses the object and doesn't track it properly. It's some kind of probability of committing gross error.

3.2 *Robot Tracking*

In this test robot was being moved slowly around the playing field. Plots of coordinates are shown on Fig. 11. Robot's position was recognized incorrectly at 8 frames out of 783 which is approximately 1%. In next test robot was moved very fast in various directions. Plots of coordinates are shown on Fig. 12. Robot's position was recognized incorrectly at 35 frames out of 476 which is approximately 7.4%. In this case the frequency of gross error occurrence is quite big, although robots cannot move itself with such big velocities, so it's the absolute upper limit of occurrence frequency of errors caused by very fast motion.

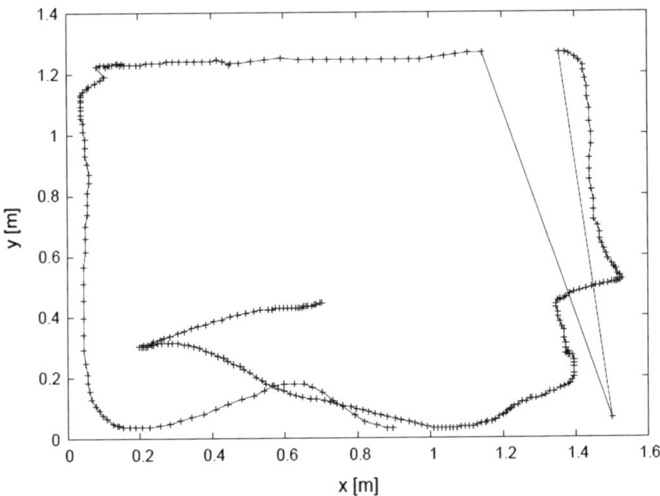

Fig. 11 A trajectory of robot slow

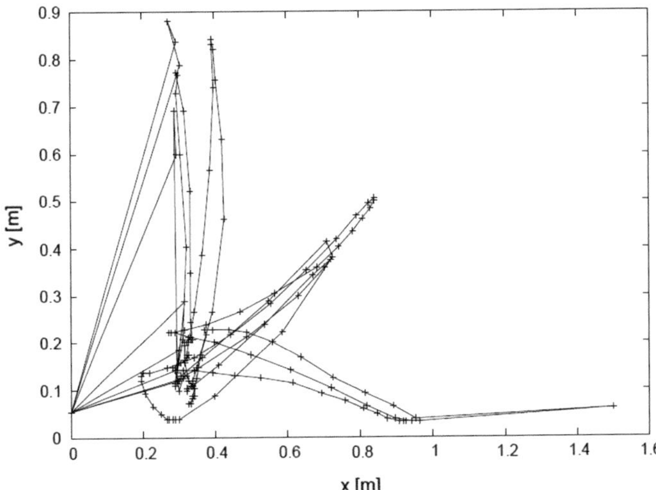

Fig. 12 A trajectory of robot fast

3.3 Tracking Accuracy

In this test robot moving with constant speed. Two general cases are present—going along straight line (both wheels rotating with the same speed) and going along a circle (wheels rotating with different speed).

For each test the theoretical curve will be fitted to results of the algorithm (samples returned by the application). Then the standard deviation of samples with respect to this curve will be presented (19).

$$S = \sqrt{\frac{1}{n}\sum_{i=1}^{n}(x_i - y_i)^2} \tag{19}$$

where x_i is certain coordinate of object returned by the system and y_i is theoretical value resulting from fitting the curve.

3.4 Straight Line

In this case both coordinates should be linearly dependent on time. For each test from this group the results are compared with the best fitting straight line. Then deviations are computed.

Motion along X axis. The robot was moving through the playing field along it's X axis. Figure 13 illustrates coordinates as a function of time. For this case $S_x \approx 0.0016\,\text{m} = 1.6\,\text{mm}$. Theoretical resolution is approximately 5 mm (this is a distance covered by one pixel). The result is better than was expected.

Motion along Y axis. The robot was moving through the playing field along it's Y axis. Figure 14 illustrates coordinates as a function of time. For this case $S_y \approx 0.0018\,\text{m} = 1.8\,\text{mm}$. This result is also better than resolution of the system (~ 5 mm).

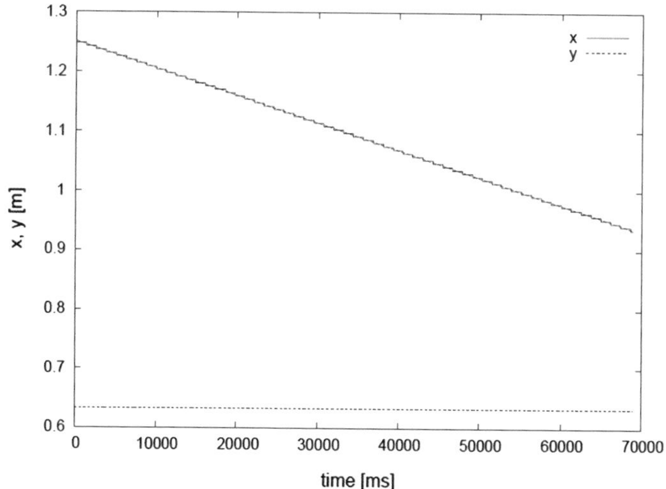

Fig. 13 x and y coordinates of robot versus time

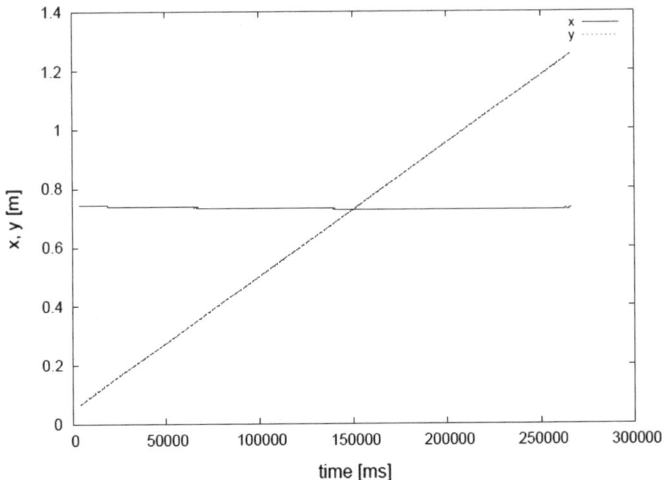

Fig. 14 x and y coordinates of robot versus time

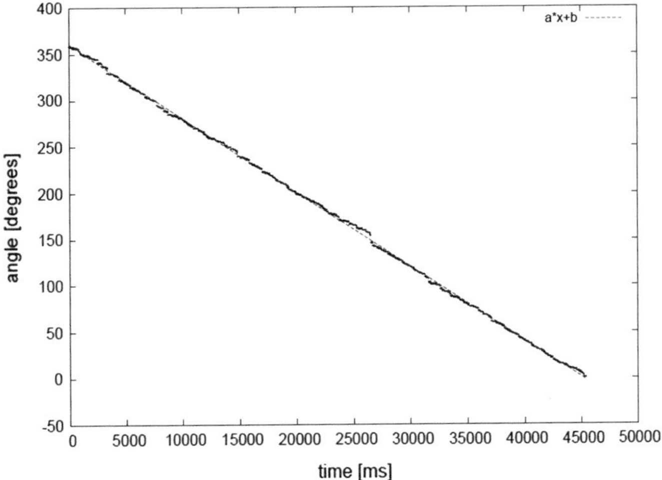

Fig. 15 Angle as a function of time

3.5 *Direction Estimation*

This section summarizes the accuracy of direction obtained by the vision system. In this case also standard deviation was used.

Slow rotation. In this experiment robot was set to rotate with constant, minimal speed (wheels were rotating with minimal speeds in opposite directions). It's angle should be linearly dependent on time. Figure 15 shows the obtained results.

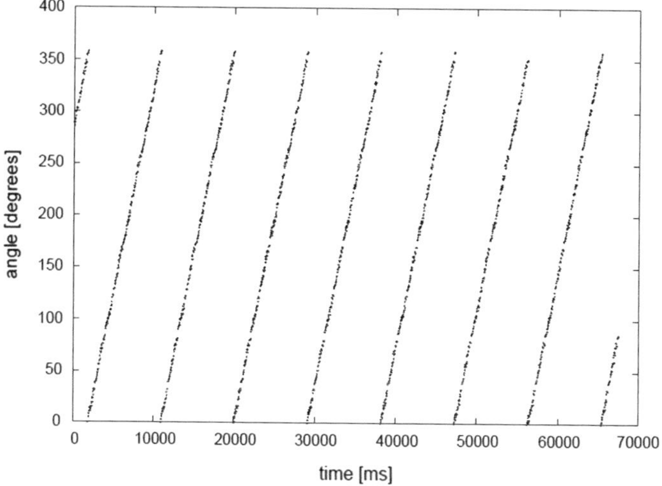

Fig. 16 Angle as a function of time in circular motion

Circular motion. In this test angle of robot going along a circle (with diameter about 0.5 m) was measured. The Fig. 16 shows a full record of several circles. For this case the standard deviation and the maximal absolute error are presented. The standard deviation is $S_\alpha \approx 2.5^\circ$ which is not a perfect result, but also not bad. The maximal deviation in this test was approximately 7.5°.

4 Conclusions

Generally goals are achieved, but as it was shown in tests it happens sometimes that object is lost by the vision system for a few frames. Fortunately it occurs rather rarely but in perfect case it should never happen.

The estimation of direction doesn't work perfectly (standard deviation is about 2°, the worst case near to 10°). Those results would be probably much better if lighting was better. The accuracy of determination of positions is better than was expected (standard deviation of order of 2 mm). The speed of processing is satisfactory (30 fps is achieved) but near to the limit. Maybe it could be slightly optimized. Small differences in lighting of playing field doesn't disturb significantly. Theoretically any marks can be used, but usage of patterns similar to other ones or to background would probably influence the quality of results. Algorithm has no problems in recovery temporary lost object unless distance limiter is turned on. This feature should be used carefully.

The new vision system can be successfully used as the part of whole system but during designing of other parts it should be kept in mind that the data get from is loaded with some small random errors and that it might happen occasionally, that recognition of one or more objects fails totally. If there is another source of data then data from vision system should be confronted with this data (e.g. robots can estimate their displacement using readings from wheels).

References

1. Skrzypczyński, P. (2005). Uncertainty models of vision sensors in mobile robot positioning. *International Journal of Applied Mathematics and Computer Science, 15*, 73–88.
2. Jaskot, K., & Łakota, T. (2016). Experimental mobile robot—software. In *Innovative Simulation Systems*, Studies in Systems, Decision and Control (Vol. 33, pp. 303–316). Springer.
3. Jaskot, K., & Łakota, T. (2016). Experimental mobile robot—software. In *Innovative Simulation Systems*, Studies in Systems, Decision and Control (Vol. 33, pp. 277–289). Springer.
4. Babiarz, A., & Jaskot, K. (2013). The concept of collision-free path planning of UAV objects. In *Advanced Technologies for Intelligent Systems of National Border Security*, Studies in Computational Intelligence (Vol. 440, pp. 81–94). Springer.
5. Iwaneczko, P., Jedrasiak, K., Daniec, K., & Nawrat, A. (2012). A prototype of unmanned aerial vehicle for image acquisition. In *Computer Vision and Graphics*. Springer.
6. Babiarz, A., Bieda, R., & Jaskot, K. (2013). A distributed control group of mobile robots in a limited area with a vision system. In *Vision Based Systems for UAV Applications*, Studies in Computational Intelligence (Vol. 481, pp. 157–175). Springer.
7. Babiarz, A., Bieda, R., & Jaskot, K. (2013). Vision system for group of mobile robots. In *Vision Based Systems for UAV Applications*, Studies in Computational Intelligence (Vol. 481, pp. 139–156). Springer.
8. Sridharan, M., & Stone, P. (2005). Autonomous color learning on a mobile robot. In *Proceedings of the Twentieth National Conference on Artificial Intelligence* (pp. 1318–1323).
9. Hartley, R., & Kang, S. B. (2005). Parameter-free radial distortion correction with centre of distortion estimation, Microsoft Research.
10. Hugemann, W. (2010). Correcting lens distortions in digital photographs. In *EVU Conference* (Vol. 19). Prague.
11. Wang, A., Qiu, T., Shao, L. (2009). A simple method of radial distortion correction with centre of distortion estimation. *Journal of Mathematical Imaging and Vision, 35*(3), 165–172. Springer Science+Business Media, LLC.
12. Parker, J. A., Kenyon, R. V., & Troxel, D. E. (1983). Comparison of interpolating methods for image resampling. *IEEE Transactions on Medical Imaging, MI-2*(1).
13. Thévenaz, P., Blu, T., & Unser, M. (2000). *Image interpolation and resampling, Handbook of medical imaging* (pp. 393–420). Orlando, FL, USA: Academic Press, Inc.
14. Miklós, P. (2004). Image interpolation techniques. In *2nd Serbian-Hungarian Joint Symposium on Intelligent Systems (SISY2004)*, Serbia and Montenegro.
15. Alexandre, L. A., & Campilho, A. C. (1998). A 2D image motion detection method using a stationary camera, RECPAD98. In *10th Portuguese Conference on Pattern Recognition*, Lisbon, Portugal.
16. Lacassagne, L., Manzanera, A., & Dupret, A. (2009). Motion detection: Fast and robust algorithms for embedded systems. In *IEEE International Conference on Image Processing (ICIP 2009)* (pp. 3265–3269).
17. Martínez-Martín, E., & del Pobil, A. P. (2012). Motion detection in static backgrounds. In Martínez-Martín, E. (Ed.) *Robust Motion Detection in Real-Life Scenarios*.
18. Widyawan, W., Zul, M. I., & Nugroho, L. E. (2012). Adaptive motion detection algorithm using frame differences and dynamic template matching method. In *The 9th International Conference on Ubiquitous Robots and Ambient Intelligence (URAI 2012)*.

19. Yong, C. Y., Sudirman, R., & Chew, K. M. (2011). Motion detection and analysis with four different detectors. In *Proceedings of the 2011 Third International Conference on Computational Intelligence (CIMSIM '11), Modelling & Simulation* (pp. 46-50). DC, USA: IEEE Computer Society Washington.
20. Gonzalez, R. C., & Woods, R. E. (2008). Intensity transformations and spatial filtering. In *Digital Image Processing* (3rd ed., Chap. 3). Pearson Prentice Hall.
21. Benezeth, Y., Jodoin, P. M., Emile, B., Laurent, H., & Rosenberger, C. (2010). Comparative study of background subtraction algorithms. *Journal of Electronic Imaging, 19*, SPIE.
22. Piccardi, M. (2004). Background subtraction techniques: A review*. In *IEEE International Conference on Systems, Man and Cybernetics* (pp. 3099–3104).

Applying Colour Image-Based Indicator for Object Tracking

Zygmunt Kuś, Jarosław Cymerski, Joanna Radziszewska and Aleksander Nawrat

1 Introduction

This article will elaborate on the idea of object tracking. All discussion and solutions will draw from the field of knowledge which deals with colour images. Using colour images offers a large scale of possibilities; however, there are numerous problems which are not present in grey scale images.

The definition of each image point in colour images domain (e.g. RGB) provides much information about a real scene observed by a camera. At the same time, the 3D size of the space of colours results in complication of calculations and interpretations. The segmentation problem for the colour images will be the first discussed problem in this paper. In general, the idea of image segmentation is presented widely in the body of the literature. The first group of the papers, which we can distinguish, concerns colour space models and pattern recognition techniques [1–3]. Segmentation on the basis of colour vector patterns [4–9], constitutes the next group in the segmentation and recognition field. Yet another aspect of the pattern recognition on the basis of colour images is the recognition of the textures as discussed in [10–13]. The last problem which is presented in [14], concerns recognition of colour characters in scene images.

Z. Kuś (✉) · A. Nawrat
Institute of Automatic Control, Silesian University of Technology,
Akademicka 16 St., Gliwice, Poland
e-mail: zkus@interia.pl

J. Cymerski
Government Protection Bureau, Warsaw, Poland
e-mail: j.cymerski@bor.gov.pl

J. Radziszewska
VR Technology, Gliwice, Poland
e-mail: j.radziszewska@vrtechnology.pl

A. Nawrat
e-mail: aleksander.nawrat@polsl.pl

A. Nawrat et al. (eds.), *Advanced Technologies in Practical Applications for National Security*, Studies in Systems, Decision and Control 106,
https://doi.org/10.1007/978-3-319-64674-9_2

In order to exploit information contained in colour images, the authors proposed the method based on clustering colours which appear in the image [16, 18]. The proposed method will operate on the space of an indicator calculated on the basis of the RGB colour definition. Therefore the proposed solution will support object recognition—object tracking [15, 17, 19] for the cases when we have efficient-enough quality colour images. The authors expect that using colour images may allow to define and analyse the properties of the tracked object with greater precision.

2 The Theoretical Basis for the Developed Method

This developed method is based on the analysis of the colour components of the RGB space. The ratios of the colour components r/g and r/b will be used to build a pattern vector which describes the features of the searched objects. The method proposed by the authors can be presented in the following steps.

I Creating the pattern vector for each of the searched objects.

(1) Let Object.jpg contain a colour image of the object defined in RGB space. We denote it as the variable $Object(i,j) \in RGB, i = 1, M_o; j = 1, N_o$; where M_o and N_o define the size of the object and, at the same time, the size of the aperture used for processing the terrain image.

For the searched object we calculate on the basis of the components (r, g, b) the representation of the object image in 2D space (L_1, L_2) where $L_1 = r/g$ and $L_2 = r/b$. In this way, we obtain a new image $NObject(i,j) \in (L_1, L_2)$.

(2) Next, we calculate for each point of the (L_1, L_2) space the number of the pixels in the object image which pixels have a pseudo-colour (L_1, L_2). The function $G_x(L_1, L_2) \in N, N$—natural numbers, defines for which L_1 and L_2 there are more or less pixels in the object image.

(3) Finally, we find three highest maxima G_x^{max} and we store their value and position on (L_1, L_2) plane—in the order from the highest to the lowest ones. In this way, we obtain the pattern vector W_{zob}, presented in (1), which defines the searched object features.

$$W_{zob} = \begin{bmatrix} G_x^{max1} & L_1^1 & L_2^1 \\ G_x^{max1} & L_1^1 & L_2^1 \\ G_x^{max1} & L_1^1 & L_2^1 \end{bmatrix} \quad (1)$$

II Calculating feature vector for all positions of the aperture in the terrain image where the object will be searched.

For each position of the aperture we follow the same as in point I. We assume that the size of the aperture is equal to the size of the object image. In this way, we obtain the feature vectors $W_z(i,j)$, presented in (2), for each position of the aperture in the terrain image with objects.

$$W_z(i,j) = \begin{bmatrix} G_x^{max1}(i,j)\ L_1^1(i,j)\ L_2^1(i,j) \\ G_x^{max1}(i,j)\ L_1^1(i,j)\ L_2^1(i,j) \\ G_x^{max1}(i,j)\ L_1^1(i,j)\ L_2^1(i,j) \end{bmatrix} \qquad (2)$$

III Calculating the distance between the pattern vector and feature vector for each object in each aperture position.

As a distance function $D(i,j)$ we assume the difference between the positions of the following maxima according to (3).

$$D(i,j) = ||W_z - W_z(i,j)|| = \sum_{x=1}^{x=3} \sqrt{[L_1^x - L_1^x(i,j)]^2 + [L_2^x - L_2^x(i,j)]^2} \qquad (3)$$

In this way, we obtain the distance function $D(i,j)$ between the feature vector $W_z(i,j)$ in a given position and pattern vector W_{zob}.

IV Calculating the position of the function *D(i, j)* minimum in the *(i, j)* plane.

The place where $D(i,j)$ assumes the minimum value we define as a place where the searched object is located. In order to present obtained results in the useful form, we define $D_1(i,j) = max(D_1(i,j)) - D_1(i,j)$. This function will be used to present obtained results. In this case the maximum of the function $D_1(i,j)$ corresponds to the place in the (i,j) plane where the object is located.

Fig. 1 The searched objects and their locations in the terrain

Fig. 2 The searched objects: **a** object 1, **b** object 2, **c** object 3, **d** object 4

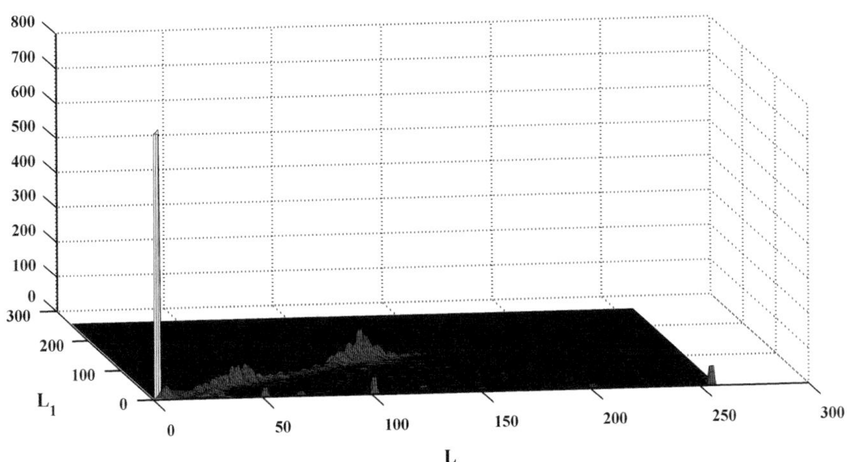

Fig. 3 Characteristics of the object 1 over the (L_1, L_2) plane

3 Examples

The functioning of the proposed method will be presented in the example of four different objects search. The objects and their locations in the terrain are presented in Fig. 1.

Figure 2 presents objects' images for which we will calculate pattern vectors. As it was described in Sect. 2, the first stage will involve calculating for each object

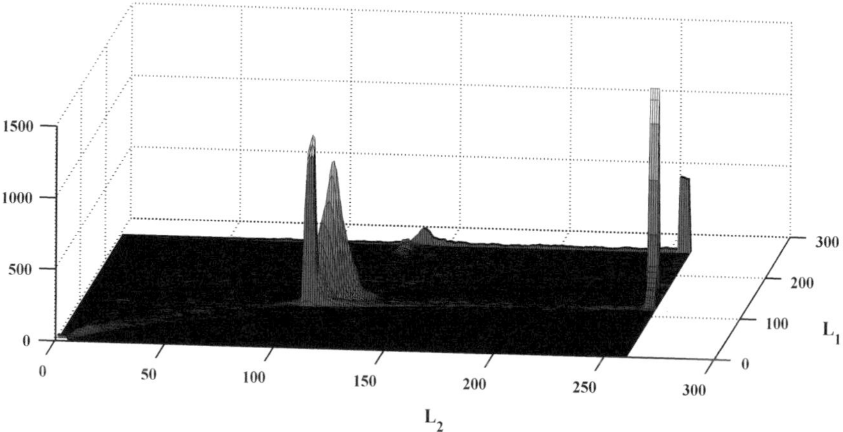

Fig. 4 Characteristics of the object 2 over the (L_1, L_2) plane

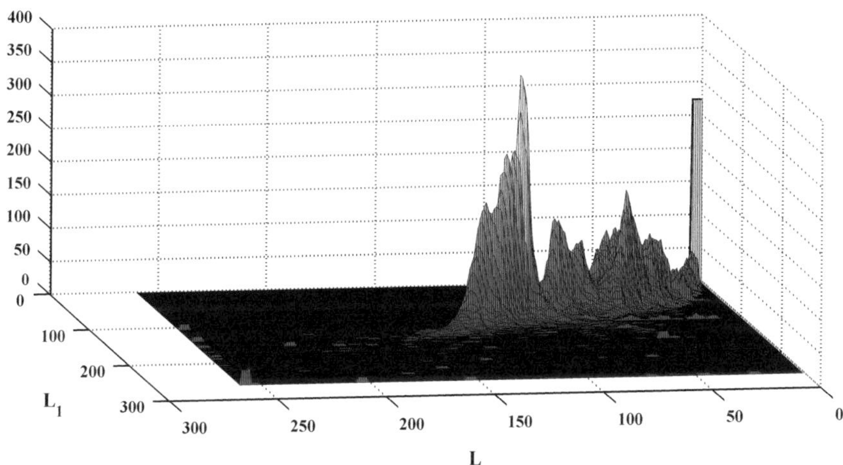

Fig. 5 Characteristics of the object 3 over the (L_1, L_2) plane

the pseudo-image *NObject*(i,j) in which each pixel is defined by two elements $L_1(i,j) = r(i,j)/g(i,j)$ and $L_2(i,j) = r(i,j)/b(i,j)$. In this way, we obtain a new image *NObject*$(i,j) \in (L_1, L_2)$. Next, we calculate the function $G_x(L_1, L_2)$ for each object. These functions were presented in Figs. 3, 4, 5 and 6. We can see that for each object we have certain values (L_1, L_2) for which there are more pixels in the object image.

In the next step, we find the positions and values of these maxima. They were graphically represented in Fig. 7.

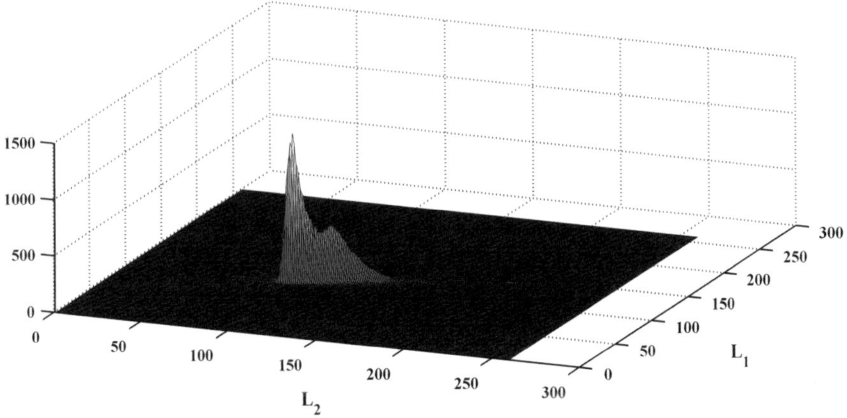

Fig. 6 Characteristics of the object 4 over the (L_1, L_2) plane

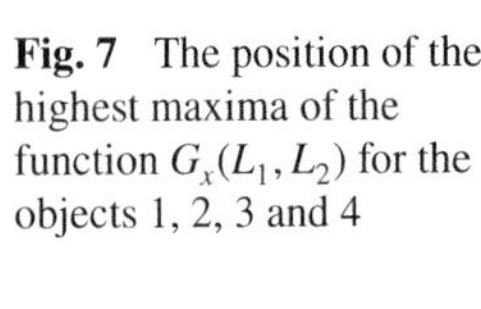
Fig. 7 The position of the highest maxima of the function $G_x(L_1, L_2)$ for the objects 1, 2, 3 and 4

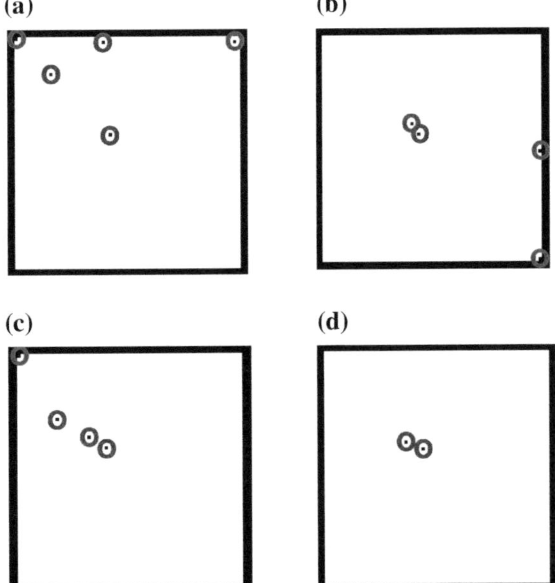

Table 1 shows numerical values of the pattern vectors for each object.

Then we calculate the distance function $D_1(i,j)$ for each object. They were presented in Figs. 8, 9, 10 and 11. We can see that there are certain maxima which define the location of the searched objects.

Table 1 The model vectors characterising 4 examined objects

Object 1	Object 2
$W_{zob} = \begin{bmatrix} 795 & 4 & 4 \\ 121 & 112 & 112 \\ 69 & 44 & 48 \end{bmatrix}$	$W_{zob} = \begin{bmatrix} 797 & 136 & 260 \\ 520 & 104 & 108 \\ 462 & 116 & 116 \end{bmatrix}$
Object 3	**Object 4**
$W_{zob} = \begin{bmatrix} 228 & 4 & 4 \\ 194 & 108 & 104 \\ 87 & 76 & 48 \end{bmatrix}$	$W_{zob} = \begin{bmatrix} 771 & 104 & 96 \\ 232 & 112 & 116 \\ 0 & 0 & 0 \end{bmatrix}$

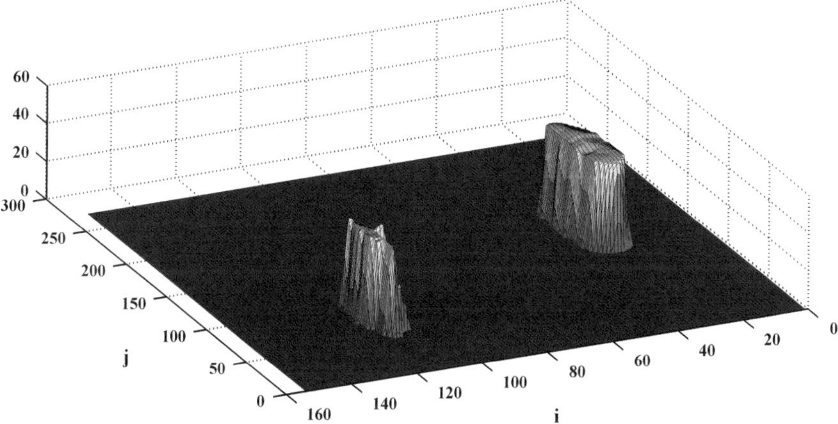

Fig. 8 Distance $D_1(i,j)$ for the object 1

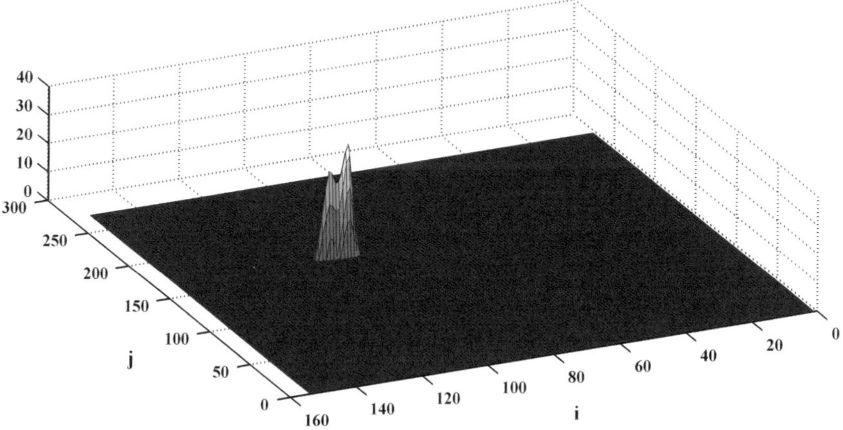

Fig. 9 Distance $D_1(i,j)$ for the object 2

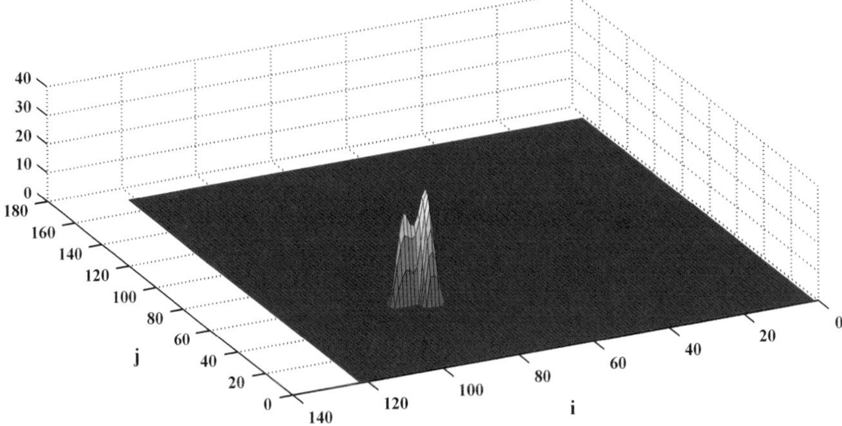

Fig. 10 Distance $D_1(i,j)$ for the object 3

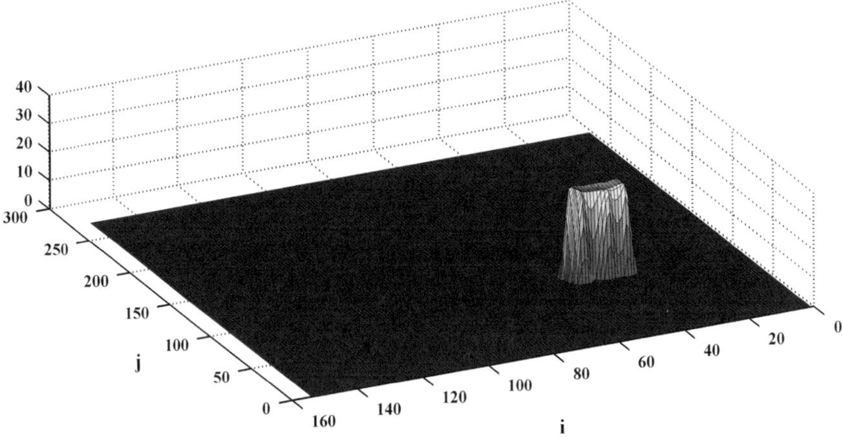

Fig. 11 Distance $D_1(i,j)$ for the object 4

The found locations of the objects are presented in Fig. 12. The contour lines, which were marked by appropriate colours, correspond to the minimal values of the distances between the pattern vector for a given object and the pattern vector for the current location of the aperture.

As we can see, the object 2 has been found correctly as well as object 3 and 4. Whereas the object's 1 location is indicated in two places. One of these places is

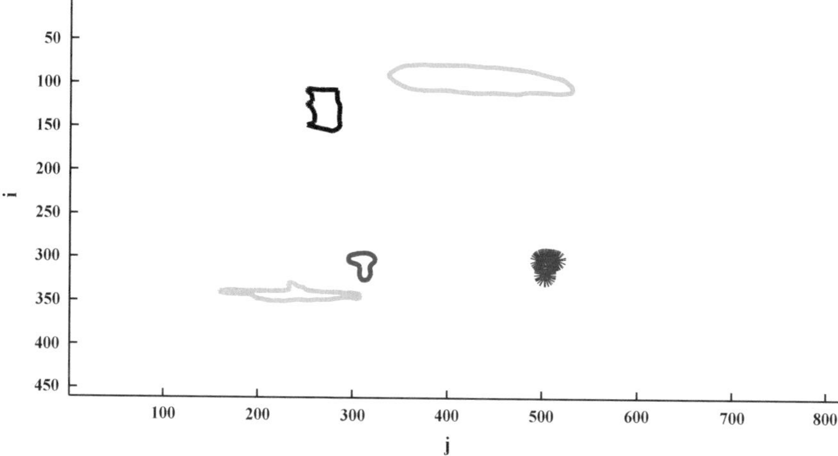

Fig. 12 The location of the searched objects. *Green*—ob.1, *Red*—ob.2, *Blue*—ob.3, *Black*—ob.4

correct. It results from the fact that the dominating colours in the object 1 are similar to the colours in the object 3. Nonetheless, the presented examples illustrate the correct functioning of the devised method in the cases when the objects has different colours.

4 Conclusion

The devised method combines both: easy calculation and the possibility of using information contained in colour images. The proposed solution was based on the calculation of an indicator which described the colour features of the object. The authors assumed that the role of this indicator would be realised by the ratio between the red and green components as well as the ratio between red and blue components. The abovementioned ratios were used to create the pattern vector. Such a defined pattern vector was used to calculate the error function and the minimum of this function indicated the object location.

The proposed approach turned out to be particularly useful in the cases when the object and terrain colours were significantly different.

This paper presented the examples of object tracking for both: the different object colour from the terrain colour and the similar object colour to the terrain colour. As expected, in the case of an object with the colour which blurs in with the background colour, the object recognition may be difficult when we base only on the object colour. However, it seems to be beneficial to use the developed method in order to enlarge features space used for the analysis of grey scale images. In this way, we may significantly enlarge the correctness of pattern recognition and object tracking.

References

1. Toh, V., Gray, A. J., Knight, C. H., & Glasbey, C. A. (2003). Comparing colour space models and pattern recognition techniques for segmentation of mammary tissue images. In *International Conference on Visual Information Engineering, 2003. VIE 2003.*
2. Ye, Z., Mohamadian, H., & Ye, Y. (2011). Comparative study of linear and nonlinear colour model identification based optimal feature extraction. In *2011 IEEE 12th International Symposium on Computational Intelligence and Informatics (CINTI).*
3. Liu, C., & Yang, J. (2009). ICA color space for pattern recognition. *IEEE Transactions on Neural Networks, 20*(2).
4. Lee, S. H., Choi, J. Y., Ro, Y. M., & Plataniotis, K. N. (2012). Local color vector binary patterns from multichannel face images for face recognition. *IEEE Transactions on Image Processing, 21*(4).
5. Castelnovi, M., Musso, P., Sgorbissa, A., & Zaccaria, R. (2003). Surveillance robotics: Analyzing scenes by colors analysis and clustering. In *2003 Computational Intelligence in Robotics and Automation. Proceedings.*
6. Pribula, V., & Canosa, R. L. (2013). LBP-inspired detection of color patterns: Multiplied local score patterns. In *Image Processing Workshop (WNYIPW).* New York: IEEE Western
7. Saigaa, D., Fedias, M., Harrag, A., Bouchelaghem, A., & Drif, M. (2011). Color space MS-based feature extraction method for face verification. In *2011 11th International Conference on Hybrid Intelligent Systems (HIS).*
8. Deb, K., Lim, H., & Jo, K.-H. (2009). Vehicle license plate extraction based on color and geometrical features. In *IEEE International Symposium on Industrial Electronics. ISIE 2009.*
9. Mindru, F., Moons, T., & Van Gool, L. (1999). Recognizing color patterns irrespective of viewpoint and illumination. In *Computer Vision and Pattern Recognition.*
10. Choi, J. Y., Ro, Y. M., & Plataniotis, K. N. (2012). Color local texture features for color face recognition. *IEEE Transactions on Image Processing, 21*(3)
11. Arivazhagan, S., & Benitta, R. (2013). Texture classification using colour local texture features. In *2013 International Conference on Signal Processing Image Processing & Pattern Recognition (ICSIPR).*
12. Porebski, A., Vandenbroucke, N., & Macaire, L. (2007). Iterative feature selection for color texture classification. In *IEEE International Conference on Image Processing, 2007. ICIP 2007* (Vol. 3).
13. Ko, T. (2006). Viewpoint-invariant and illumination-invariant classification of natural surfaces using general-purpose color and texture features with the ALISA dCRC classifier. In *Applied Imagery and Pattern Recognition Workshop, 2006. AIPR 2006. 35th IEEE.*
14. Wakahara, T. (2008). Figure-ground discrimination and distortion-tolerant recognition of color characters in scene images. In *19th International Conference on Pattern Recognition, 2008. ICPR 2008.*
15. Jedrasiak, K., Bereska, D., & Nawrat, A. (2013). The prototype of gyro-stabilized UAV gimbal for day-night surveillance. *Advanced technologies for intelligent systems of national border security* (pp. 107–115). Berlin, Heidelberg: Springer.
16. Ryt, A., Sobel, D., Kwiatkowski, J., Domzal, M., Jedrasiak, K., & Nawrat, A. (2014). Real-time laser point tracking. In *International Conference on Computer Vision and Graphics* (pp. 542–551). Springer International Publishing.

17. Daniec, K., Iwaneczko, P., Jedrasiak, K., & Nawrat, A. (2013). Prototyping the Autonomous Flight Algorithms Using the Prepar3Dő Simulator. In *Vision based systems for UAV applications* (pp. 219–232). Springer International Publishing.
18. Nawrat, A. (2008). *Jedrasiak* (pp. 69–76). Fast colour recognition algorithm for robotics, Problemy Eksploatacji: K.
19. Sobel, D., Jedrasiak, K., Daniec, K., Wrona, J., Jurgas, P., & Nawrat, A. (2014). Camera calibration for tracked vehicles augmented reality applications. In *Innovative control systems for tracked vehicle platforms* (pp. 147–162). Springer International Publishing.

Image Processing in Thermal Cameras

Tomasz Sosnowski, Grzegorz Bieszczad and Henryk Madura

1 Introduction

In modern security systems more and more commonly thermal camera are used exploiting infrared radiation imaging to perimeter observation and thread detection, especially when there is a need to proceed in limited visibility conditions of complete darkness. Implementation of cameras in security system causes a substantial increase of information fed to security system operator. Thermal image processing system can help the security system operator by enhancing the relevant information in the image, which enables to discern important image details [1]. More advanced imaging systems detect threats automatically and present information about that fact directly to the operator. To perform such tasks, special algorithms are implemented for detection and tracking of objects in the image. On the grounds that, the detection algorithms used for tracking and vision systems operating in the visible light cannot be directly used for the analysis of thermal images [2], special methods have been developed for detection and tracking of objects on the thermal image.

T. Sosnowski (✉) · G. Bieszczad · H. Madura
Military University of Technology, Institute of Optoelectronics, Warsaw, Poland
e-mail: tomasz.sosnowski@wat.edu.pl

G. Bieszczad
e-mail: grzegorz.bieszczad@wat.edu.pl

H. Madura
e-mail: henryk.madura@wat.edu.pl

A. Nawrat et al. (eds.), *Advanced Technologies in Practical Applications for National Security*, Studies in Systems, Decision and Control 106,
https://doi.org/10.1007/978-3-319-64674-9_3

2 The General Structure of the Infrared Camera—Algorithms and Methods for Image Processing in Infrared Cameras

Thermal cameras are more and more often used as the observation device in security systems for perimeter protection, military systems for object detection, identification and tracking, in pollution detection and many more. In such systems, it is important to process infrared information that the resulting image is faithfully corresponding to the observed situation. More and more common use of infrared cameras as surveillance equipment means that they should be as simple to use as possible. This forces the need for implementation of automatic thermal image processing and analysis methods. These methods allow to simplify the operation of the camera by means of automatic adjustment of operating parameters of the thermal imager. The methods used should also allow the work of the infrared camera not only as a tool to support observation, but also for detecting and identifying emerging objects and phenomena. As used in the device processing method is dependent on the particular application and the type of the analyzed information [1], therefore they cannot be universal or selected once and for all. Moreover such automatic systems performing processing and image analysis must have a relatively small size and low power consumption. In general an electronic system in the thermal camera can be divided in three essential modules [3–6]:

- focal plane array module,
- control and digital processing module,
- imaging module.

Simplified schematic of electronic system is shown on Fig. 1.

In detector array module there is a high performance analog to digital converter that allows transforming analog signals form detectors to digital form possible to process in control and image processing module. In detector array module there are also sophisticated power sources for powering and biasing internal circuitry of the array and special filters that enables noise immunity.

Control and image processing unit sends synchronization and control signals responsible for proper readout from the detector array module [6, 7]. For example if the camera has to work in broad range of temperatures without temperature stabilization, the control unit has to control the biasing voltages of the array.

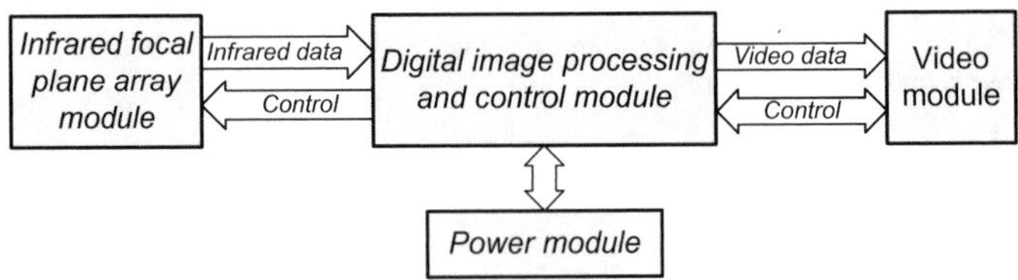

Fig. 1 Schematic of electronic modules in microbolometric thermal camera

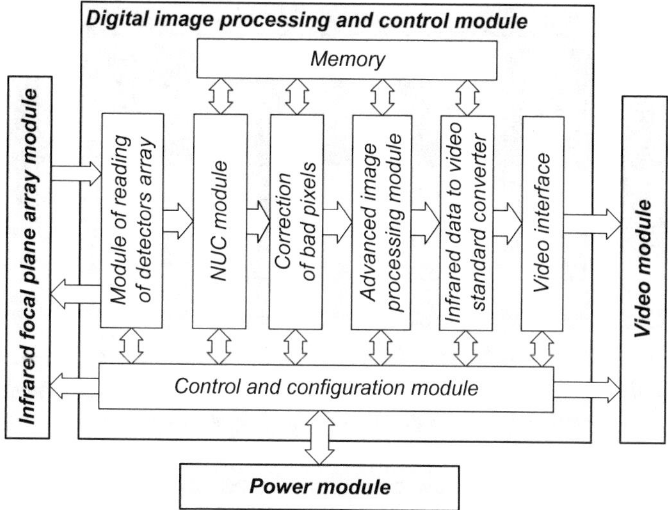

Fig. 2 Functional block diagram of control and image processing module

Biasing voltages are largely responsible for the sensitivity of the detectors on the focal plane array and for maintaining an adequate common mode level of the output signal.

Control and image processing unit is responsible for configuration of all other modules and image processing of collected infrared image data. The main tasks performed by the image processing and control module are the following: control of focal plane array to read the values of all detectors in the matrix, nonuniformity correction, correction of signal from defective detectors and generation of data for the display module. The block diagram of control and image processing module is shown on Fig. 2.

The main problem of hardware implementation of complex image processing algorithms such as image enhancement algorithm, is their high computational cost. The implementation of complex algorithm by means of software in a general-purpose processor usually does not give satisfactory results. For complex image processing algorithms implemented in real time Application Specific Integrated Circuits are often used. However, due to a predefined architecture typically they have limited functionality and have a relatively long development time and considerable price. The alternative is to use reconfigurable computing architecture based on programmable devices [4].

The use of programmable system is a flexible and powerful solution that allows the execution of complex operations of image processing in real time. Data processing modules organized in pipeline, usually exchange data using standardized data bus, for example with VideoBus [5, 6]. This allows to swap the order of operations performed by image processing modules without interfering with the overall pipeline system. Therefore, the control and digital image processing module

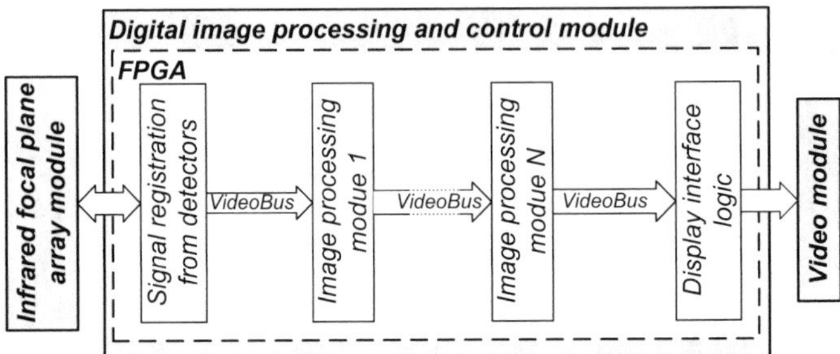

Fig. 3 Block diagram of image processing chain implementation in FPGA

is often built on the basis of two basic systems: programmable system e.g. the FPGA and microcontroller circuit. The programmable system performs the processing of image data, which requires significant computing power, the microprocessor performs all activities related to the control and other activities that demand relatively small computational power. A block diagram of image data processing modules implementation in the FPGA is shown in Fig. 3.

Generally speaking, image processing in the thermal imaging camera can be divided into three groups of methods and algorithms. The first group of algorithms and processing methods are the algorithms and methods that are necessary for the operation of the thermal imager. This group includes such processing as nonuniformity correction of detector responses in FPA [7, 8] and algorithms for replacing signal from faulty detectors. The second group of algorithms are the algorithms used to improve image quality, in order to enable and facilitate the interpretation of the thermal visualization by the operator or the vision system. The third group of image processing algorithms are the methods of data analysis for the automatic detection and tracking of objects in an image and interpret the scene (i.e. Machine vision).

3 Basic Thermal Image Processing Algorithms

A first group of algorithms and processing methods are the algorithms and methods that derive from the working principles of the infrared detector array, and the physical laws relating to infrared radiation. The main processing algorithms is a correction of response nonuniformity of each detector in the array and the detection and replacement of damaged detectors. There are many algorithms for detectors response nonuniformity. The most commonly used and popular methods include one-point and two-point correction methods [7, 8].

On Fig. 4 is schematically shown the nonuniformity correction algorithm of the FPA with the two point method. Regarding the replacement of defective detectors, the procedure is divided into two phases. The first phase involves detection and localization of the faulty detector. Second phase involves interpolation of the signal

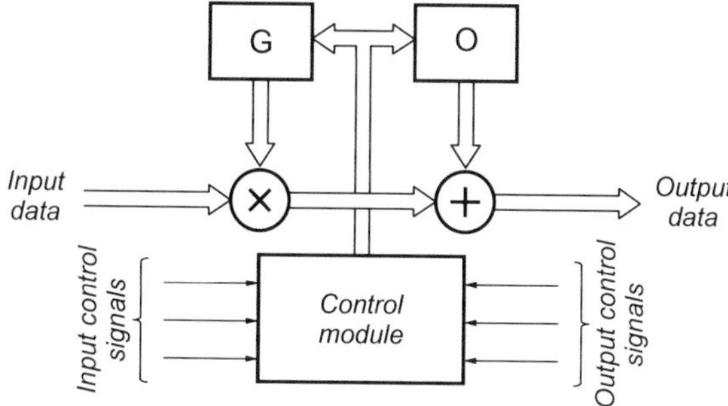

Fig. 4 Block diagram of hardware nonuniformity correction module

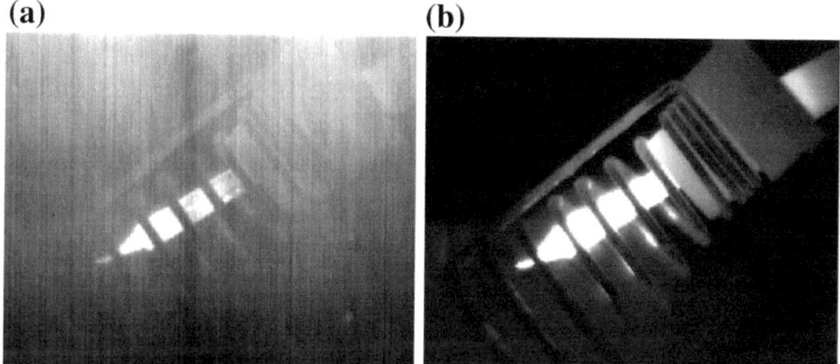

Fig. 5 Exemplary thermal images before (**a**) and after (**b**) nonuniformity correction

for signal replacement from faulty detector. Generally, one can distinguish the following manifestations of damaged detector:

- detector returns a value outside the dynamic range of the readout circuit,
- detector returns a constant value, independent of the incident radiation (does not respond to changes in incident radiation),
- detector has too low or too high sensitivity to incident radiation,
- detector is characterized by high noise, independent of the incident radiation,
- detector flashes, that means it significantly alters its value with a very low frequency in range of 1 Hz or less (this symptom occurs in cooled detectors only).

Algorithms and methods for detecting malfunctioning detectors are based on the causes of faults in the detectors. The most commonly used methods for detecting defective sensors include: offset criterion, noise criterion, sensitivity criterion, and algorithms for blinking pixels detection. On Fig. 5 the effect of non-uniformity correction operation and removal of defective detectors is shown.

4 Initial Image Processing—Image Enhancement

In general, thermal image is a visualization of the infrared radiation emitted by the observed object and its surroundings. Visualization of the thermal scene is significantly different from the image recorded by the video camera. It follows that the thermal image is often difficult in interpretation. Proper interpretation of the thermal image is associated with proper understanding of infrared radiation properties as well as specific object properties and surrounding scenery. Throughout the process of perception (analysis and interpretation of the image) an important role plays not only objective factors like the emissivity of the object, but also subjective factors ex. human perception of visual information. Unfortunately, the optimum parameters setting of infrared camera alone (sharpness, temperature range) does not guarantee a correct detection and interpretation [2, 9, 10].

Therefore, in a thermal imaging camera, special image processing is performed, for example to ensure automatic dynamic range control of the entire image or selected area. The primary purpose of techniques to improve the quality of the image (image enhancement) is so that the image obtained as a result was more suitable for a particular application, not only to make it more pleasing for human observer. Differences between visible spectrum video and thermal image causes that adaptation of image processing techniques is more challenging and demands custom solutions [2, 9, 11].

The most common algorithms to improve image quality are: image contrast modification, sharpening, removing geometry distortions, smoothing and denoising etc. Described in the article methods to improve the infrared image quality comprises in three categories: context-free methods (also called point methods), contextual image processing, methods for histogram modification.

4.1 Context-Free (Point) Image Processing Methods

The most common Operation in the early stages of image processing is the point (context-free) processing. It is a fundamental and at the same time the simplest operation performed in image processing. Point image processing is typically implemented in hardware using a look-up-table (LUT) processing. LUT processing performs pixel value transformation using an array of data contained in the memory. A diagram of LUT technique implementation is presented on Fig. 6.

A typical example is the LUT operation is correction of brightness and contrast of the infrared image. Change in the brightness or contrast corresponds to an adequate change in values within the LUT (Fig. 7).

Figure 7a shows the changes taking place in the table when modifying the image contrast. Increasing the contrast causes the angle of the slope to increase, what means that the differences in the levels of brightness of the individual pixels in the output image are also increased. The maximum number of brightness levels of the

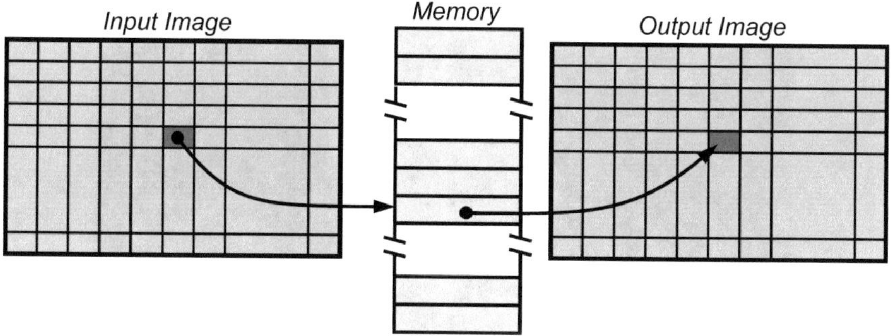

Fig. 6 The implementation scheme of "look-up-table"

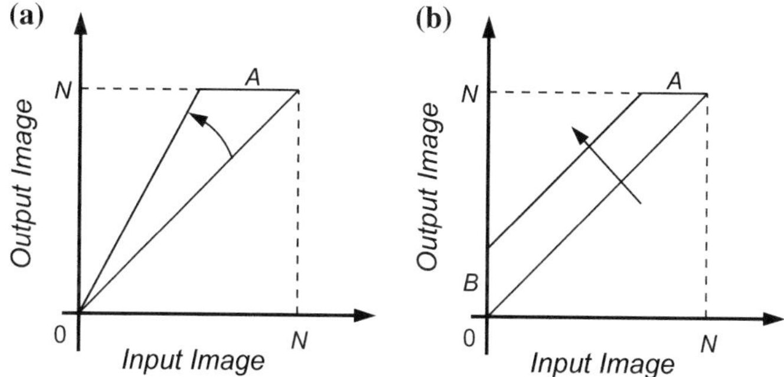

Fig. 7 Modifying the brightness (**b**) and contrast (**a**) of the image using the LUT

input and output is constant and equals N. For this reason, increasing the contrast is followed by cutting the dynamic range of input stream as shown on the picture by section A. All levels of brightness of the original image in range indicated by section A results in the truncated level of brightness equaling N. Increasing the brightness of the image corresponds to the translation of conversion function in the vertical direction as shown on Fig. 7b. This is equivalent to adding certain value to every cell in LUT. In this case the cut off segments also occurs. Some of the input brightness levels will not be present in the final image. Absent levels not represented in final image are indicated by section B. Nature of the functions carried out by the LUT values can be closely matched to the specific application. For example, we may care about increasing the contrast only in the central region of the brightness levels. On Fig. 8 is an example illustration of the LUT operation result to normalize the images from the infrared camera. Applied operation has enabled a significant increase in image contrast.

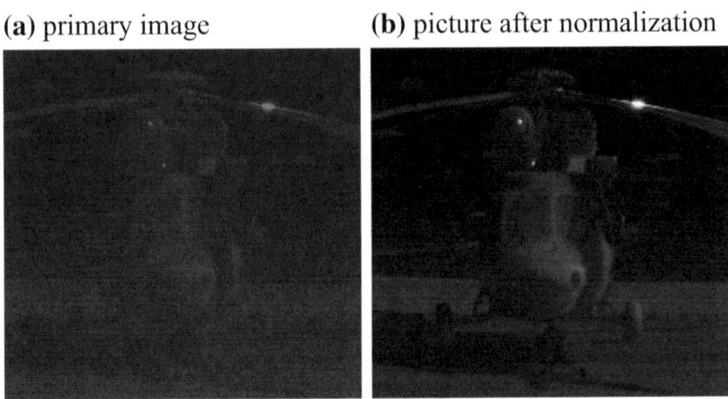

Fig. 8 Example application of LUT operation to normalize the image

4.2 Contextual Image Processing

Contextual image processing involves replacement of original value of the pixel with value calculated based on the context (environment) of this point. In other words, the new value is calculated based on the neighborhood of the original pixel. The function realizing contextual image processing to be described by the equation:

$$L^{'}(m,n) = \frac{1}{\sum\limits_{i,j \in K} w(i,j)} \sum_{i,j \in K} L(m-i, n-j)w(i,j) \tag{1}$$

where: $L(m,n)$—function representing input values, $L^{'}(m,n)$—function representing output values of the image, $w(i,j)$—weighting factors (kernel of the transform function). Kernel shape is generally square, with an odd number of cells on each side. Based on the kernel coefficients and neighborhood pixels, a new value is calculated. Although the shape of the mask can be any in practice most commonly used mask size 3 × 3.

For realizing functions of the context image processing in real-time, the hardware dedicated processing unit can be used. Dedicated real time processing unit consist of delay lines (buffers), to ensure a continuous flow of the stream of data, and in the same time simultaneous access to neighborhood of actually processed pixel. This allows to perform corresponding arithmetic operation in one batch. Processing module for the kernel size of 3 × 3 is schematically shown in Fig. 9. Contextual image processing module opens possibility to perform vast amounts of image processing methods only by changing applied kernel coefficients. The most commonly used methods with application of contextual image processing are among others: low-pass filtering, high pass filtering, edge detection, feature detection, denoising etc.

The low pass filter elements of the image passes a low frequency signals, while suppressing or blocking elements with a high frequency. These filters are often used

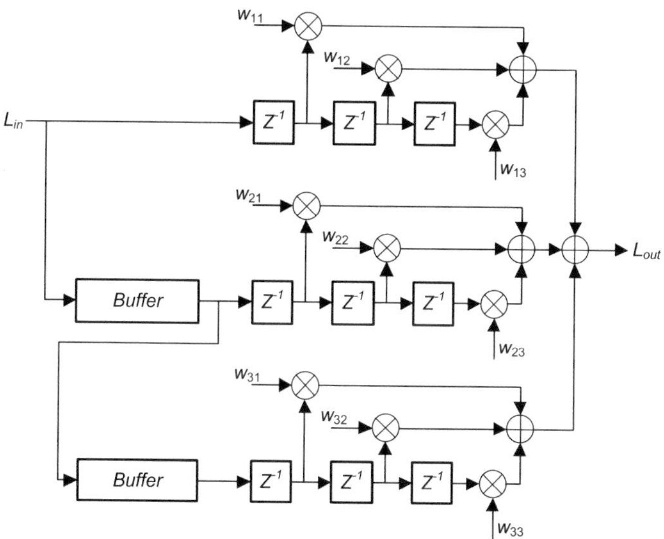

Fig. 9 Block diagram of the convolution module for kernel size of 3 × 3

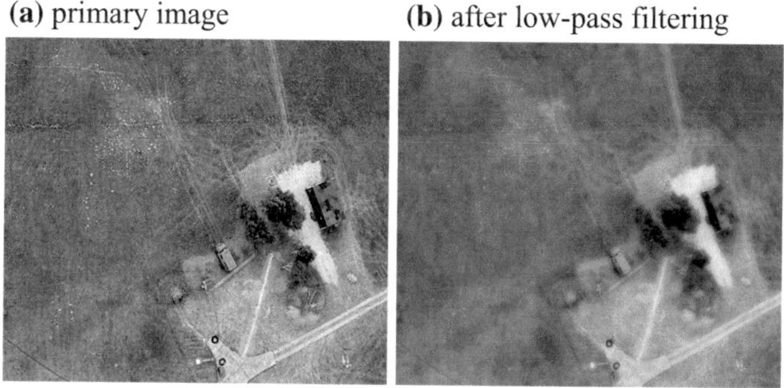

Fig. 10 Example application of low-pass filtering in order to remove single point interference

to reduce noise in the image, on the other side they causes reduction of fine details in the image. On Fig. 10 an exemplary effect of the low pass filter is shown. Used filter is intended to remove image noise.

High Pass Filtering is used to strengthen the high frequency information present in the image, while preserving the integrity of the low-frequency information. High Pass Filtering is important when objects in the image has to be highlighted or identified.

A high-pass filter increases the sharpness of the image but on the other hand the negative effect occurs from amplification of the noise. On Fig. 11 are shown the examples of high pass filtering.

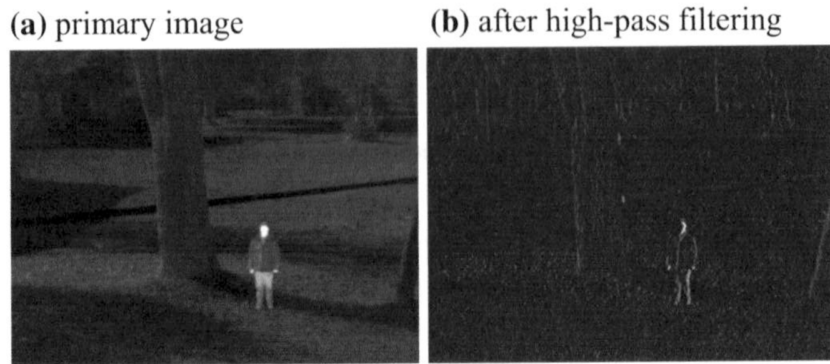

Fig. 11 An example of the of high-pass filtering application to highlight objects in the image

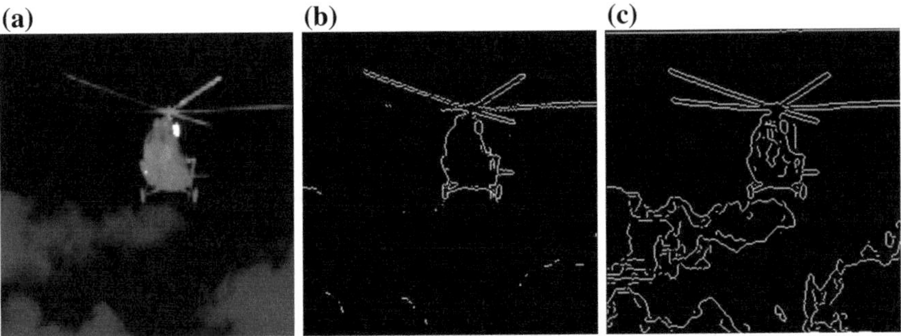

Fig. 12 Edge detection in the image: **a** original image, **b** Sobel filter, **c** Canny filter

Filters used for edge detection are sometimes called contour detectors. These filters are commonly used for the classification of shapes of objects in the image. They operate on the principle of gradient detection. The gradient is defined as the spatial difference of brightness in the image. Gradient achieves the highest value in places where there are the biggest changes in the brightness between adjacent pixels.

Laplacian operators are often used for edge detection. In comparison with other methods, they are omnidirectional. On Figs. 12 and 13 there is an illustration of edge detection and enhancement.

4.3 Histogram Modification

Histogram of brightness levels is the statistical distribution of the occurrence of a pixels of each brightness level in the image. In the picture we are dealing with both a finite number of pixels and a finite number of brightness levels, implies that the histogram is a discrete function. Histogram modification comprises the operations performed on the image in order to obtain the required shape of the histogram.

Fig. 13 Modification of the thermographic image by edge sharpening

One of the basic and most common methods of this type is the histogram equalization. Histogram equalization technique involves such modification of the source image so that the resulting histogram is as flat as possible. Assuming that the histogram represents brightness levels occurrence probability p(i) then we can calculate the cumulative distribution D(i) of the probability distribution using the formula:

$$D(n) = \sum_{i=1}^{n} p(i) \tag{2}$$

Based on the cumulative distribution we obtain an intermediate LUT, which is the function for converting brightness levels in the current image to new levels of brightness. LUT table calculation can be expressed by the formula:

$$i' = \mathrm{int}\left[\frac{D(i) - D(0)}{D(0)}(K-1)\right] \tag{3}$$

where: *i*—the index of the brightness of the original image, *i'*—the brightness of the resulting image, *D*(*i*)—cumulative distribution for *i*-th brightness level, *K*—number of brightness levels, int—rounding operator. Histogram equalization can be used to highlight details that are barely visible in the image because of the low contrast, but it should be noted that this is not an universal method and for some histogram shapes it does not give satisfactory results.

The module for histogram calculation is one of the more complex modules in digital imaging, however the principle of operation of the module alone is fairly simple and involves counting the occurrences of each pixel value. The implementation of incrementing operation consists of three basic steps: read the memory location, increment the value, and then store the resulting values back to memory. The system calculates the histogram has one cycle of computing operations to perform three operations. Therefore, the implementation requires a relatively fast memory of considerable size. Schematically, the histogram calculation module is shown in Fig. 14. On Fig. 15 is an example of the application of histogram equalization operation of the thermal image of the helicopter.

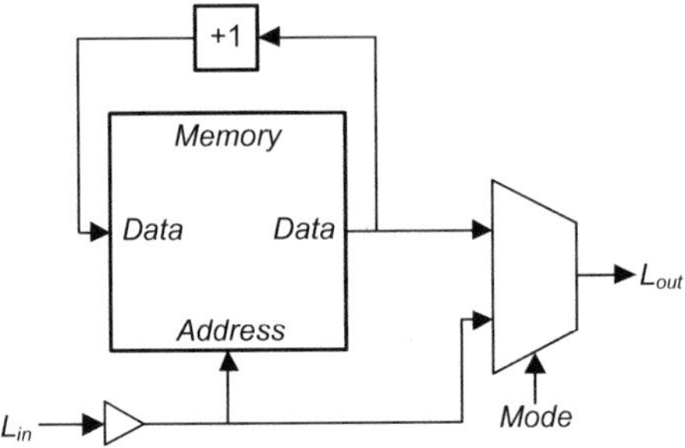

Fig. 14 The block diagram of the module that calculates histogram

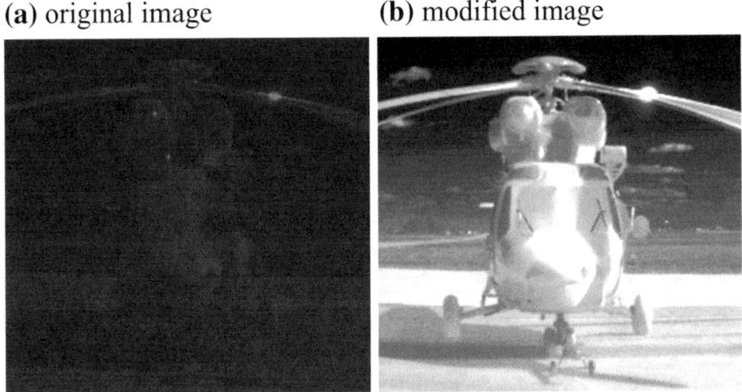

Fig. 15 Example applications of the histogram equalization operation

5 Detection, Recognition and Object Tracking Algorithms

Image analysis is very often used in the detection, classification, recognition and identification of moving objects and targets. Automatic target recognition (ATR) system should be designed to help analyze the image by the analyst by increasing the amount of data that he can process. I also should help by increasing the possible time to spent on desired region of interest ROI by automatic preselection of more interesting objects. It makes possible to reduce the number of analysts and accelerates the analysis of the situation. In general, we require algorithms that allow ATR to distinguish targets from ordinary objects and allow to distinguish between different classes of targets. An important requirement for ATR system is to show exact location and orientation of the moving targets. Generally, in the ATR system the objective is implemented by the sequence of operations

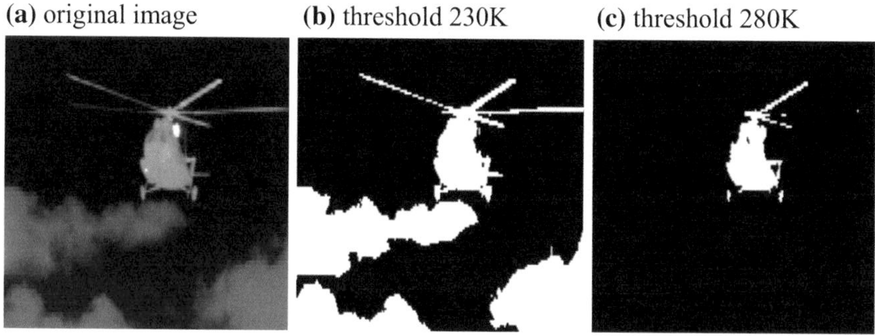

Fig. 16 Infrared image binarization **a** for different threshold temperature: 230 K (**b**), 280 K (**c**)

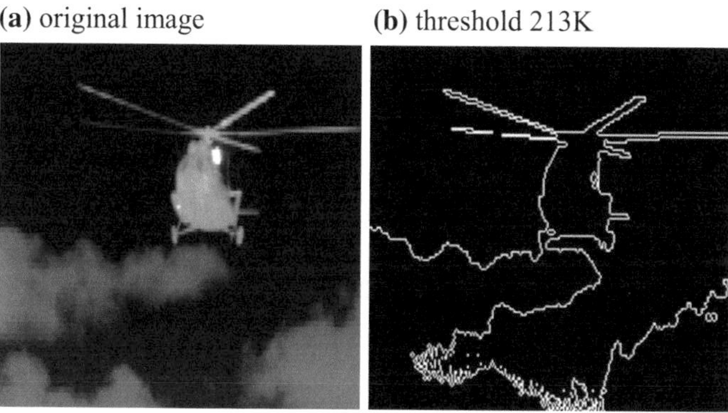

Fig. 17 Detection of closed areas—temperature threshold 213 K

consisting of: detection, classification, recognition and identification. In preprocessing the methods of edge detection are commonly used to detect spatial discontinuity. It is quite an effective way to get a quick effect by indicating interesting elements in the image without significant degradation of image. Edge detection than can help operator to identify shape of the object.

For objects detection one can use some simple image processing methods such as: binarization, edge detection, skeletonization, and morphological transformations such as erosion dilation composed together to form more complex algorithms. Examples of use of the above operations for the thermographic images were presented in the Figs. 16, 17, 18 and 19.

Then, on the basis of images resulting from pre-processing operations, potential objects has to be separated. Separation involves the division of the image into fragments corresponding to different detected objects.

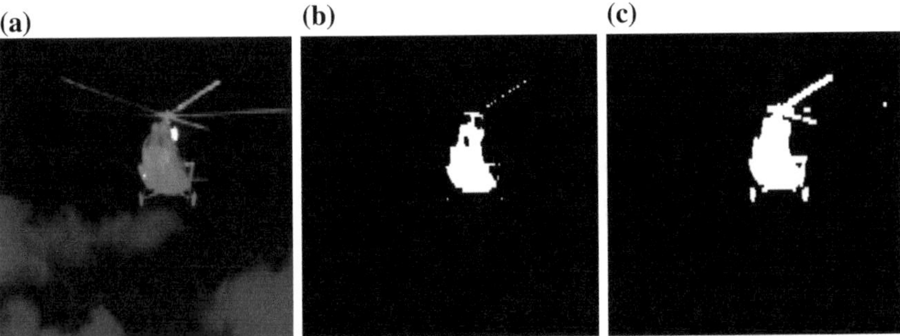

Fig. 18 Examples of use in the thermographic image (**a**) morphological operators like erosion (**b**) and dilatation (**c**)

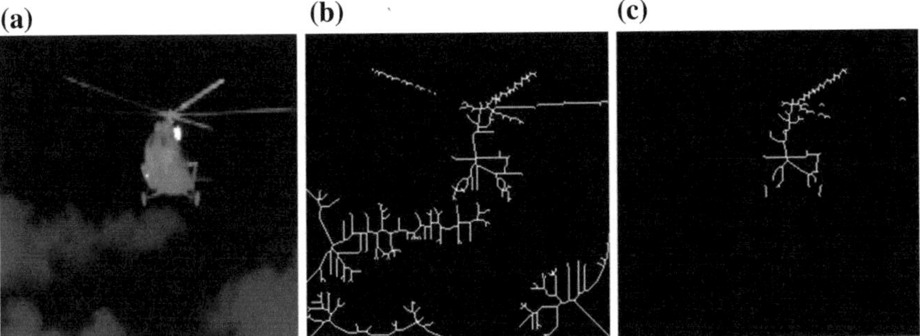

Fig. 19 Infrared image skeletonization **a** for different threshold temperature: 240 K (**b**), and 280 K (**c**)

5.1 Object Detection Method

Object detection involves performing a series of operations on the thermal image to highlight the image interesting, from the point of view of its appointed task. The developed method for objects detection performs the task by recording the thermal image and calculating the reference image. Then a comparison is made between newly acquired thermographic image and the reference image. Object detection is then performed by discriminating changes from the background image. Simplified diagram of the algorithm to detect objects in the infrared image is shown in Fig. 20.

According to the algorithm, the first step in the analysis is the determination of the initial reference picture. Determination of the reference image f^* s based on averaging a certain number of images for which there is no object detected in accordance with formula:

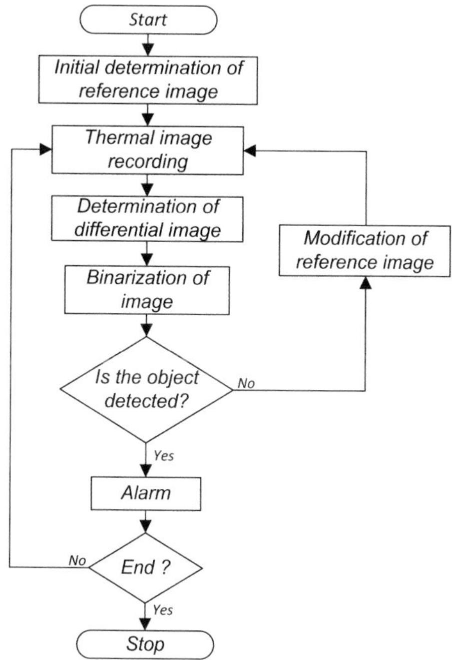

Fig. 20 Simplified diagram of the algorithm to detect objects in the infrared image

$$f^*(m,n) = \frac{1}{K}\sum_{k=1}^{K} f_k(m,n). \tag{4}$$

where $f_k(m,n)$, $f^*(m,n)$—pixel value respectively of the source image and the resulting image placed in m–th row and n–th column in k–th picture in the sequence.

After determination of the reference image, algorithm goes to determination of the different image Δf for each newly recorded image. A differential image is calculated according to the equation:

$$\Delta f(m,n) = \left|f(m,n) - f^*(m,n)\right|. \tag{5}$$

The next operation is to check whether the determined value of a differential image for each pixel falls into the detection threshold. If the calculated value of the differential image is greater than a predetermined detection threshold L then the object is detected. If the value of the differential image is below the detection threshold, the detection did not occurs and the recorded image is used to adaptively modify the reference image.

Probability of detection (P_w) and probability of false alarm (P_f) was determined in experiment along with their relationship to threshold level. Minor threshold level cause the background elements in the image to be classified as the detected object. With high detection threshold some elements of an object are not detected what

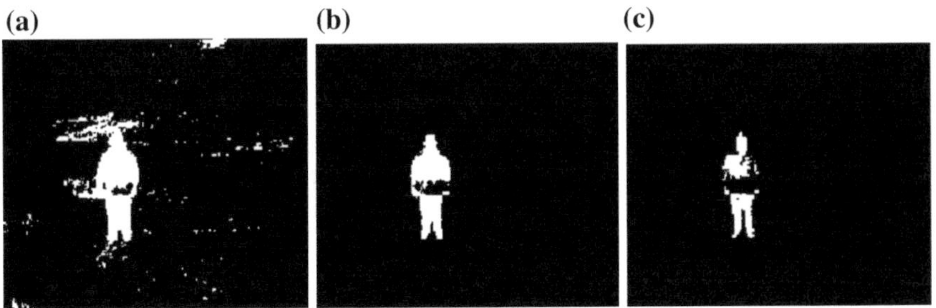

Fig. 21 Objects detection results for threshold values of 1.0 °C (**a**), 3.3 °C (**b**), 6.0 °C (**c**)

Table 1 Summary of the probability of object detection (P_w) and false alarm probability (P_f) at different distances from the object and different values of the proportionality coefficient α_P

Proportionality coefficient	Distance 160 m		Distance 50 m		Distance 20 m	
	P_f	P_w	P_f	P_w	P_f	P_w
0.50	0.063	0.888	0.385	0.794	0.110	1
0.55	0.063	0.833	0.308	0.692	0.110	1
0.60	0.063	0.770	0.077	0.666	0.110	1
0.65	0	0.580	0	0.590	0	1

causes that interesting object is distorted and blurred. Exemplary results of detection for different values of the detection threshold are shown in Fig. 21.

The most important factor influencing the high probability of detection achieved with simultaneous low probability of false alarm is the right choice of detection threshold. To get the best performance the detection threshold L is adaptively determined according to the following formula:

$$L = \alpha_P \cdot \Delta f_{MAX} \tag{6}$$

where: α_P—proportionality coefficient, Δf_{MAX}—maximum value of the pixel in differential image.

Proper selection of the proportionality factor is essential to correctly identify the potential location of an object, and extract more of its components and thereby reduce the number of false alarms. The results of research carried out for the set of the objects and different proportionality factor is shown in Table 1.

5.2 *Objects Recognition by Means of Radial Shape Function*

One of the object recognition methods is an algorithm using the so-called. radial shape function [12]. This method has some similar properties to the human vision

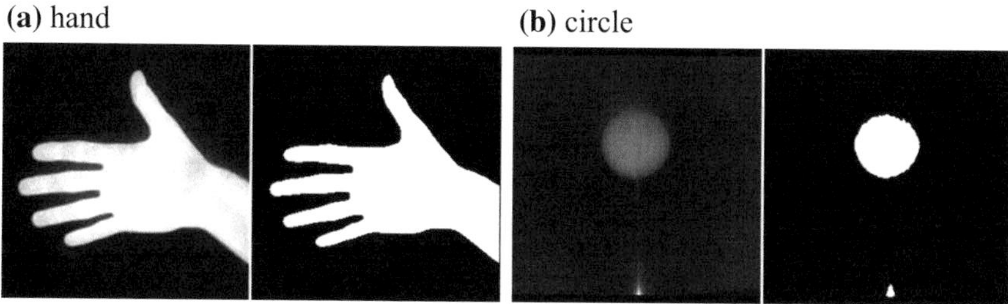

Fig. 22 Thermovision image (on *left*) binarized and filtered result (on *right*) for two objects of different shape

and pattern recognition capabilities. The fundamental advantages of the method include immunity to distance, location and size change of the recognized object. The effectiveness of shape recognition is independent of the shape rotation. Moreover, this method has a relatively high recognition rate with low computational demands. This makes if a good candidate to embed in infrared cameras for image analysis in real-time.

The first operation in is binarization carried out in such a way that the detected object is represented by a value of one (white) and the background by zero (black). After binarization some distortions may occur specially on the edges of objects. In order to eliminate the noise there has been erosion operation applied. Sample results of such preprocessing are shown in Fig. 22.

The main part of the algorithm involves determination of a function describing the shape of the object called the radial shape function (RSF). First stage of RSF determination consists of describing the object in polar coordinates. Center of polar coordinate system is set to the statistical center of gravity of the object. According to the theory of statistical moments coordinates of the center of gravity (centroid) can be described as [12]:

$$x = \frac{M_{1,0}}{M_{0,0}}, \quad y = \frac{M_{0,1}}{M_{0,0}}$$

where:

$$M_{p,q} = \sum_{x=0}^{N-1} \sum_{y=0}^{M-1} x^p \cdot y^q \cdot f(x,y),$$

$f(x,y)$—binary image value at the point (x, y).

After calculating the position of the center of the coordinate system, the radial shape function f_{RSF} is being calculated. This is done by determining the coordinates of points belonging to the edges of the object in polar coordinates according to equation:

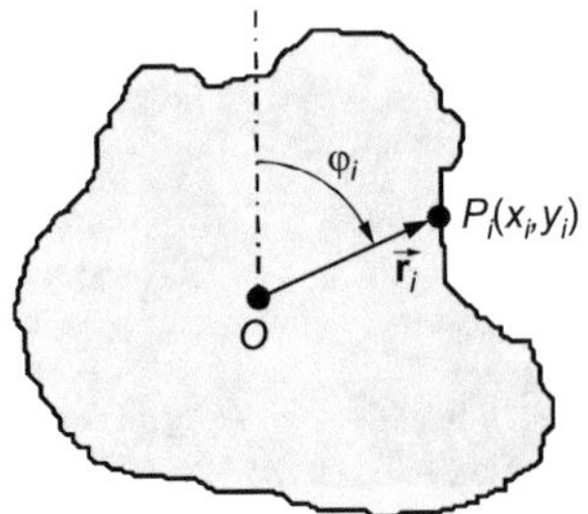

Fig. 23 Method of determining the polar coordinates for successive points on the objects edge

$$f_{RSF}(\varphi) = |\vec{\mathbf{r}}_i| = \sqrt{x_i^2 + y_i^2} = \frac{x_i}{\cos\varphi_i} \tag{7}$$

where: (x_i, y_i) are the coordinates of successive points on the edge of the object in relation to the center of gravity, while φ is the angle between point on the edge and the vertical axis. Determined in this way radial shape function expresses the distance between the center of gravity of the object and its edge as a function of angle. The described method of radial shape function determination is illustrated in Fig. 23.

In the next step of the object recognition algorithm the previously determined radial shape function is being low pass filtered. This operation is introduced to eliminate noise caused by the irregular edges of the object and noise. Then the radial shape function is subjected to a discrete Fourier transform [12] according to the equation:

$$F_{RSF}(\Phi_k) = \left| \sum_{n=0}^{N-1} f_{RSF}(\varphi_n) e^{-j2\pi kn/N} \right| \tag{8}$$

The obtained spectrum of radial shape function F_{RSK} is normalized to the mean value which can be written using the equation:

$$F^*_{RSF}(\Phi_k) = \frac{F_{RSF}(\Phi_k)}{F_{RSF}(0)} \tag{9}$$

Based on the obtained in the above manner normalized spectrum the similarity score is determined. Similarity score ω_k is defined as sum of differences between individual spectral components F^*_{RSF} and the spectrum of reference object W_k:

$$\omega_k = \sum_i \left| F^*_{RSFi} - W_{ki} \right| \tag{10}$$

After determining the similarity scores for all reference vectors, system is taking the decision to recognize the object. Classification decision is made by selecting object pattern with smallest similarity score. In the Table 2 there are values of the

Table 2 The values of similarity scores for selected objects obtained during the algorithm testing

Pattern	Object on the picture			
	Circle	Hand	Square	Triangle
Circle	**0.37258**	6.1515	0.68999	2.7068
Hand	4.6372	**0.90317**	3.0745	2.6486
Square	0.85206	5.338	**0.6428**	2.6556
Triangle	3.3157	3.3243	2.3643	**0.8632**

similarities shown, obtained for the chosen objects. Similarity scores that determined the recognition of the object has been highlighted in the table.

5.3 *Object Tracking Algorithm*

After successful target detection and recognition, the system can track the movement of the object of interest. Object tracking is an image processing procedure of finding chosen object on the following frame using knowledge about its position in previous frames. This means that the object tracking method differs from previously described image processing techniques by the fact that they analyze multiple frames in time. Among many tracking algorithms one can distinguish methods with different levels of complexity, different computational demands and varying execution time. During the development of the method we have focuses on the tracking algorithms, which enables real time operation and can be implemented in a low power digital signal processor or FPGA. Having all this in mind the Sum-of-the-Squared Differences algorithm was chosen [13].

Gradient based methods like Sum-of-Squared-Differences [14, 15] localize targets by analyzing differences between consequent frames. Finding target movement is performed by searching minimum of cost function in space and time. Cost function in this approach is a sum of squared differences. Sum of squared differences coefficient is a measure of difference between two fragments of images. Both fragments of images should have equal size. Assuming, that the two fragments are $(2h + 1)$ by $(2h + 1)$ in size and that they centers have the coordinates (x, y) and (u, v) respectively, the SSD coefficient can be calculated according to the following relation:

$$SSD = \sum_{\substack{i \in (-h, h), \\ j \in (-h, h)}} \left\{ [f_{k-1}(x+i, y+j) - f_k(u+i, v+j)]^2 \right\} \quad (11)$$

i, j—point coordinates with respect to the centers of compared fragments.

If we assume that the tracked object is present on the f_{n-1} frame and is centered around (x, y) coordinates, then finding this object on the consecutive frame means, that the point (u, v) is to be found, for which the SSD coefficient has minimal value.

Then the point (u, v) becomes the center of the tracked object on consecutive f_n frame. The search for the minimal value of SSD coefficient is performed in the neighborhood of the past location of tracked object. The general idea of finding an object using the Sum-of-the-Squared Differences method is presented in Fig. 24.

Fragment containing found object became a new model for further search in following frame. Replacing old model with the fragment of picture containing a newly found one is called the model update. In traditional version of algorithm the model update is made every frame.

This approach can lead to some undesired effects. In case where object tracking will fail for one frame, algorithm will forget the object. After that, tracking of this object will not be possible, because the proper model has become obsolete automatically. That is why SSD algorithm has low long time reliability. Noise in one frame only, can cause the whole algorithm to fail in further frames. That is why there is a need to develop some new special routines to make this algorithm immune to partial or full occlusions, noise and changes of objects appearance. To distinct reliable position estimation from noisy one the special SSDVar coefficient was developed:

$$SSDVar = \sum_{i \in \langle -h, h \rangle, j \in \langle -h, h \rangle} \left[(SSD[i,j] - \min(SSD[i,j]))^2 \right] \tag{12}$$

This coefficient can be used to evaluate the quality of objects localization. High SSDVar indicates that the minimum of SSD coefficient was clear. Low SSDVar indicates that the estimated localization is not so reliable. This coefficient was used to determinate the model update procedure. When the SSDVar parameter was higher than arbitrary threshold, model update procedure is made like in traditional version of SSD algorithm. When SSDVar is lower than the threshold, the model update procedure is skipped. This prevents the algorithm to forget the object model when the new estimation of object localization is unreliable.

Exemplary plots of SSD coefficient map for disrupted and non-disrupted object is shown in Fig. 25.

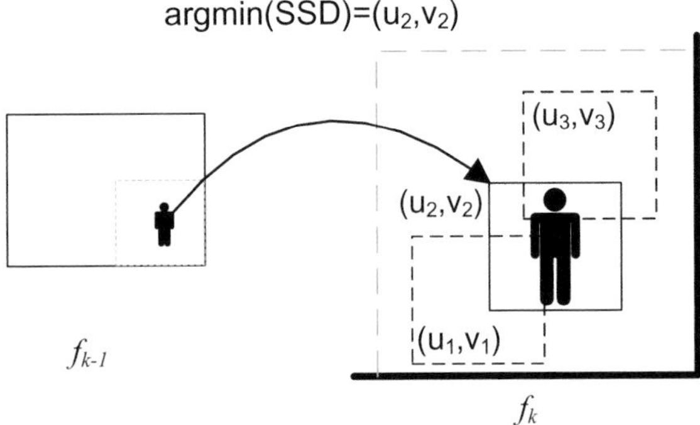

Fig. 24 Localizing the object by finding the area for which the SSD coefficient is the smallest

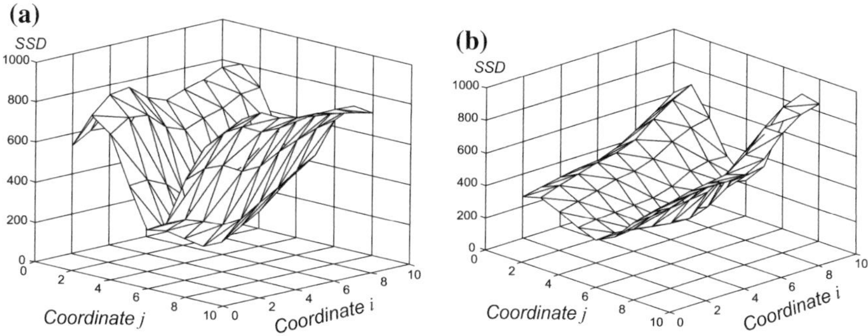

Fig. 25 Charts of SSD values in the neighborhood of the object when the object is undisturbed (**a**) and, when disturbed (**b**)

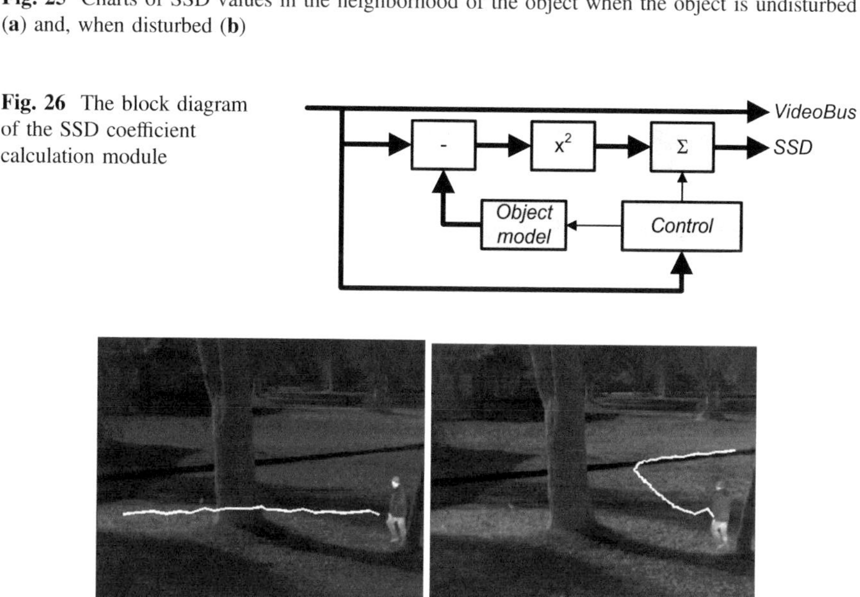

Fig. 26 The block diagram of the SSD coefficient calculation module

Fig. 27 Obtained trajectory of objects using developed tracking method in the infrared image

Object tracking algorithm according to the proposed method is quite computationally complex and requires relatively high computing power. Furthermore, it is desirable that tracking module introduced the shortest possible delay in the signal processing path. One way to speed up the operation of the system is to provide specialized hardware module performing the calculation for the proposed algorithm. That is why the data processing algorithms have been implemented and tested in the FPGA structure. This approach provides high performance thanks to multiple acceleration techniques like pipelining and parallelism. A simplified diagram of a module for calculating the SSD coefficient for the data stream is presented in Fig. 26.

The results of obtained trajectory of objects using developed tracking method in the infrared image is shown in Fig. 27.

6 Conclusion

The article presents the general structure of a thermal imaging camera and shows the image processing operations performed in the observation of infrared cameras that are used in security systems, detection and recognition. Provided examples of processing operations for thermal images and their implementations in the system addresses specifics of infrared image processing that differs from commonly used visual image processing. High processing power demands for image processing techniques used in real-time infrared cameras demands usage of hardware accelerated modules especially when infrared camera has to be used in security and control applications, where data latency is a critical issue.

Presented by the authors developed methods for detecting and tracking objects on the thermographic images has been a subject to experimental verification. Based on these results, it turned out that a detection method is characterized by a high probability of detection at a relatively low probability of false alarm. Results of operation were dependent on applied internal parameters so the adaptive functions were described like adaptive threshold of detection. Developed method of object tracking allows effective determination of the trajectory of moving objects in thermographic images. Method was particularly robust to low image resolution typical for infrared cameras and blurred edges of objects. When implemented in a digital system based on the structure of the programmable FPGA both methods are performed in real time. The combination of both methods in one system allows for effective support for decision-making.

References

1. Gonzalez, R. C., & Woods, R. E. (2002). *Digital image processing* (2nd ed.). Prentice-Hall.
2. Holst, G. C. (1998). *Testing and evaluating of infrared imaging systems*. SPIE Optical Engineering Press.
3. Sosnowski, T., Orżanowski, T., Kastek, M., & Chmielewski, K. (2007). Digital image processing system for thermal cameras. In *Advanced Infrared Technology and Applications AITA 9*, Leon, 8–12.10.2007.
4. Sosnowski, T., Bieszczad, G., Kastek, M., & Madura, H. (2010). Digital image processing in high resolution infrared camera with use of programmable logic device. *Proceedings of SPIE, 7838,* 78380U.
5. Bieszczad, G., Sosnowski, T., Madura, H., Kastek, M., & Bareła, J. (2010). Adaptable infrared image processing module implemented in FPGA. *Proceedings of SPIE, 7660,* 76603Z.
6. Bieszczad, G., Sosnowski, T., Madura, H., Kastek, M., & Bareła, J. (2011). Image processing module for high-speed thermal camera with cooled detector. *Proc. SPIE, 8012,* 80120L.
7. Orżanowski, T., Madura, H., Kastek, M., & Sosnowski, T. (2008). Nonuniformity correction algorithm for microbolometer infrared focal plane array. In M. Strojnik (Ed.), *Advanced infrared technology and applications 2007* (pp. 263–269). Leon: Mexico.

8. Krupiński, M., Bieszczad, G., Sosnowski, T., Madura, H., & Gogler, S. (2014). Non-uniformity correction in microbolometer array with temperature influence compensation. *Metrology and Measurement Systems, XXI*(4), 709–718.
9. Dulski, R., Powalisz, P., Kastek, M., & Trzaskawka, P. (2010). Enhancing image quality produced by IR cameras. *Proceedings of SPIE, 7834,* 783415.
10. Dulski, R., Madura, H., Piatkowski, T., & Sosnowski, T. (2007). Analysis of a thermal scene using computer simulations. *Infrared Physics and Technology, 49*(3), 257–260.
11. Accetta, J. S., & Shumaker, D. L. (1993). *The infrared and electro-optical systems handbook.* Bellingham, WA: Ann Arbor MI and SPIE Press.
12. Bieszczad, G. (2005). Metoda rozpoznawania wzorców w obrazie graficznym, VI Międzynarodowa Konferencja Elektroniki I Telekomunikacji Studentów i Młodych Pracowników Nauki, SECON 2005, 8–9.11.2005 Warszawa.
13. Bieszczad, G., & Sosnowski, T. (2008). Real-time mean-shift based tracker for thermal vision. In *Quantitative InfraRed and Thermography QIRT 2008 Conference.*
14. Hager, G., & Belhumeur, P. (1998). Efficient region tracking with parametric models of geometry and illumination. *IEEE Transactions on Pattern Analysis and Machine Intelligence, 20*(10), 1025–1039.
15. Venkatesh Babu, R., Patrick, P., & Patrick, B. (2007). Robust tracking with motion estimation and local Kernel-based color modeling. *Image and Vision Computing, 25*, 1205–1216.

Nystagmus Detection System

Robert Bieda, Krzysztof Jaskot and Jan Łazarski

1 Introduction

Eye tracking is the process of location either point of gaze (where one is looking) or pupil position relative to the head [1].

Currently there is a few eye tracking techniques. The three most common techniques are:

- Videooculography (VOG),
- Electrooculography (EOG),
- Infrared Reflection Devices.

Videooculography is a method of recording eye movement across video frames. This is effective and non-invasive technique popular in Medical Researches [2, 3].

Electrooculography technique is based on measuring skin potentials. By placing electrodes around the eye, it is possible to measure small differences in the skin potential that correspond, among others, to eye movement [4–6].

Infrared techniques generally works by quantifying the difference between the amount of infrared light reflected by the sclera between a sensor (phototransistor) pair. Infrared is used so that one can test subjects in darkness [7].

In this paper we focused on Gaze Tracking using Videooculography. Gaze reflects our attention and intention. This means that information about gaze direction can be very useful in human computer interaction such as device control by disabled

R. Bieda · K. Jaskot (✉) · J. Łazarski
Institute of Automatic Control, Silesian University of Technology,
Akademicka 16, 44-100 Gliwice, Poland
e-mail: krzysztof.jaskot@polsl.pl

R. Bieda
e-mail: robert.bieda@polsl.pl

J. Łazarski
e-mail: jan.lazarski@gmail.com

A. Nawrat et al. (eds.), *Advanced Technologies in Practical Applications for National Security*, Studies in Systems, Decision and Control 106,
https://doi.org/10.1007/978-3-319-64674-9_4

patients. It is also important in medical diagnostics and scientific researches. Nowadays there are also other application of this technology connected with studies about costumer preferences like market research or users interactions with websites.

Despite growing popularity of eye tracking, this technology is still very expensive. In this work we present one of the currently cheapest commercial gaze tracking system called myGaze developed by Visual Interaction. System contain four elements: Device, Programmable interface to provide access to the device (API), Software which use the API to interact with the Device and Eye Tracking Server which collects the data from the Device, and provide data via API [8].

myGaze like almost every gaze tracking system need personal calibration at the beginning. The procedure requires from one to nine markers on screen. Calibration is very time-consuming but it is necessary due to differences in eye size (about 10% individual difference in radius) and structure (even small differences in fovea position matters) between individuals [9].

In this paper we present our own gaze tracking application created in LabVIEW programming environment and based on myGaze Eye Tracking System. National Instruments LabVIEW is a graphical development environment used to design flexible and scalable measurement and control systems [10].

In this paper we describe all parts belonging the test stand. Then we focus on details regarding to myGaze Tools Palette creation process. Next we describe workflow of integration with myGaze Eye Tracking Server and details about our system main application. After that we test potential usefulness of created system in Human-Computer Interface and Medical Researches. And then we also describe algorithm used to calculate factors of optokinetic nystagmus. At the end of this paper we describe safety aspect of myGaze technology.

2 Test Stand

The main part of created system is Eye Tracking Device. This device work in binocular mode with sampling rate 30 Hz. This device is very precise, gaze position accuracy equals 0.5°. There is also no problems with glasses and lenses. Device is able to track gaze position if distance between patients head and device is bigger than 50 cm and smaller than 75 cm [8]. This system do not require special additional light source or any tools to stiffen the head. But because results from created system are compared with results from other VOG Eye Tracking System described in [11] both of them are used. Results from created system are also compared with results from EOG Eye Tracking System called "Biopack" described in [11, 12], test stand contains this EOG System.

Test stand (Fig. 1) contains myGaze Eye Tracking Device, infra red light source, simple headrest and camera which belongs to referenced Eye Tracking System, and EOG "Biopack" System.

Fig. 1 Test stand

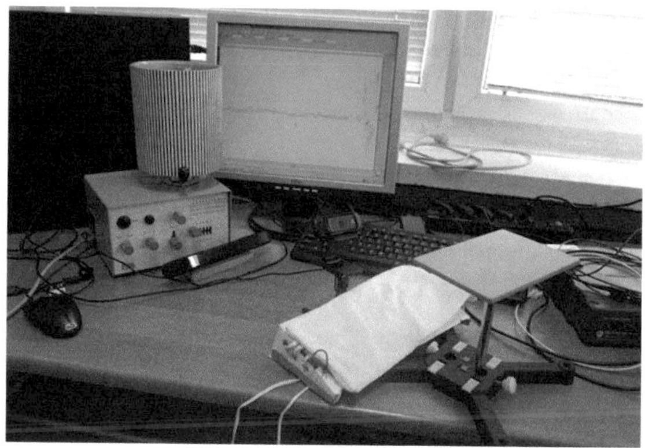

3 myGaze API LabVIEW Tools Palette

To use in our LabVIEW application programmable interface (API) from myGaze system we create new tools palette in LabVIEW by importing myGaze API Static Library and creating structures containing temporary data. New palette contain twelve data controls and forty eight functions.

Figure 2 shows the workflow of normal interaction with myGaze Eye Tracking Server. Figure contains four main parts: connection, calibration, processing and disconnection.

At first part we start the myGaze Eye Tracking Server, then we establishes a connection to Server and we defines the myGaze monitor attached geometry.

At second part we set the calibration and validation parameters, then we start calibration and validation procedure. During calibration and validation person have to track calibration point. Then if calibration accuracy is satisfying it is possible to start processing stage.

At third part we acquire, store and process data from myGaze Eye Tracking Device.

At fourth part we close the connection with myGaze Eye Tracking Server.

Calibration procedure works in three modes. First requires 0 markers and second 1 marker on the screen. This modes uses default parameters and can generate errors. Third requires five markers and this mode (Fig. 3) is the most precise mode.

Block Diagram and Front Panel of test application to interact with myGaze Server (Figs. 2, 4 and 5).

4 Using myGaze System for Gaze Tracking

To test system usefulness in Human-Computer Interaction we take a look at statistical accuracy factors for five marker positions. Factors are shown in Table 1. Screen

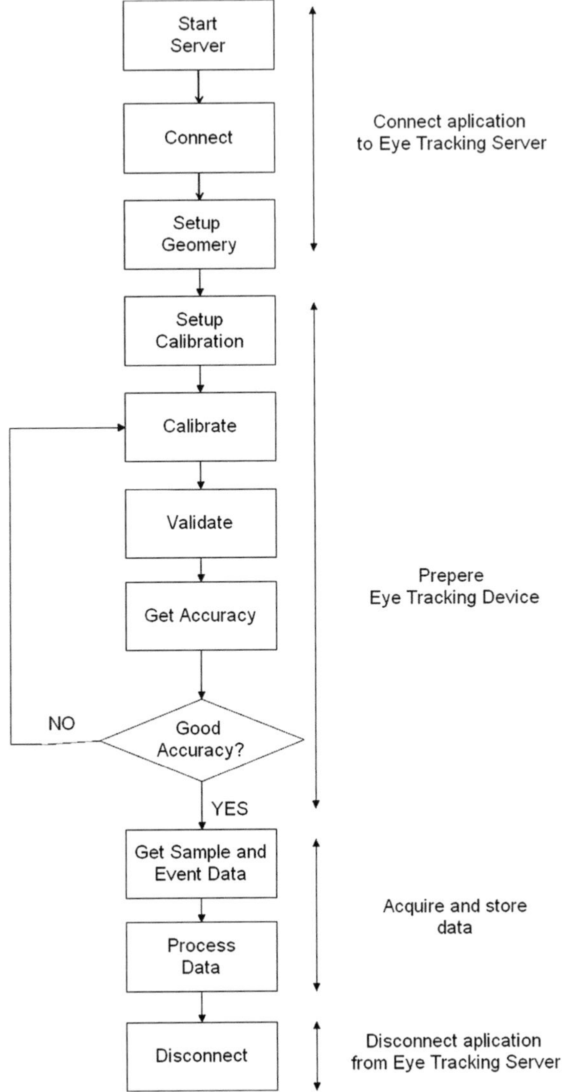

Fig. 2 Connection with myGaze server—workflow

resolution is 1360 × 768. Figure 6 show gaze positions on screen during accuracy test. We calculate following statistical factors:

- F_1—mean gaze position on screen, (X [px]; Y [px])

$$F_{1x} = \sum_{i=1}^{n} \frac{x_g(i)}{n}, \quad F_{1y} = \sum_{i=1}^{n} \frac{y_g(i)}{n} \tag{1}$$

Fig. 3 myGaze calibration markers

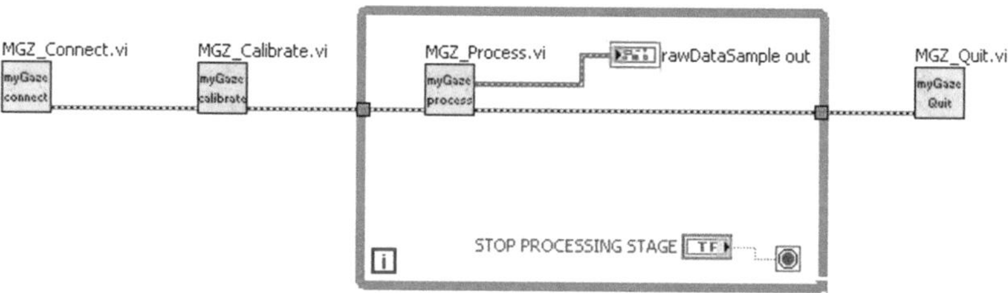

Fig. 4 Test application block diagram

- F_2—mean distance between marker position and successive gaze positions, [px] [mm]

$$F_2 = \sum_{i=1}^{n} \frac{\sqrt{((x_m - x_g(i))^2 + (y_m - y_g(i))^2)}}{n} \tag{2}$$

- F_3—standard deviation of distance between marker position and successive gaze positions, [px] [mm]

$$F_3 = \sqrt{(\frac{\sum_{i=1}^{n}(\sqrt{((x_m - x_g(i))^2 + (y_m - y_g(i))^2)} - F_2)}{n})} \tag{3}$$

- F_4—distance between marker position and mean gaze position, [px] [mm]

$$F_4 = \sqrt{((F_{1X} - x_M)^2 + (F_{1Y} - y_M)^2)} \tag{4}$$

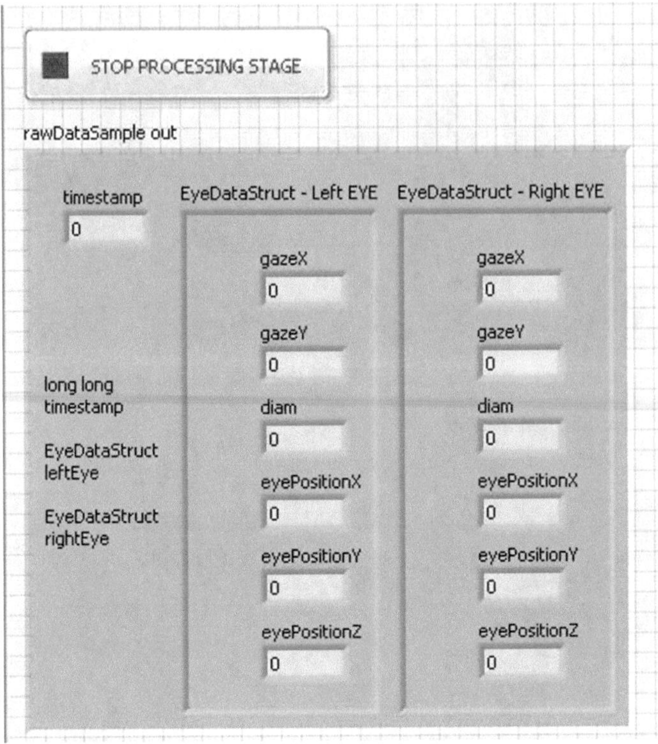

Fig. 5 Test application front panel

where,
x_m, y_m—marker coordinates
x_g, y_g—gaze coordinates
n—number of gaze points for single marker.

Unfortunately myGaze library do not provide functions to get eye rotation angles relative to the Ficks axis [13], so factors are represented in point of gaze screen pixel values. To convert pixels into millimeters we assume that single pixel have 0.26458 [mm] width.

Mean distance between marker position and gaze position is always lower than one centimeter. Also the standard deviation factor is small, it never exceed tree millimeters. This means that this system is enough precise to be used in some Human-Computer Interact Applications.

First experiment prove that analyzed system can be used to targeting areas on screen by gaze fixation. But will this system be useful if we want to record and analyze fixation point changing process? During next test patient was alternately looking at two markers. First one was on the left side of screen and the second was on right side.

Table 1 Gaze accuracy test results—statistical factors

Marker	Marker position (X [px]; Y [px])			F_1 (X [px]; Y [px])		
1	[680; 385]			[685; 380]		
2	[80; 90]			[95;102]		
3	[1265; 90]			[1257; 123]		
4	[1265; 680]			[1242; 708]		
5	[80; 680]			[88; 688]		
Marker	F_2[px]	F_3[px]	F_4[px]	F_2[mm]	F_3[mm]	F_4[mm]
1	3.580	1.813	0.748	0.947	0.479	0.198
2	26.661	4.613	19.792	7.054	1.220	5.236
3	17.378	5.725	15.071	4.597	1.514	3.987
4	11.796	8.703	10.282	3.121	2.302	2.720
5	15.052	5.664	13.741	3.982	1.498	3.635

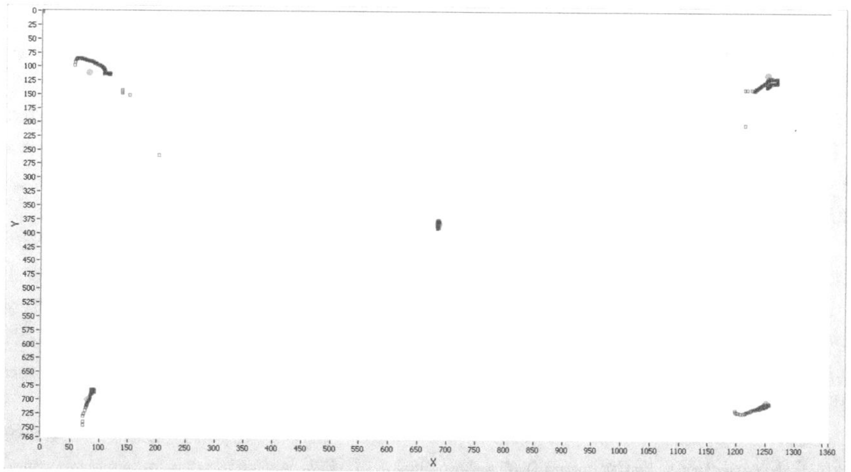

Fig. 6 Gaze accuracy test results—test board; *Green*—markers, *Red*—gaze points on screen

Results form system are compared with results from electrooculograph. First graph present horizontal coordinate of gaze on screen acquired by created system. Second graph shows horizontal coordinate of eye position acquired by electrooculograph.

Signals shapes are the almost identical which means that created system shows in approximation gaze trajectory. But first graph is smoothed, this graph do not contain information about small noises or little eye oscillations. This myGaze feature was very helpful during previous test but during present one, it causes loss of information (Fig. 7).

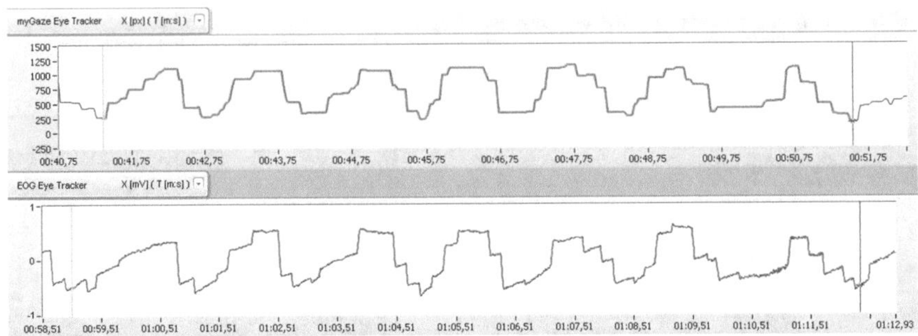

Fig. 7 Horizontal gaze coordinate graphs

5 Gaze Tracking in Medical Researches

Eye movement is very powerful diagnosis tool. It can help with laryngology, neurology and ophthalmology diseases. But information must be precise and free from noise. Number of samples per second is also important and depends on analyzed disease [14, 15].

Our system is operating at 30 Hz. So to test usefulness of myGaze in Medical Researches we analize eye movement during optokinetic stimulation (video sequence with moving vertical stripes on the monitor screen). This stimulation provoke optokinetic nystagmus characterized by fast eye oscillations [14, 15].

Generally rhythmic, oscillating motions of the eyes are called nystagmus. Nystagmus is not necessarily connected with health problems. For diagnostic purposes it can be induced. Optokinetic nystagmus, caloric nystagmus, and positional nystagmus fit into this category. In this paper we focus on optokinetic nystagmus (OKN). It is a normal eye reaction occurring when looking at objects passing across the field of vision, as in viewing from a moving vehicle [14].

Standard optokinetic nystagmus signal of gaze horizontal coordinate is shown on Fig. 8. Single beat contain fast and slow phase [15].

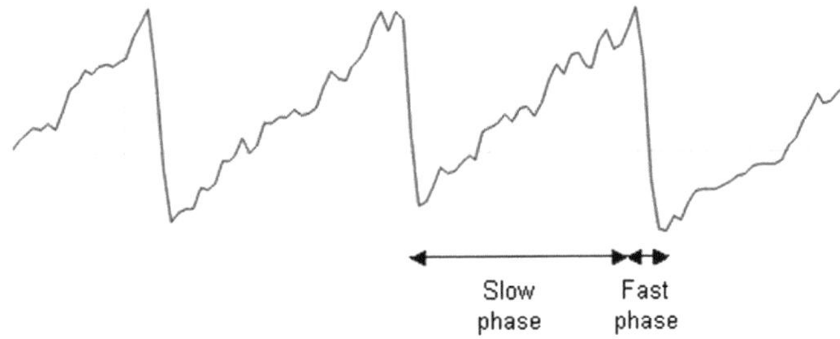

Fig. 8 Optokinetic nystagmus signal

6 Using myGaze System During Optokinetic Stimulation

During first experiment we acquire optokinetic signal with typical amplitude, but during second one we acquire signal with ten times smaller amplitude.

Results from first test (Fig. 9) are compared with results from videooculography described in [11]. First graph present horizontal coordinate of gaze on screen acquired by created myGaze system. Second graph shows horizontal coordinate of eye position acquired by videooculography.

Results from first test include information about the reason of eye movement. Signal is smoothed but in this case this fact do not prevent from calculating factors of this signal.

Results from second test are compared with results from "Biopack" electrooculograph described in [11, 12]. First graph present horizontal coordinate of gaze on screen acquired by created system. Second graph shows horizontal coordinate of eye position acquired by electrooculograph (Fig. 10).

On second graph it is possible to see very small eye oscillations, but first graph is too smoothed and signal information about nystagmus is impossible to interpret.

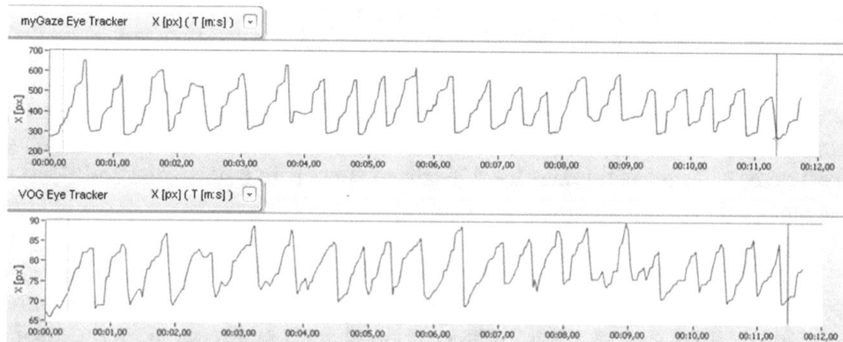

Fig. 9 Acquired signals with typical amplitude

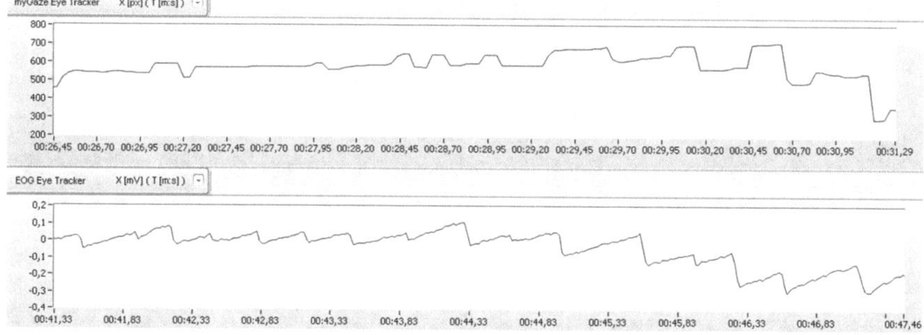

Fig. 10 Acquired signals with ten times smaller amplitude

Both experiments prove that created system can not be used in more advanced Medical Research tests due to myGaze binocular mode and smoothing feature. But this system can be useful for preliminary stage of Medical Tests.

7 Beats Separation Algorithm

For more accurate assessment we will calculate optokinetic signal factors. To do that we create algorithm which first separates signal into single beats, then calculate values of durations and amplitudes for fast and slow phases for each beat and finally calculate statistical values of OKN signal factors. For tests we use signal (Fig. 11) acquired by created system based on myGaze.

First we must devide signal (Fig. 11) which contain horizontal coordinates of gaze into individual separate beats. To do that we use the wavelet-based methods. All procedure is presented on Fig. 12.

Wavelet transforms are multiresolution representations of signals. Wavelets are families of functions $\Psi_{s,t}(x)$

$$\Psi_{s,t}(x) = \frac{1}{\sqrt{|s|}}\Psi\left(\frac{x-t}{s}\right) \tag{5}$$

generated from a single base wavelet $\Psi(x)$ by dilations and translations where s is the dilation (scale) parameter, and t is the translation parameter. Wavelets must have mean zero, and the useful ones have localized support in both spatial and Fourier domains [16].

Algorithm steps:

Step I: Noise reduction using wavelet transform. Used function "WA Denoise" applies the wavelet transform to noisy data, then it obtains detail and approximation coefficients. Next applies soft threshold to coefficients and reconstruct the coefficients after thresholding and transforms them back into the original domain [17]. Denoise effect is presented on Fig. 13.

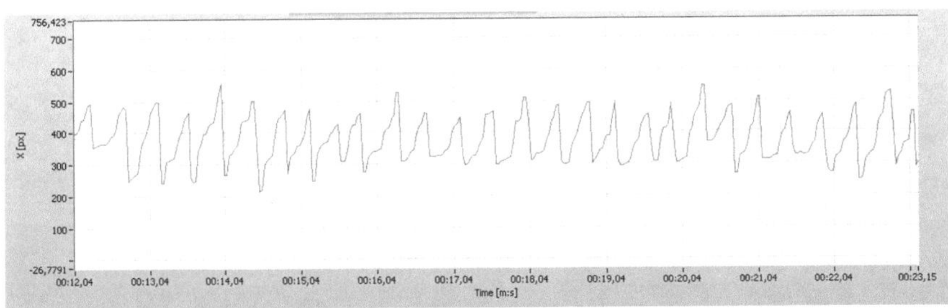

Fig. 11 Test signal

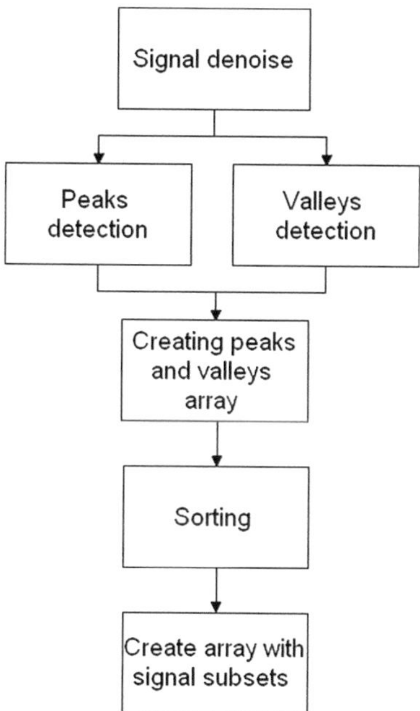

Fig. 12 Beats separation algorithm diagram

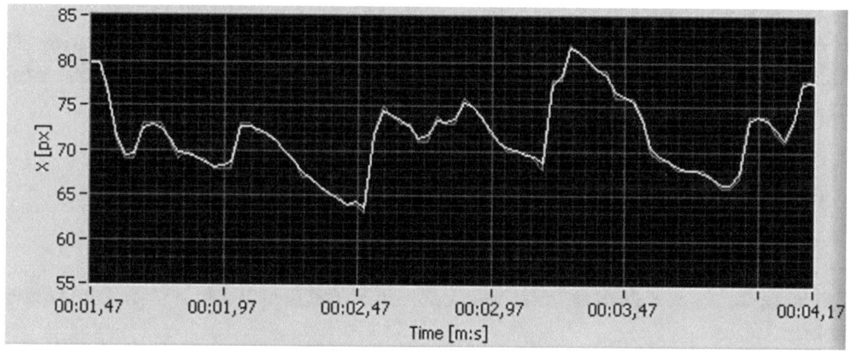

Fig. 13 Part of test signal (Fig. 11); *Red*—before noise reduction, *White*—after noise reduction

Step II: Peaks and valleys detection using wavelet transform. Used function "WA Multiscale Peak Detection" executes the following steps to detect peaks [17] (Fig. 14).

1. Calculates the undecimated wavelet transform (UWT) level based on the width and the UWT level based on the threshold frequency, separately. This VI chooses the greater UWT level from the two results to control the decomposition level.
2. Performs UWT on the input signal by using the biorthogonal wavelet and the greater UWT level calculated in step 1.

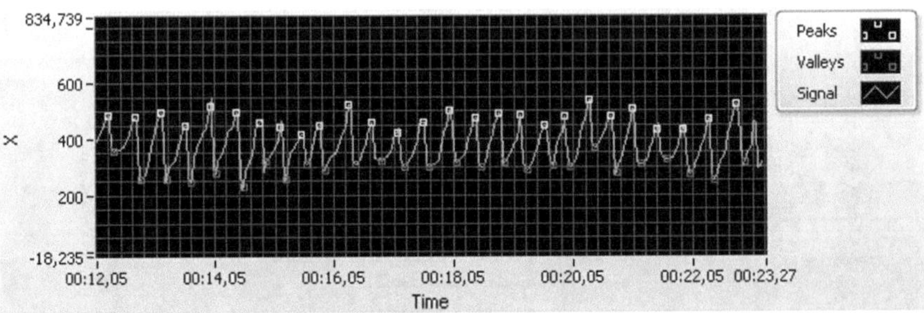

Fig. 14 Detected peaks and valleys on test signal

3. Searches for zero-crossing points in the detail coefficients at all levels.
4. Selects the zero-crossing points at the largest scale as the coarse estimation of the real peaks.
5. Searches the finer scale for the corresponding nearest zero-crossing point for each detected point.
6. Repeats step 5 until this VI reaches the finest scale, which is the first level.

Step III: Creating common array for peaks and valleys by adding peaks and valleys data into new array.

Step IV: Incrementally sort data in array in order to time of occurrence using bubble sort algorithm.

Step V: Creating new array with signal subsets. In this step into new array we added successive signal parts located between sequential points from sorted array created in step III.

8 Fundamental OKN Nystagmus Signal Factors

Fundamental factors of gaze horizontal coordinate OKN nystagmus signal are shown on Fig. 15.

Using array with signal subsets we calculate duration and amplitude for every fast and slow phase.

We get slow phase duration by subtracting successive peak time of occurrence from previous volley time of occurrence. Fast phase duration we get by subtracting successive volley time of occurrence from previous peak time of occurrence. Figure 16 illustrate successive duration values for test signal.

Similarly slow phase amplitude we get by subtracting successive peak position from previous volley position and counting absolute value. Fast phase amplitude we get by subtracting successive volley position from previous peak position and counting absolute value. Figure 17 illustrate successive amplitude values for test signal.

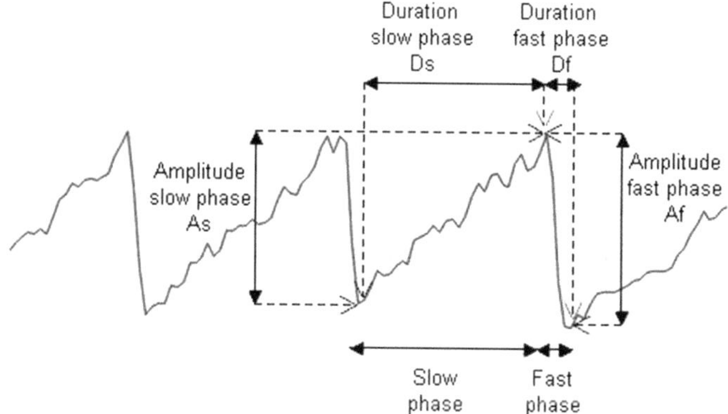

Fig. 15 Fundamental factors of OKN nystagmus signal

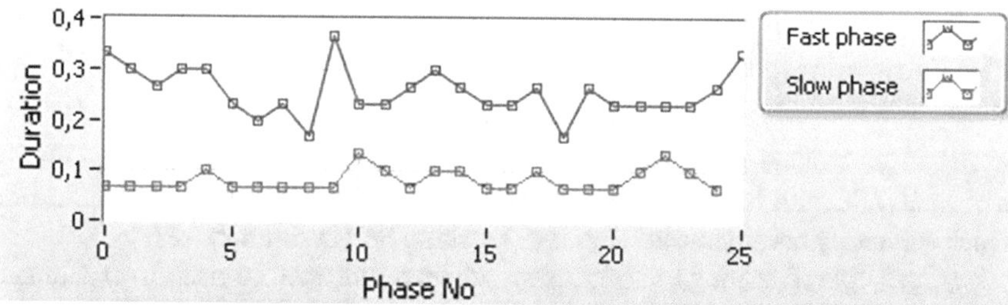

Fig. 16 Test signal duration values

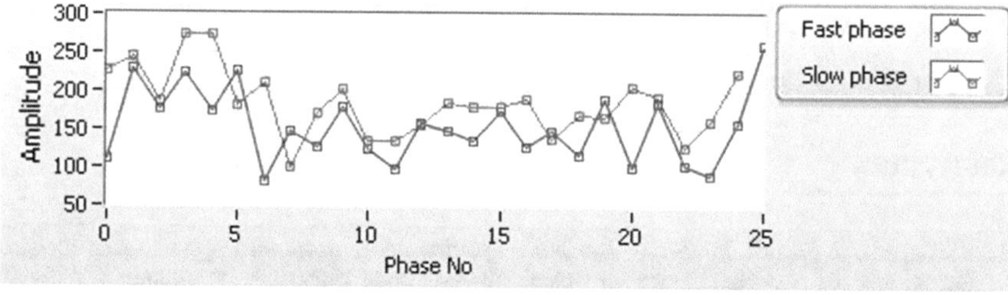

Fig. 17 Test signal amplitude values

For test signal mean slow phase duration equals 0.248 [s], and mean fast phase duration equals 0.080 [s]. This means that fast phase is approximately three times faster than slow phase. Mean slow phase amplitude equals 147.685 [px], and mean fast phase amplitude equals 182.656 [px]. Difference between them is less than 15% of their sum. Standard deviation of slow phase duration (0.047 [s]) and fast phase duration (0.021 [s]) are less than 35% of mean values. Also standard deviation of slow phase amplitude (47.122 [px]) and fast phase amplitude (43.135 [px]) are less than 35% of mean values. This informations may mean that myGaze system record a valid OKN nystagmus signal.

9 Safety Aspect

According to the manufacturers specifications [8] myGaze Eye Tracking Device meets the standard IEC 60601 (International electrotechnical commission) [18]. This standard refers to safety requirements of all electrical, electronic and related technologies. This norm was applied to European law in year 2012. This standard contains such norms as norm IEC30301-1-2 for electromagnetic compatibility. Requirements refers to safety rules and electrical parameters adjusting. This norm nowadays is very often required if device is designed for medical purposes. myGaze Device also meets the standard EN 55011 (European Norm) [19]. This norm refers to industrial and scientific devices used in medicine. This means that myGaze Eye Tracking Device complies safety standards required by European law and should not be harmful.

10 Conclusion

myGaze Eye Tracking System is safe, stable and easy to calibrate. myGaze library is compatybile with LabVIEW environment and it is possible to create application which can easily communicate with Eye Tracking Server through API.

myGaze Eye Tracker is very precise and for sure can be usefull in Human-Computer Interface or to obserwate gaze trajectories during slow eye movement.

Because of myGaze System smoothing feature and low working frequency (30 Hz) this device can not be used in advanced and precise researches. But in initial stage of Medical Test this system could be usefull.

References

1. Cieśla, M., & Kozioł, P. (2012). *Eye pupil location using webcam*. Department of Physics, Astronomy and Applied Computer Science, Jagiellonian University, Reymonta 4, Krakow, Poland.
2. Bieda, R., Jaskot, K., Jaworski, M., Jaworski, P., & Wyględowska-Promieńska, D. (2015). System INTEGRA. *Przegląd Elektrotechniczny, 91*(9), 235–241.
3. Gans, R. E. (2001). Video-oculography: A new diagnostic technology for vestibular patients. *The Hearing Journal* (Lippincott Williams & Wilkins, Inc.), *54*(5).
4. Binias, B., Palus, H., & Jaskot, K. (2015). Real-time detection and filtering of eye movement and blink related artifacts in EEG. In *20th International Conference on Methods and Models in Automation and Robotics (MMAR)* (pp. 903–908), Międzyzdroje, Poland.
5. Binias, B., Palus, H., & Jaskot, K. (2016). Real-time detection and filtering of eye blink related artifacts for brain-computer interface applications. In *4th International Conference on Man-Machine Interactions, Advances in Intelligent Systems and Computing* (Vol. 391, pp. 281–290). Springer.
6. Morimoto, C. H., & Mimica, M. R. M. (2005). Eye gaze tracking techniques for interactive applications. *Computer Vision and Image Understanding, 98*, 4–24.

7. Hain, T. Eye movement recording devices. Retrieved February 20, 2015, from http://www.dizziness-and-balance.com/practice/eyemove.html.
8. myGaze System Manufacturer Website (Visual Interaction GmbH). Retrieved February 20, 2015, from http://mygaze.com.
9. Ohno, T., Mukawa, N., & Yoshikawa, A. (2002). *FreeGaze: A gaze tracking system for everyday gaze interaction*. NTT Communication Science Laboratories, NTT Corporation.
10. National Instruments Corporation Website. Retrieved February 20, 2015, from http://poland.ni.com.
11. Łazarski, J. (2014). *Support system for the diagnosis of eye diseases*. Master thesis, Silesian University of Technology, Gliwice, Poland.
12. Biopack Systems, Inc. http://www.biopac.com/.
13. Zatsiorsky, V. M. (1998). *Kinematics of human motion*. Human Kinetics.
14. Dell'Osso, L. F. & Daroff, R. B. (2006). *Eye movement characteristics and recording techniques*. Duane's Ophthalmology, Lippincott Williams & Wilkins.
15. Swart, W. (2011). *Nystagmus and eye reflex sensor*. Department of Mechanical and Mechatronic Engineering, University of Stellenbosch.
16. Xu, Y., Weaver, J. B., Healy, D. M. Jr., & Lu, J. (1994). Wavelet transform domain filters: A spatially selective noise filtration technique. *IEEE Transactions On Image Processing, 3*(6).
17. LabVIEW 2010 Advanced Signal Processing Toolkit Documentation. http://zone.ni.com/reference/en-XX/help/371419D-01/lvwavelettk/waveletapplicationpal/.
18. International Elektrotechnical Commission Website. http://www.iec.ch.
19. European Standard Website. http://rfemcdevelopment.eu/index.php/en/emc-emi-standards/en-55011-2009.

Weighted Pattern Vector for Object Tracking with the Use of Thermal Images

Zygmunt Kuś, Joanna Radziszewska and Aleksander Nawrat

1 Introduction

The images obtained from an camera equipped in infrared sensor can display invisible thermal energy emitted from objects. It may be very useful in the low visibility conditions as well as in the cases when we need to get information about thermal features of the objects. Progress in the production of thermal cameras causes that thermal images became more and more common images in adverse visibility conditions. There are situation in which visible image provides the information which allows to distinguish two objects with the same thermal images. In other cases the visible images are the same for different objects whereas we can see some differences in the thermal images. Complementary information of the objects may be obtained when we use both thermal and visible images. Thermal imaging technology has been widely used in many applications. Because of the thermal features of the human being we may find a lot of applications include detection, tracking and identification of people [1–10]. Thermal characteristic may be helpful in the recognition of or tracking such objects as vehicles on the road [11, 12] or animals which can appear on the road [13].

Non-contact malfunction identification and a diagnosis of equipment [14, 15] or for inspecting the production process [16] may be provided also with using thermal images. The images fusion task appears when one has to solve problems which one

Z. Kuś (✉) · A. Nawrat
Silesian University of Technology, Institute of Automatic Control,
Akademicka 16 St., Gliwice, Poland
e-mail: zkus@interia.pl

J. Radziszewska
VR Technology, Gliwice, Poland
e-mail: j.radziszewska@vrtechnology.pl

A. Nawrat
e-mail: aleksander.nawrat@polsl.pl

A. Nawrat et al. (eds.), *Advanced Technologies in Practical Applications for National Security*, Studies in Systems, Decision and Control 106,
https://doi.org/10.1007/978-3-319-64674-9_5

encounters when we use image processing systems in such applications as: image recognition [17], an unmanned aerial vehicle used for image acquisition [18], object tracking [19] or images fusion [20].

In the following paper, we will consider the problem of object tracking and pattern recognition in the conditions of not distorted visibility of the object [21, 22]. We assume that both visual and thermal images are good quality images.

2 The Method of Determining the Values of Weights for the Visual and Thermal Parts of the Features Vector

This paper will consider the cases in which the visual image does not have to be destroyed or noised by fog or smoke. Nevertheless, some important features of the tracked object are not visible in the visual image. One of the examples of such a situations is a tracked car with either turned on or turned off engine. Another example may be the recognition of an element with some specific features [23, 24]. This element is located among other elements which have the same visual image; however, the thermal image allows us to recognise these specific features. The abovementioned problems result in the necessity of using thermal images in the objects recognition process.

The proposed method consists of three stages. These stages result from the dynamics of the object tracking process. The first stage is based on the dynamical features vector and it assumes that we provide object tracking with the use of visual and thermal images. The second stage assumes that there is an initial state in which we know the location of the tracked object in both: visual and thermal images. In such a case we can precisely define the region of the image corresponding to the object. It enables us to calculate the pattern vector during object tracking and correct the connection between the object image and its pattern vector [25–27].

It may be obtained in a manual mode when the system operator decides about the moment of the pattern vector change or in an automatic mode when the automatic algorithm causes that a pattern vector follows the changes of the object image. The last stage conducts the recognition of an object on the basis of the pattern vector prepared in the previous stages.

The changes of the pattern vector are realised as the changes of weights with which the visual and thermal pattern vectors are used to calculate the distance.

Figure 1 presents the proposed algorithm.

Section 3 presents the examples of object tracking in the case when we take into consideration the thermal features of the tracked car.

3 Examples

The general idea consists in this that we possess the certain additional knowledge about the object and we try to represent this knowledge in the form of weights with which the visual and thermal pattern vectors are used to calculate the distance.

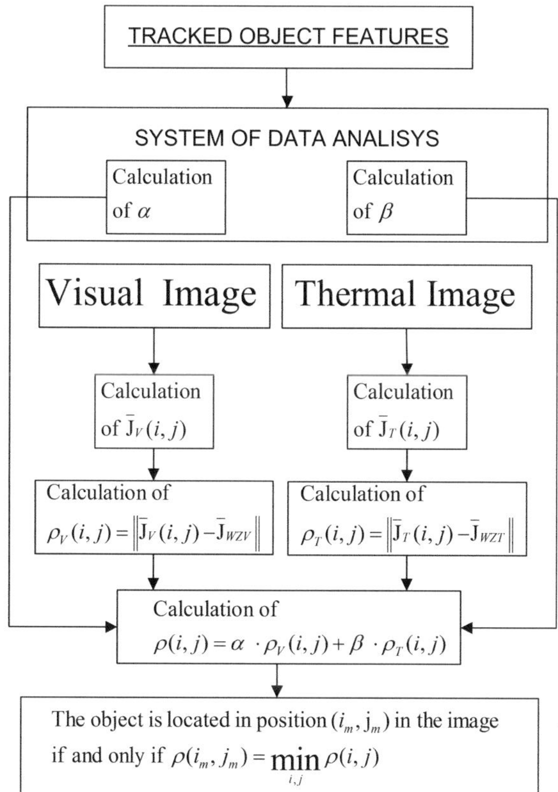

Fig. 1 The algorithm of pattern vector modification and object recognition

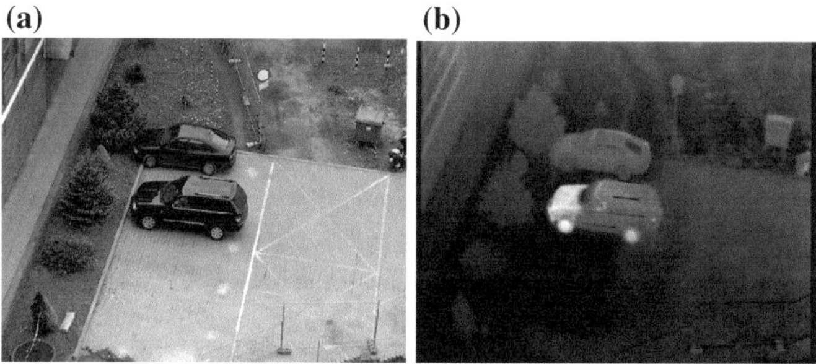

Fig. 2 The visual image (**a**), thermal image (**b**) for the experiment 1

Experiment 1

The first experiment illustrates the situations when the visible images of the different objects are very similarly whereas the thermal images contain the information which allows to distinguish these objects. The Fig. 2 presents visual and thermal images of the terrain with two cars. An off-road vehicle was the tracked object.

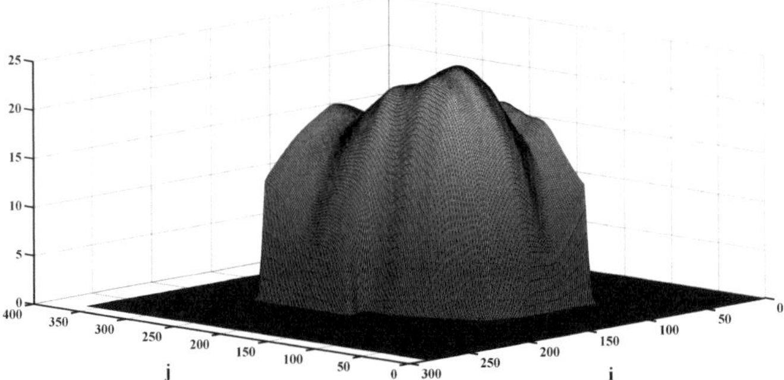

Fig. 3 The shape of the reversed distance between the object pattern vector and feature vector calculated for the experiment 1. Distance ρ calculated for $\alpha = 0$ and $\beta = 1$

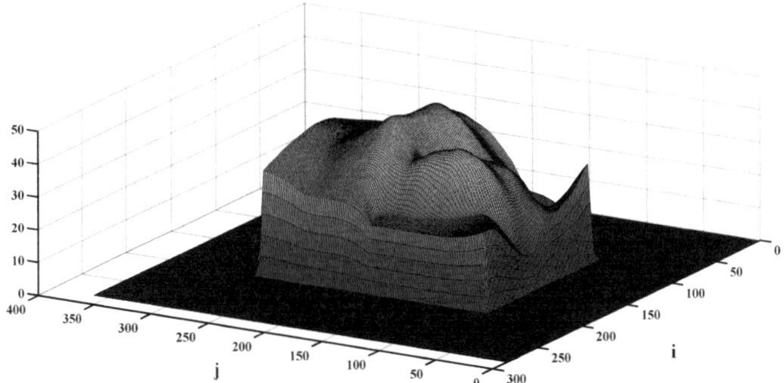

Fig. 4 The shape of the reversed distance between the object pattern vector and feature vector calculated for the experiment 1. Distance ρ calculated for $\alpha = 1$ and $\beta = 1$

Figure 3 presents the shape of the reversed distance between the object pattern vector and feature vector calculated for the experiment 1. Distance ρ in this case was calculated for $\alpha = 0$ and $\beta = 1$. The maximum of the shape corresponds to the place where the tracked object is located. We can state that there is unequivocal maximum which allows to locate the object what was presented in Fig. 5.

Figure 4 presents the shape of the reversed distance between the object pattern vector and feature vector calculated for the experiment 1. Distance ρ in this case was calculated for $\alpha = 1$ and $\beta = 1$. The maximum of the shape corresponds to the place where the tracked object is located. We can see that the visual image in this case causes that few similar maxima appear in the shape of distance function. It results in some problems with object recognition what may be observed in Fig. 6.

The Fig. 5 presents results of the objects recognition on the basis of the thermal image of terrain. It corresponds to the distance ρ calculated for $\alpha = 0$ and $\beta = 1$. The green colour denotes isoline corresponding to the minimum distance to the pattern of the object.

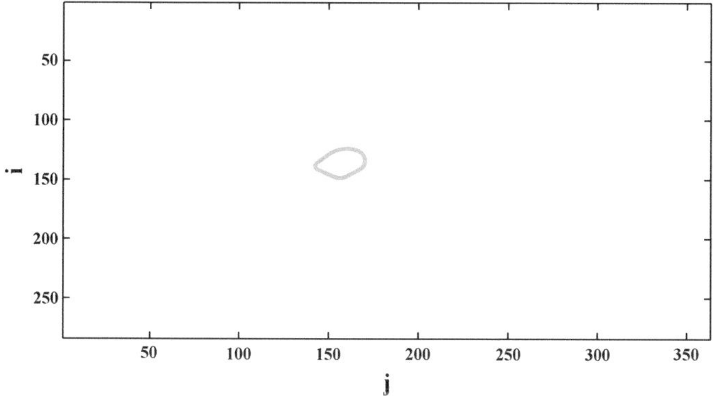

Fig. 5 Object location calculated in experiment 1 for $\alpha = 0$ and $\beta = 1$

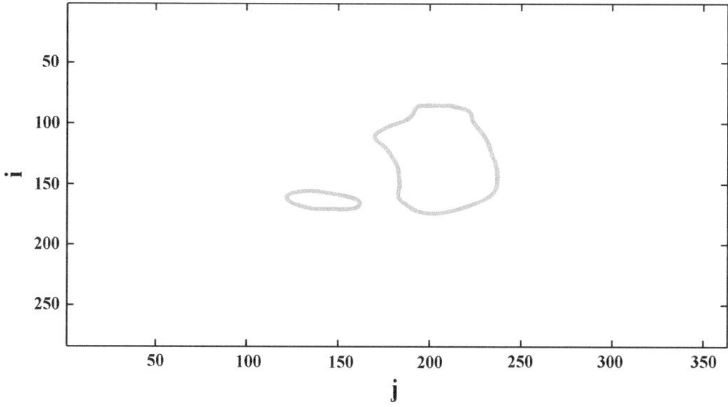

Fig. 6 Object location calculated in experiment 1 for $\alpha = 1$ and $\beta = 1$

The Fig. 6 presents results of the objects recognition on the basis of the visual and thermal images of terrain. It corresponds to the distance ρ calculated for $\alpha = 1$ and $\beta = 1$. The green colour denotes isoline corresponding to the minimum distance to the pattern of the object.

Finally, we can state that sometimes we have better information when we have less information.

Experiment 2

The second experiment illustrates the situations when both visible and thermal images of the different objects are similarly. In this case it is not easy which image gives us better information. However shapes of distance function ρ presented in Figs. 8, 9 and 10 shows that thermal image gives more unequivocal information about maximum localisation.

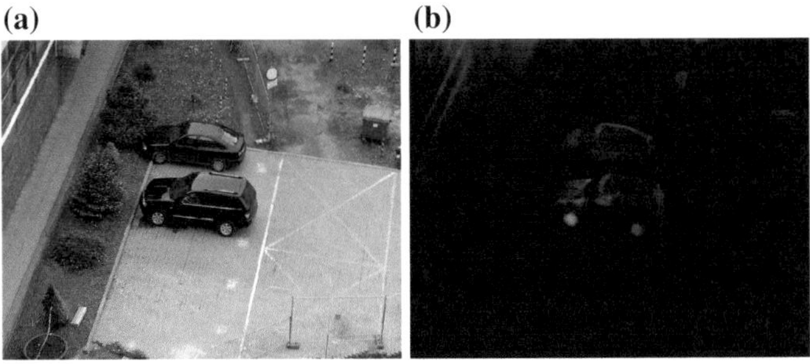

Fig. 7 The visual image (**a**), thermal image (**b**) for the experiment 2

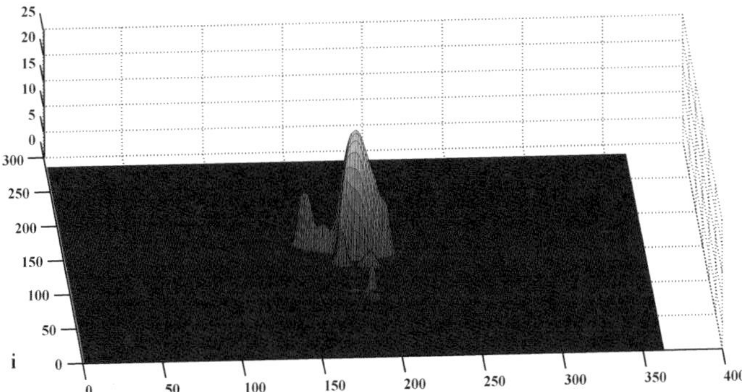

Fig. 8 The shape of the reversed distance between the object pattern vector and feature vector calculated for the experiment 2. Distance ρ calculated for $\alpha = 0$ and $\beta = 1$

The Fig. 7 presents visual and thermal images of the terrain with two cars. An off-road vehicle was the tracked object.

Figure 8 presents the shape of the reversed distance between the object pattern vector and feature vector calculated for the experiment 2. Distance ρ in this case was calculated for $\alpha = 0$ and $\beta = 1$. The maximum of the shape corresponds to the place where the tracked object is located.

Figure 9 presents the shape of the reversed distance between the object pattern vector and feature vector calculated for the experiment 2. Distance ρ in this case was calculated for $\alpha = 1$ and $\beta = 0$. The maximum of the shape corresponds to the place where the tracked object is located.

Figure 10 presents the shape of the reversed distance between the object pattern vector and feature vector calculated for the experiment 2. Distance ρ in this case was calculated for $\alpha = 1$ and $\beta = 1$. The maximum of the shape corresponds to the place where the tracked object is located.

The Fig. 11 presents results of the objects recognition on the basis of the thermal image of terrain. It corresponds to the distance ρ calculated for $\alpha = 0$ and $\beta = 1$. The green colour denotes isoline corresponding to the minimum distance to the pattern of the object.

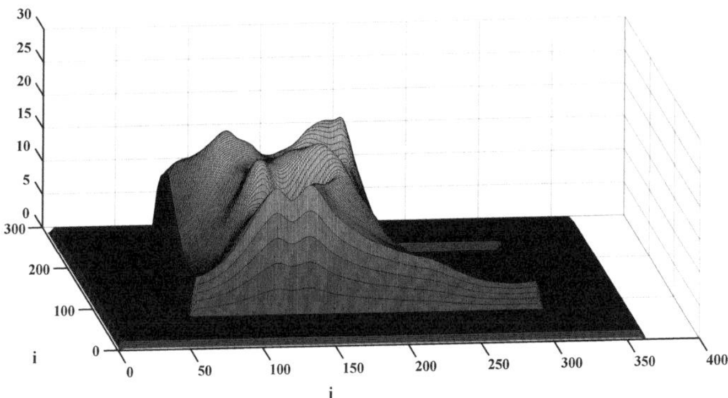

Fig. 9 The shape of the reversed distance between the object pattern vector and feature vector calculated for the experiment 2. Distance ρ calculated for $\alpha = 1$ and $\beta = 0$

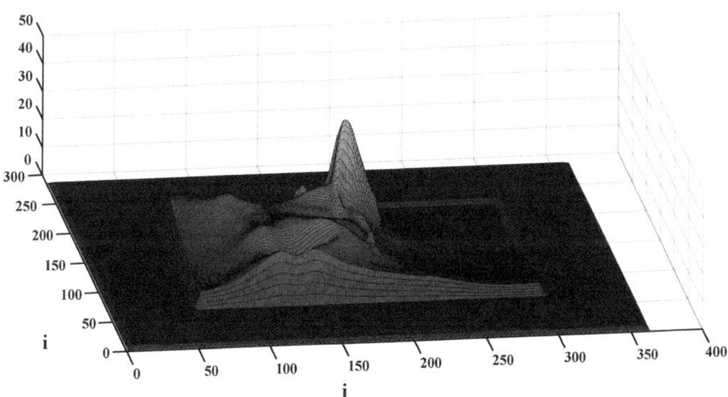

Fig. 10 The shape of the reversed distance between the object pattern vector and feature vector calculated for the experiment 2. Distance ρ calculated for $\alpha = 1$ and $\beta = 1$

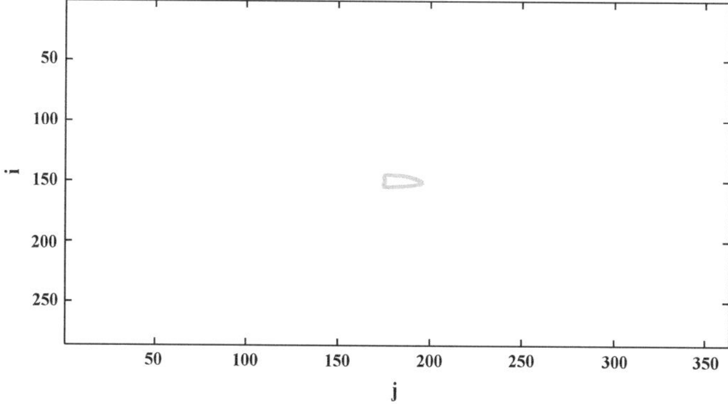

Fig. 11 Object location calculated in experiment 2 for $\alpha = 0$ and $\beta = 1$

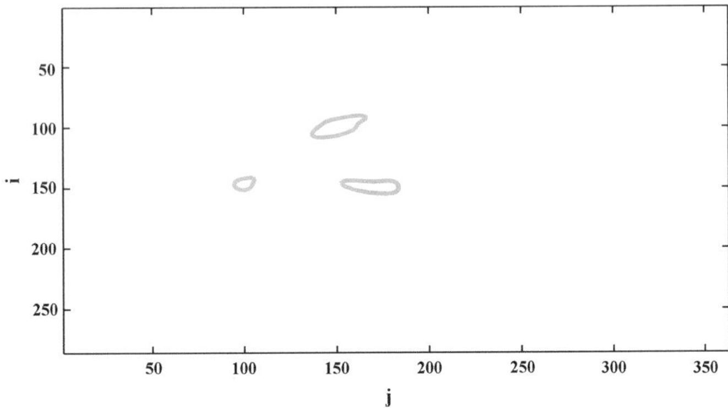

Fig. 12 Object location calculated in experiment 2 for $\alpha = 1$ and $\beta = 0$

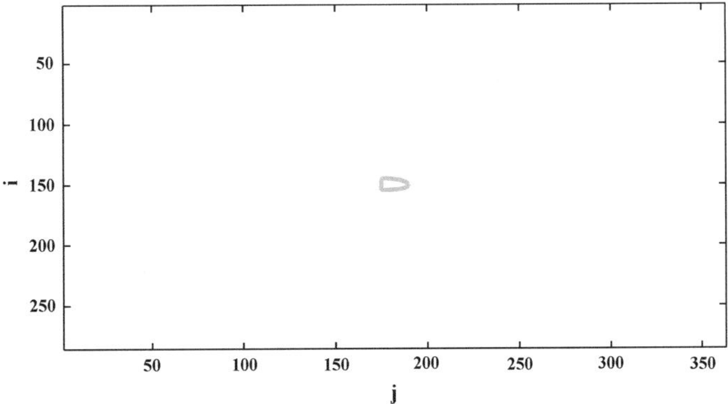

Fig. 13 Object location calculated in experiment 2 for $\alpha = 1$ and $\beta = 1$

The Fig. 12 presents results of the objects recognition on the basis of the visual image of terrain. It corresponds to the distance ρ calculated for $\alpha = 1$ and $\beta = 0$. The green colour denotes isoline corresponding to the minimum distance to the pattern of the object.

The Fig. 13 presents results of the objects recognition on the basis of the visual and thermal images of terrain. It corresponds to the distance ρ calculated for $\alpha = 1$ and $\beta = 1$. The green colour denotes isoline corresponding to the minimum distance to the pattern of the object.

According to figures from experiment 1 and 2 we may state that correct stetting values of coefficients α and β allows to obtain the shape of distance ρ with more distinguished maximum. It results in the greater accuracy of finding the maximum.

4 Conclusions

The authors presented in the paper the method of object tracking in a case when we have preliminary knowledge of an object. Object tracking on the basis of visual and thermal images allows to use this knowledge to choose what kind of information we want to use. We assumed that object tracking may be conducted with the pattern vector modification basing on this additional knowledge. The authors assumed that we acquire the visual and thermal images in each moment of object tracking. The pattern vector and current feature vector for an image of a given type were used to compute the distance between the object pattern vector and feature vector calculated for a given location of the aperture. It was calculated for both: the visual and thermal images. The pattern vectors of the tracked object based on the visual and thermal images were computed separately. This fact allowed to steer of influence of these vectors on recognition process by setting proper values of some coefficients. These coefficients decides which part of the pattern vector (visual, thermal or both) was finally used to find object location. Presented examples illustrated the cases when the object may have the same visual images and different thermal images what may be helpful in pattern recognition task.

References

1. Kang, B.-D., Jeon, K.-H., Kyoung, D., Kim, S.-H., Hwang, J.-H. (2012). Multiple human body tracking based on normalized cross-correlation of average histogram using the fusion of color and thermal image sensor. In *RO-MAN 2012*. IEEE. doi:10.1109/ROMAN.2012.6343785.
2. Padole, C. N. ; Alexandre, L. A. (2010) Wigner distribution based motion tracking of human beings using thermal imaging. In *2010 IEEE Computer Society Conference on Computer Vision and Pattern Recognition Workshops (CVPRW)*. doi:10.1109/CVPRW.2010.5543226.
3. Ciric, I., Cojbasic, Z., Nikolic, V., Antic, D. (2013). Computationally intelligent system for thermal vision people detection and tracking in robotic applications. In *2013 11th International Conference on Telecommunication in Modern Satellite, Cable and Broadcasting Services (TELSIKS)* (Vol. 02). doi:10.1109/TELSKS.2013.6704447.
4. Stolkin, R., Rees, D., Talha, M., & Florescu, I. (2012). Bayesian fusion of thermal and visible spectra camera data for mean shift tracking with rapid background adaptation. In *Sensors*. IEEE. doi:10.1109/ICSENS.2012.6411350.
5. Treptow, A., Cielniak, G., & Duckett, T. (2005) Active people recognition using thermal and grey images on a mobile security robot. In *2005 IEEE/RSJ International Conference on Intelligent Robots and Systems. (IROS 2005)*. doi:10.1109/IROS.2005.1545530.
6. Choi, E.-J., & Park, D.-J. (2010). Human detection using image fusion of thermal and visible image with new joint bilateral filter. In *2010 5th International Conference on Computer Sciences and Convergence Information Technology (ICCIT)*. doi:10.1109/ICCIT.2010.5711182.
7. Li, Z., Wu, Q., Zhang, J., & Geers, G. (2011). SKRWM based descriptor for pedestrian detection in thermal images. In *2011 IEEE 13th International Workshop on Multimedia Signal Processing (MMSP)*. doi:10.1109/MMSP.2011.6093800.
8. Wang, W., Zhang, J., & Shen, C. (2010). Improved human detection and classification in thermal images. In *2010 17th IEEE International Conference on Image Processing (ICIP)*. doi:10.1109/ICIP.2010.5649946.

9. Szczodrak, M., & Szwoch, G. (2013). An approach to the detection of bank robbery acts employing thermal image analysis. In *Signal Processing: Algorithms, Architectures, Arrangements, and Applications (SPA)*.
10. Zhang, G., Shen, T., Liu, W., & Yuan, H. (2010). The application of dual thresholds binarization algorithm for pedestrian segmentation in thermal imaging. In *2010 International Conference On Computer and Communication Technologies in Agriculture Engineering (CCTAE)* (Vol. 1). doi:10.1109/CCTAE.2010.5544733.
11. Woeber, W., Szuegyi, D., Kubinger, W., Mehnen, L. (2013). A principal component analysis based object detection for thermal infra-red images. In *2013 55th International Symposium ELMAR*.
12. Sangnoree, A., & Chamnongthai, K. (2009). Method, Robust, & for analyzing the various speeds of multitudinous vehicles in nighttime traffic based on thermal images. In *Fourth International Conference on Computer Sciences and Convergence Information Technology, ICCIT '09*. doi:10.1109/ICCIT.2009.186.
13. Zhou, D., Wang, J. (2011). Identification of deer in thermal images to avoid deer-vehicle crashes. In *2011 International Conference on Electronics and Optoelectronics (ICEOE)* (Vol. 3). doi:10.1109/ICEOE.2011.6013376.
14. Huang, S.-Y., Mao, C.-W., Cheng, K.-S. A VQ-based approach to thermal image analysis for printed circuit boards diagnosis. *IEEE Transactions on Instrumentation and Measurement, 54*(6). doi:10.1109/TIM.2005.858546.
15. Zhang, W. (2010). Remote malfunction diagnosis system based on infrared thermal imaging and RIA. In *2010 Symposium on Photonics and Optoelectronic (SOPO)*. doi:10.1109/SOPO.2010.5504340.
16. Ginesu, G., Giusto, D. D., Margner, V., Meinlschmidt, P. Detection of foreign bodies in food by thermal image processing. *IEEE Transactions on Industrial Electronics, 51*(2). doi:10.1109/TIE.2004.825286.
17. Jedrasiak, K., & Nawrat, A. (2009). Image recognition technique for unmanned aerial vehicles, computer vision and graphics. Lecture notes in computer science (Vol. 5337, pp. 391–399). Springer.
18. Iwaneczko, P., Jedrasiak, K., Daniec, K., & Nawrat, A. (2013) A prototype of unmanned aerial vehicle for image acquisition. *Computer vision and graphics*. Lecture notes in computer science (Vol. 7594, pp. 87–94). Springer.
19. Nawrat, A., Jedrasiak, K. (2009). SETh system spatio-temporal object tracking using combined color and motion features. In *9th WSEAS International Conference on Robotics, Control and Manufacturing Technology*, Hangzhou, China.
20. Jedrasiak, K., Nawrat, A., Daniec, K., Koteras, R., Mikulski, M., & Grzejszczak, T. (2013). A prototype device for concealed weapon detection using IR and CMOS cameras fast image fusion. In *Computer vision and graphics*. Lecture notes in computer science (Vol. 7594, pp. 423–432). Springer.
21. Ryt, A., Sobel, D., Kwiatkowski, J., Domzal, M., Jedrasiak, K., & Nawrat, A. (2014). Real-time laser point tracking. In *International Conference on Computer Vision and Graphics* (pp. 542–551). Springer International Publishing.
22. Nawrat, A., & Jedrasiak, K. (2008). Fast colour recognition algorithm for robotics. *Problemy Eksploatacji*.
23. Josinski, H., Switonski, A., Jedrasiak, K., Kostrzewa, D. (2012). Human identification based on gait motion capture data. In *Proceedings of the 2012 International MultiConference of Engineers and Computer Scientists*.
24. Switonski, A., Josinski, H., Jedrasiak, K., Polanski, A., & Wojciechowski, K. (2010). Classification of poses and movement phases, ICCVG 2010. Lecture notes in computer science. Springer.
25. Bereska, D., Daniec, K., Jedrasiak, K., & Nawrat, A. (2013). Gyro-stabilized platform for multispectral image acquisition. In *Vision based systems for UAV applications* (pp. 115–121). Springer International Publishing.

26. Daniec, K., Iwaneczko, P., Jedrasiak, K., & Nawrat, A. (2013). Prototyping the Autonomous Flight Algorithms Using the Prepar3D Simulator. In *Vision based systems for UAV applications* (pp. 219–232). Springer International Publishing.
27. Sobel, D., Jedrasiak, K., Daniec, K., Wrona, J., Jurgas, P., & Nawrat, A. (2014). *Camera calibration for tracked vehicles augmented reality applications*. Springer International Publishing: Innovative Control Systems for Tracked Vehicle Platforms.

Pixel Classification for Skin Detection in Color Images

Bartosz Binias, Mariusz Frąckiewicz, Krzysztof Jaskot and Henryk Palus

1 Introduction

Segmentation of color image regions that contain skin pixels is a very important and challenging task of modern image processing. The general objective of described problem is to return the output image, whose every pixel was classified as either representing skin or not [1, 2]. Such information can be then used in various applications in computer vision [3, 4]. Important and interesting tasks, where skin segmentation is required, are automatic face [5–7] or gesture detection [8–10]. Additionally, many of the most effective filters of adult-only content are based on the information from the segmented skin regions [11, 12]. Image coding using regions of interest as presented in [13], is another very important example of the application of skin segmentation algorithms.

The most common approach to skin detection problem is pixel-wise, color-based classification [4]. In such methods each pixel is being classified independently from its neighbours only on the basis of its color features [14, 15]. Such discrimination between skin and non-skin pixels can be performed using skin color model represented either as a set of rules or thresholds or derived from used machine learning algorithm [4]. However, it must be noted that basing solely on the color information may not be sufficient for the task. It is a well known fact, that most popular color spaces such as the *RGB* and *HSV* or perceptually uniform *CIELab* suffer from many shortcomings (for more details see [16]). In order to improve the quality of pixel-based skin detection many approaches have been proposed. Among these worth mentioning are texture-based methods, adaptation techniques and spatial analysis. A very detailed description of this techniques can be found in [4].

In this research a direct, pixel-based approach was chosen. The main purpose was to evaluate the performance in this specific task of two very popular machine

B. Binias (✉) · M. Frąckiewicz · K. Jaskot · H. Palus
Silesian University of Technology, Akademicka 16,, 44-100Gliwice, Poland
e-mail: Bartosz.Binias@polsl.pl

A. Nawrat et al. (eds.), *Advanced Technologies in Practical Applications for National Security*, Studies in Systems, Decision and Control 106,
https://doi.org/10.1007/978-3-319-64674-9_6

learning classifiers: Regularized Logistic Regression and Artificial Neural Network with Regularization trained with Backpropagation [17]. The research focused mostly on the aspect of model's error evaluation and parameter tuning. Developing new algorithms [18–21] and computer simulation [22] are classical means to achieve progress of the technology.

2 Data Description

Data used in this research was 'Skin Segmentation Dataset' provided for the UCI Machine Learning Repository [23]. The dataset consists of 50859 examples marked as skin samples and 194198 non-skin samples. Available features are pixel's values in B, G and R channels, coded with 8 bits. The skin dataset was collected by random sampling of these values from images of various gender, age and race groups obtained from Color FERET Image Database and PAL Face Database from Productive Aging Laboratory.

All 245057 available examples were randomly divided into three subsets. The training set was created by randomly selecting 60% of all skin samples and 60% of samples from the other class. That way an original proportion between both classes has been sustained. Analogously, the cross-validation and test sets were separated from the remaining examples. Each of these consisted of 20% of all skin and non-skin examples, sampled without repetition.

3 Description of Algorithms

3.1 Logistic Regression with Regularization

The goal of the regression is finding a set of parameters $\Theta \in \mathbb{R}^{n+1}$ (where n goes for the number of features and additional dimensionality stems from the bias term) that minimizes cost function presented in Eq. 1 [24]:

$$J(\Theta) = \frac{1}{M} \sum_{i=1}^{M} \left(h_\Theta(x^{(i)}) - y^{(i)} \right)^2 \tag{1}$$

where M is the total number of training examples, defined by input variables (features) $x \in \mathbb{R}^{M \times (n+1)}$ and output variables $y \in \mathbb{R}^{n+1}$ (labels). Because of the dimensionality of a single example $x^{(i)} \in \mathbb{R}^{n+1}$ the vectorized notation of the hypothesis $h_\Theta(x^{(i)})$ can be written as follows (we assume that $x_0 = 1$ for each example):

$$h_\Theta(x^{(i)}) = \Theta_0 x_0 + \Theta_1 x_1 + \cdots + \Theta_n x_n = \Theta^T x^{(i)} \tag{2}$$

For the binary classification task the preferred output of the hypothesis function would be either 0 or 1. To enable that a slight modification of the hypothesis function is required. For that a sigmoid (logistic) function is used. The improved hypothesis is presented in Eq. 3:

$$h_\Theta(x) = \frac{1}{1 + e^{-\Theta^T x}} \tag{3}$$

In order to assign bigger penalization to predictions that differ highly from the required output hypothesis $h_\Theta(x)$ should be additionally logarithmized. To avoid problems with algorithm overfitting the training data the regularization of Θ parameters (apart from bias-related Θ_0) [24] was introduced in the form of λ multiplier. The final form of the minimized cost function of the Regularized Logistic Regression method used for the binary classification is presented in Eq. 4:

$$J(\Theta) = -\left[\frac{1}{M}\sum_{i=1}^{M}\left(y^{(i)}log\left(h_\Theta(x^{(i)})\right) + (1 - y^{(i)})log\left(1 - h_\Theta(x^{(i)})\right)\right)\right] \tag{4}$$

Finding optimal parameters Θ can be performed iteratively with the use of gradient-based numerical optimization techniques such as Gradient Descent or Conjugate Gradient. For such methods to work the derivative of cost function with respect to each parameter must be calculated and provided. However, it must be noted that the Θ_0 parameter should not be regularized. Therefore, the rule for upgrading this parameter in iteration $p + 1$ presented in Eq. 5 does not take the regularization term into account.

$$\Theta_0^{(p+1)} = \Theta_0^{(p)} - \alpha\frac{1}{M}\sum_{i=1}^{M}\left(h_\Theta(x^{(i)}) - y^{(i)}\right)x_0^{(i)} \tag{5}$$

For other parameters Θ_j, where $j = \{1, 2, \dots, n\}$, the formula for finding their improved values in new iteration $p + 1$ using Gradeint Descent based method with step α is presented in Eq. 6:

$$\Theta_j^{(p+1)} = \Theta_j^{(p)} - \alpha\left[\frac{1}{M}\sum_{i=1}^{M}\left(h_\Theta(x^{(i)}) - y^{(i)}\right)x_j^{(i)} + \frac{\lambda}{M}\Theta_j^{(p)}\right] \tag{6}$$

3.2 *Artificial Neural Network Model with Regularization*

A typical Artificial Neural Network consists of structures known as layers [24]. Among them distinguished is the *input layer* and the *output layer*. Other ones are remotely referred to as *hidden layers*. Each layer is constructed of basic calculation units called *neurons*. If layer j has s_j neurons and layer $j + 1$ has s_{j+1} units then the weights of connections between neurons in particular layers $\Theta^{(j)} \in \mathbb{R}^{s_{j+1} \times (s_j + 1)}$. The process of Neural Network's output calculation is known as the *Forward Propagation* where the vector of neuron's activations in layer j $a^{(j)} \in \mathbb{R}^{s_j+1}$ (with added bias) is being calculated as was presented in Eq. 7. In first, input layer we treat inputs as activation, as in $a^{(1)} = x^{(i)}$.

$$a^{(j+1)} = g(\Theta^{(j)} a^{(j)}) \tag{7}$$

In proposed model a sigmoid activation function $g(\Theta, a)$ was used for each neuron (Eq. 8).

$$g(\Theta, a) = \frac{1}{1 + e^{-\Theta^{(j)T} a^{(j)}}} \tag{8}$$

The cost function minimized by the Artificial Neural Network with Regularization of weights is presented in Eq. 9, where K stands for the number of classes, L is the total number of layers in the network [24].

$$J(\Theta) = -\Big[\frac{1}{M}\sum_{i=1}^{M}\sum_{k=1}^{K}\big(y_k^{(i)} log\big(h_\Theta(x^{(i)})\big)_k + (1 - y_k^{(i)}) log\big(1 - h_\Theta(x^{(i)})\big)\big)_k\Big] + J_{reg}(\Theta) \tag{9}$$

where $J_{reg}(\Theta)$ is the regularization term (Eq. 10) [24].

$$J_{reg}(\Theta) = \frac{\lambda}{2M}\sum_{l=1}^{L-1}\sum_{i=1}^{s_l}\sum_{j=1}^{s_{l+1}}\big(\Theta_{ji}^{(l)}\big)^2 \tag{10}$$

The most popular procedure of Artificial Neural Networks training is the *Backpropagation* algorithm. The detailed description of the algorithm can be found in [25].

4 Error Model Evaluation

Before the selection of the best parameters for the method it is important to properly evaluate its error on training and cross-validation datasets. Such effort can help to determine whether the model is capable of explaining the variance in the data properly without overfitting to the training examples. In this research, the base input of classification algorithm's consisted of three features related to each pixel's (example's) value in one of the *RGB* color space channels. All features were scaled to the range [0, 1]. The reason for that operation is the use of the Nonlinear Conjugate Gradient based method for optimization of algorithms' cost functions. Providing that features are on the similar scale improves convergence of this method.

4.1 Learning Curves

In order to determine whether any of the proposed models struggles with high bias or variance problem the adequate learning curves were calculated as the function of classification error's dependence on the number of samples. High bias errors on training and cross-validation sets would converge with the increase of samples provided for training. However, at some point they will both set on a relatively high value. Such behaviour indicates that examined model does not explain the classes sufficiently. Models with high variance are characterized by low error achieved during classification of training set and higher number of misclassified samples in cross-validation set. The reason for that is the overfitting of the model to the training dataset. Formula used for error calculation is presented in Eq. 11.

$$J_{err}(\theta) = \frac{1}{m}\sum_{i=1}^{m}(h_{\theta}(x^{(i)}) - y^{(i)})^2 \tag{11}$$

Learning curves for both models are presented in Figs. 1 and 2. They were calculated on the basis of the performance of classifier trained on the increasing number of $m \leq M$ samples. Because error of testing set classification was measured only on m examples it tends to increase with m. The size of cross-validation dataset remained unchanged during the whole testing.

The learning curves calculated for the Regularized Logistic Regression classifier are presented in Fig. 1. It can be noted that the increase of training samples does not reduce the classification error of cross-validation data. Additionally, both error functions converge to a relatively high value. These characteristics are a clear indicators that the model suffers from high bias problem. This means that it is not capable of properly describing the classes present in the data.

Analysis of the curves presented in Fig. 2 shows that the Artificial Neural Network model does not suffer from high bias or variance. After the sufficient number of training samples is provided it can be observed that errors calculated on training and cross-validation sets converge to common value. This proves that proposed model generalizes well on the distinct features of the data. Additionally, low values of errors acquired for higher numbers of samples indicate that it is able to properly explain the differences between classes.

4.2 Tuning of Regularized Logistic Regression Model

Error analysis for Logistic Regression Model performed in the previous subsection revealed that it suffers from high bias. The best solution of this problem would be to add more features, that would help to better describe the data. In this research an addition of some higher order polynomial features (up to third degree) was proposed. Assuming that the feature vector for i-th example (with n features and unitary

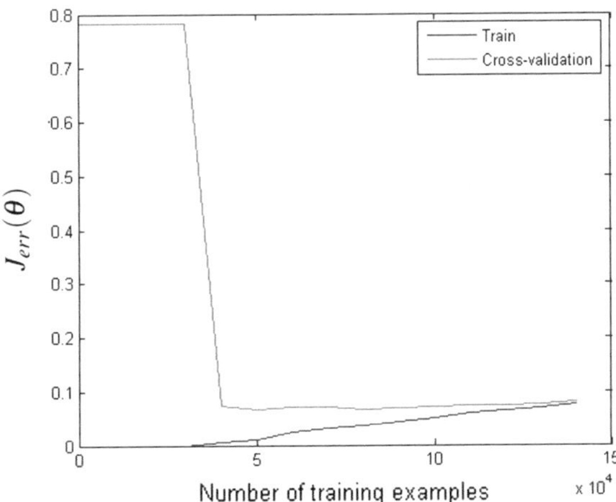

Fig. 1 Learning curve of Regularized Logistic Regression model ($\lambda = 0$)

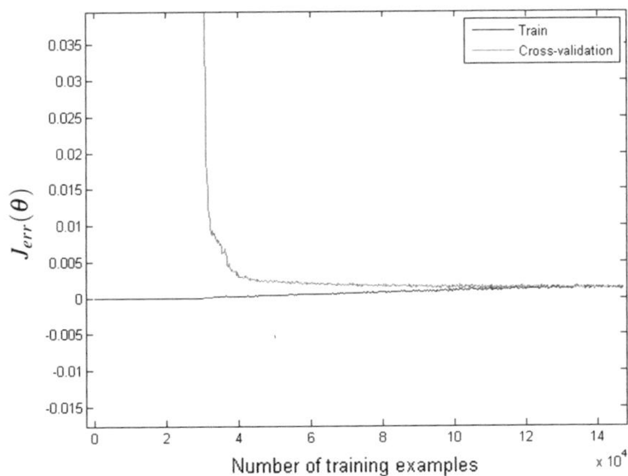

Fig. 2 Learning curve of Artificial Neural Network Model ($\lambda = 0$)

bias x_0) is of form $x^{(i)} = [x_0^{(i)} x_1^{(i)} \ldots x_n^{(i)}]^T$, then addition of higher order would mean providing all the combination of features (except for bias) up to required order, i.e. $x_1x_2, x_2^2, x_1x_2x_3, x_3^2x_1, etc$. Presented in Fig. 3 are the learning curves calculated for the extended set of features. Achieved results prove that the extension of feature vector is the solution of high bias problem for this case. For the final tuning of the parameter the best weight of the regularization term was selected.

Finally, an additional tuning of the regularization weight was applied. Basing on the results presented in Fig. 4 $\lambda = 3$ was chosen as the best regularization parameter for the classification of testing data.

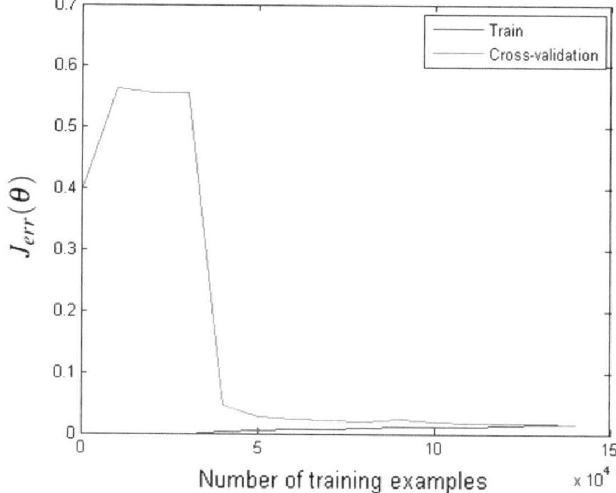

Fig. 3 Learning curve of Regularized Logistic Regression model with polynomial features of up to third order ($\lambda = 0$)

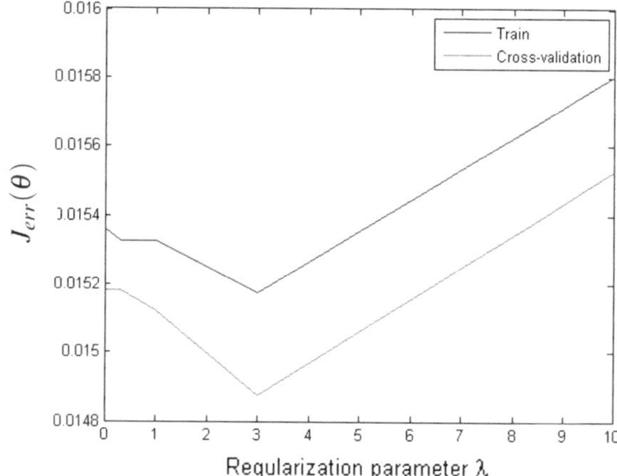

Fig. 4 Validation curve of Regularized Logistic Regression model presenting influence of regularization parameter λ on classification error

4.3 Tuning of Artificial Neural Network Model

The Artificial Neural Network Model trained with Backpropagation did not exhibit any signs of high bias or variance despite it being very basic with only one hidden layer. In order to find the best number of neurons in hidden layer model's classification error was calculated for training and cross-validation datasets with regularization parameter $\lambda = 1$. Thanks to such high regularization weight, achieved results were oriented mostly towards bias, not variance validation. Achieved validation curve is presented in Fig. 5.

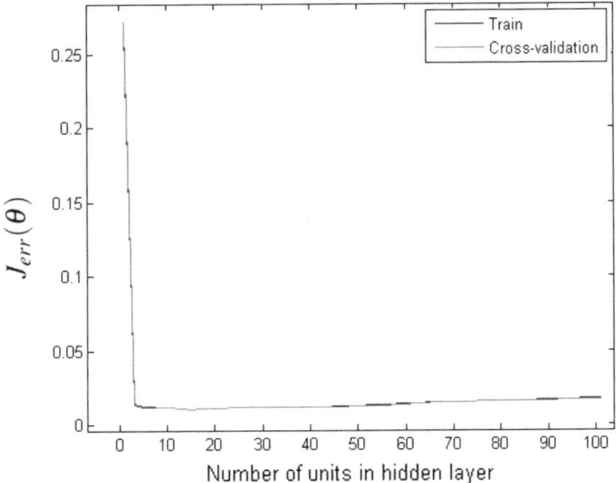

Fig. 5 Validation curve of ANN presenting the influence of hidden layer neurons on classification error ($\lambda = 1$)

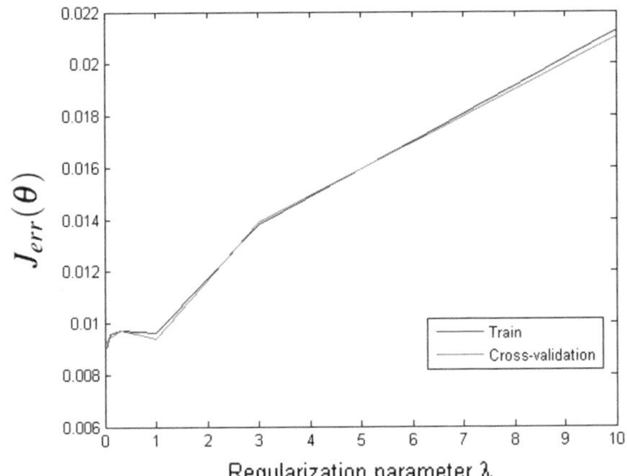

Fig. 6 Validation curve of ANN (with 15 neurons in hidden layer) presenting influence of regularization parameter λ on classification error

According to achieved results the best number of hidden neurons is 15. For that value the additional tuning was performed in order to select appropriate regularization parameter λ. Basing on the results presented in Fig. 6 $\lambda = 0.001$ was chosen as the best regularization parameter for the classification of testing data.

5 Results

Both models with the best settings selected in Sect. 4 were used for the classification of test data. However, it must be noted that after a careful evaluation of both model's errors it was determined that the Artificial Neural Network classifier proved to perform significantly better. For each model a confusion matrix was created on both cross-validation and test dataset. Because in this case proportion between classes is skewed evaluating algorithms only on the basis of their accuracy would not be meaningful. Having that in mind, classifier's precision, recall and specificity were also calculated. Precision refers to a number of True Positive cases over all cases classified as Positive (here as skin samples). To describe the ratio between True Positives and actual Positive examples, the Recall can be used. Specificity is the equivalent of the Recall for the Negative class. In Table 1 presented are results achieved for cross-validation and test sets for both proposed models.

Examination of the results achieved by the Artificial Neural Network model shows that it's performance on the testing and cross-validation datasets is nearly perfect. Very high values of all used measures are the indicators that not only the accuracy of the model is high, but also the general quality of classification. On the other hand, the performance of the Logistic Regression is less stable. Although it achieved comparable accuracy and slightly higher recall, it is significantly less precise. The lacking precision of skin segmentation algorithms is often listed as one of the most important issues of pixel-based detectors.

To additionally test the performance of the Artificial Neural Network model it was tested and visually examined on the real images that did not belong the 'Skin Segmentation Dataset' provided for the UCI Machine Learning Repository. In Figs. 7 and 8 presented are results of segmentation of the images from PUT database. The PUT face database [26] contains high resolution color face images of 100 people acquired on a uniform background under controlled lighting conditions.

Analysis of the segmentation results proves the high quality of the developed classifier. The visual examination of the output image reveals that all the most important information as well as some really deep details of the original image were captured, thus providing a very satisfactory, precise and accurate skin segmentation result.

Table 1 Performance of both proposed classifiers evaluated on cross-validation and test data

Model	Logistic regression		Artificial Neural Network	
Data	CV	Test	CV	Test
Accuracy	0.9849	0.9831	0.9990	0.9987
Precision	0.9321	0.9248	0.9967	0.9947
Recall	1.0000	1.0000	0.9987	0.9989
Specificity	0.9809	0.9787	0.9991	0.9986

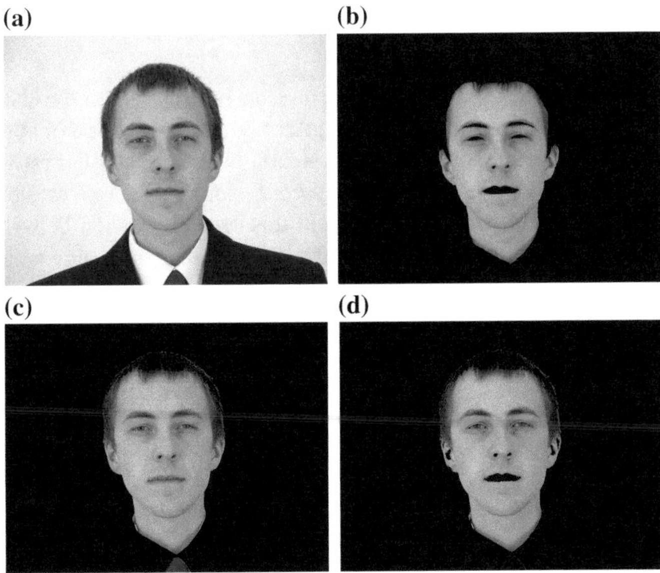

Fig. 7 Results of skin segmentation performed on real image: **a** original image, **b** manually segmented ground truth image, **c** result of segmentation with Regularized Logistic Regression, **d** result of segmentation with Artificial Neural Network

It must be noted that the algorithm allowed the occurrence of some False Positive classification in the right side of the image. The most problematic regions of the image are related to areas where hair connects with skin. Because color features of this regions correspond belong to the developed skin model it is impossible to avoid classifying them as Positive cases. However, such inaccuracies can be easily removed with the use of some image postprocessing techniques mentioned in the Sect. 1 of this paper. To furtherly evaluate a quality of the segmentation the Jaccard similarity coefficient [27] was calculated for both images. The obtained values were 0.897 for Regularized Logistic Regression and 0.920 for Artificial Neural Network model for the first image and respectively 0.943 and 0.948 for the second image. The Jaccard index is a statistic commonly used for the evaluation of the similarity and diversity of two sets. For a full similarity the Jaccard index would take the value of 1. High values achieved for Artificial Neural Network classifier correspond with visual evaluation of the results.

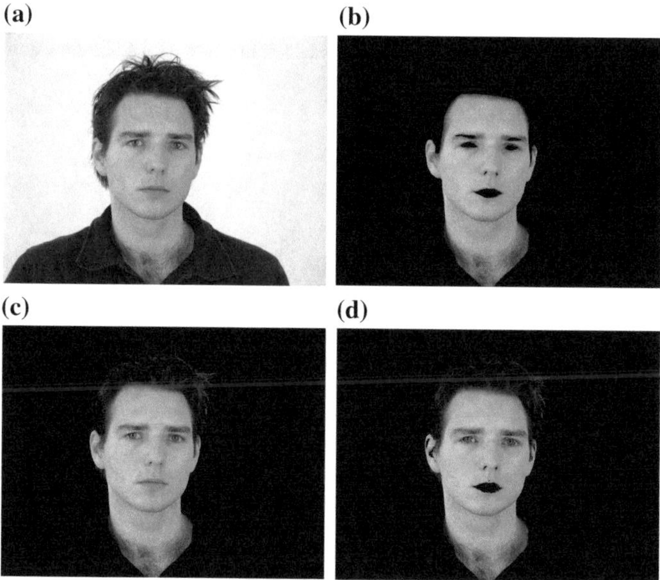

Fig. 8 Results of skin segmentation performed on real image: **a** original image, **b** manually segmented ground truth image, **c** result of segmentation with Regularized Logistic Regression, **d** result of segmentation with Artificial Neural Network

6 Conclusions

In this paper the performance of two prominent classification algorithms was evaluated and compared in the task of skin pixel detection. It was proven that the exploratory data analysis approach can produce some highly satisfying results without the need of referring to some image specific methods, especially during low-level processing.

Achieved results indicate that the Artificial Neural Network is capable of extracting some hidden features and structures in the data. This can be best observed when used classifiers are compared. The Regularized Logistic Regression's performance was poor until some additional features were created and provided as extra inputs to the classifier. At the same time the Artificial Neural Network (with only one hidden layer) performed significantly better using only basic features. This holds an important advantage over approaches where features must be designed by the data scientist during model creation. The reason for that is, that such automatic feature extraction is not constrained by the invention of the designer.

The further improvements to the method can still be applied in order to achieve even better results. An addition of different features derived from other known color spaces could prove to be a valuable extension of the method. Additionally, some advanced postprocessing like texture-based models, adaptation techniques and spatial analysis could be introduced to the proposed algorithm.

Acknowledgements This work was supported by Polish Ministry for Science and Higher Education under internal grant BKM/514/RAu1/2015/t-21 for Institute of Automatic Control, Silesian University of Technology, Gliwice, Poland.

References

1. Kakumanu, P., Makrogiannis, S., & Bourbakis, N. (2007). A survey of skin-color modeling and detection methods. *Pattern Recognition*, *40*(3), 1106–1122.
2. Khan, Rehanullah, Hanbury, Allan, Stöttinger, Julian, & Bais, Abdul. (2012). Color based skin classification. *Pattern Recognition Letters*, *33*(2), 157–163.
3. Phung, S. L., Bouzerdoum, A., & Chai, D. (2005). Skin segmentation using color pixel classification: Analysis and comparison. *IEEE Transactions on Pattern Analysis and Machine Intelligence*, *27*(1), 148–154.
4. Kawulok, M., Nalepa, J., & Kawulok, J. (2014). Skin detection and segmentation in color images. In M. Emre Celebi & B. Smolka (Eds.), *Advances in low-level color image processing* (Vol. 11, pp. 329–366). Lecture notes in computational vision and biomechanics. Netherlands: Springer.
5. Kawulok, M. (2005). Application of support vector machines in automatic human face recognition. *Medical Informatics & Technology (MIT)*, *9*, 143–150.
6. Chaves-González, Jose M., Vega-Rodríguez, Miguel A., Gómez-Pulido, Juan A., & Sánchez-Pérez, Juan M. (2010). Detecting skin in face recognition systems: A colour spaces study. *Digital Signal Processing*, *20*(3), 806–823.
7. Hajraoui, Abdellatif, & Sabri, Mohamed. (2014). Face detection algorithm based on skin detection, watershed method and gabor filters. *International Journal of Computer Applications*, *94*(6), 33–39.
8. Kawulok, M. (2008). Dynamic skin detection in color images for sign language recognition. *Proceedings of the ICISP, LNCS*, *5099*, 112–119.
9. Grzejszczak, T., Kawulok, M., & Galuszka, A. (2016). Hand landmarks detection and localization in color images. *Multimedia Tools and Applications*, *75*(23), 16363–16387.
10. Daniec, K., Jedrasiak, K., Nawrat, A., & Bereska, D. (2013). Gyro-stabilized platform for multispectral image acquisition. In A. Nawrat & Z. Kuś (Eds.), *Vision based systems for UAV applications* (pp. 115–121). Springer.
11. Zarit, B. D., Super, B. J., & Quek, F. K. H. (1999). Comparison of five color models in skin pixel classification. In *International Workshop on Recognition, Analysis, and Tracking of Faces and Gestures in Real-Time Systems, 1999. Proceedings* (pp. 58–63). IEEE.
12. Lee, J. S., et al. (2007). Naked image detection based on adaptive and extensible skin color model. *Pattern Recognition*, *40*(8), 2261–2270.
13. Chen, M. J., Chi, M. C., Hsu, C. T., & Chen, J. W. (2003). ROI video coding based on H.263+ with robust skin-color detection technique. *IEEE Transactions on Consumer Electronics*, *49*(3), 724–730.
14. Chai, D., & Bouzerdoum, A. (2000). A Bayesian approach to skin color classification in YCbCr color space. In *TENCON 2000. Proceedings* (Vol. 2, pp. 421–424). IEEE.
15. Palus H. (2006). Color image segmentation: Selected techniques. In R. Lukac & K. N. Plataniotis (Eds.), *Color image processing: Methods and applications* (pp. 103–128). Boca Raton: CRC Press.
16. Palus, H. (1992). Colour spaces in computer vision. *Machine Graphics and Vision*, *1*(3), 543–554.
17. Al-Mohair, K., & Suandi, S. A. (2012). Human skin color detection: A review on neural network perspective. *International Journal of Innovative Computing, Information and Control*, *8*(12), 8115–8131.

18. Świtoński, A., Josiński, H., Jędrasiak, K., Polański, A., & Wojciechowski, K. (2010). Classification of poses and movement phases. *Computer Vision and Graphics*, 193–200.
19. Ryt, A., Sobel, D., Kwiatkowski, J., Domzal, M., Jedrasiak, K., & Nawrat, A. (2014, September). Real-time laser point tracking. In *International Conference on Computer Vision and Graphics* (pp. 542–551). Springer: Cham.
20. Sobel, D., Jędrasiak, K., Daniec, K., Wrona, J., Jurgaś, P., & Nawrat, A. M. (2014). Camera calibration for tracked vehicles augmented reality applications. In *Innovative control systems for tracked vehicle platforms* (pp. 147–162). Springer International Publishing.
21. Jedrasiak, K., & Nawrat, A. (2008). Fast colour recognition algorithm for robotics. *Problemy Eksploatacji, 3*, 69–76.
22. Daniec, K., Iwaneczko, P., Jędrasiak, K., & Nawrat, A. (2013). Prototyping the autonomous flight algorithms using the Prepar3D® simulator. In *Vision based systems for UAV applications* (pp. 219–232). Springer International Publishing.
23. Bhatt, R., & Dhall, A. (2010). Skin Segmentation Dataset, UCI Machine Learning repository.
24. Du, K.-L., & Swamy, M. N. S. (2013). *Neural networks and statistical learning*. Springer Science & Business Media.
25. Werbos, P. J. (1994). *The roots of backpropagation: From ordered derivatives to neural networks and political forecasting* (Vol. 1). Wiley.
26. Kasinski, A., Florek, A., & Schmidt, A. (2008). The PUT face database. *Image Processing and Communications*, *13*(3–4), 59–64.
27. Jaccard, P. (1912). The distribution of the flora in the Alpine Zone. *New Phytologist*, *11*(2), 37–50.

Part II
Design of Control Algorithms for Unmanned Mobile Robots

Ensuring the safety is every day more closely related to the practical use of unmanned mobile robots. They are commonly used wherever hidden action is necessary or action is associated with a high probability of injury or significant losses. Combat units and emergency services are equipped with all sorts of "drones" and robots for sapper operations. All these objects regardless of the degree of autonomy and the type require increasingly sophisticated control systems. This chapter is devoted to the development of control systems dedicated for unmanned vehicles.

For example, the proposed set of articles can trace the whole spectrum of issues related to the discussed topic. Ranging from control systems dedicated to the object with a single drive after a multi-level microprocessor control systems designed for UAV and solutions based on multi-agent environment.

Certainly it will be interesting to read the results of works that were devoted to the practical implementation of the speed control algorithm designed for UAV type SEPL. An important advantage of this study is to compare the results obtained in computer simulations of the results obtained during the actual test flight of the object.

Equally interesting are the considerations devoted to a modified method of synthesis BLT control system designed to dampen noise in the control helicopter UAVs. The authors focus on the lowest level of control that represents the various stages of the method and the results they obtained synthesizing control of cruise linear climb, and the angular velocity of the object around its axis. According to the authors modified method BLT offers greater opportunities for tuning PI the tuning process may take into account different requirements for the control system.

The works presented in the next article focuses on multi-level control system using the fusion of data coming from the vision system and the navigation system odometry. In order to increase the reaction rate of the experimental part of the mobile robot control system functions was transferred to the onboard system, the master system has been designed to solve global tasks.

Another article is devoted to the multi-agent environment, in which each agent tries to reach individual target regardless of its definition and possible conflict with another agent. The proposed method of finding a solution combines the advantages. However, since the solution in the general case is incomplete, authors propose a method to solve the problem by finding a plan by using multiple elements of hierarchical game theory.

The last article is devoted to the very substantial issue of controlling a swarm of unmanned objects. The main objectives and the implementation of a system for prototyping of control algorithms swarm using a simulation environment was presented. It has been shown that the use of a swarm of UAVs can effectively accelerate tasks required to execute within the area. For instance imaging reconnaissance could be executed through scattering of the main task into the partial tasks performed by the individual objects in the swarm.

Combining Data from Vision and Odometry Systems for More Accurate Control of Mobile Robot

Robert Bieda, Krzysztof Jaskot, Tomasz Łakota and Karol Jędrasiak

1 Introduction

Mobile mini-robots constitute an important equipment for research of control algorithms. During last decade, many constructions for this purpose were developed. One of examples is the series of Khepera robots, which are widely used at many universities in robotics courses. It is commercial product, with emphasis put on high modularity [1]. Very different approach is presented in the work [2], where the cheap, compact mobile robot is described. There are also many constructions intended to be used in robo soccer competitions [3].

This paper describes control algorithms implemented in the experimental mobile robot [4, 5]. It is assumed that general motion planning is performed by the master control system, and only some basic tasks are realized by the robot itself. They are mostly tasks, that use data from encoders, so by implementing them in the on-board system, they can react much faster than it would be, if current encoder readings was sent to the master system, and decision was taken there. Since most of tasks need knowledge about robot position and/or orientation, the algorithms that allow to estimate robot's pose are also described here.

The overall structure of the system is presented in the Fig. 1. Each green block denotes some algorithm. The arrows show direction of data flow between them. The Table 1 shows the same structure in a form of stack (position estimation is not shown there). Each layer is responsible for only one kind of tasks and has contact only with

R. Bieda · K. Jaskot (✉) · T. Łakota · K. Jędrasiak
Institute of Automatic Control, Silesian University of Technology,
Akademicka 16, 44-100 Gliwice, Poland
e-mail: krzysztof.jaskot@polsl.pl

R. Bieda
e-mail: krzysztof.jaskot@polsl.pl

K. Jędrasiak
e-mail: karol.jedrasiak@polsl.pl

A. Nawrat et al. (eds.), *Advanced Technologies in Practical Applications for National Security*, Studies in Systems, Decision and Control 106,
https://doi.org/10.1007/978-3-319-64674-9_7

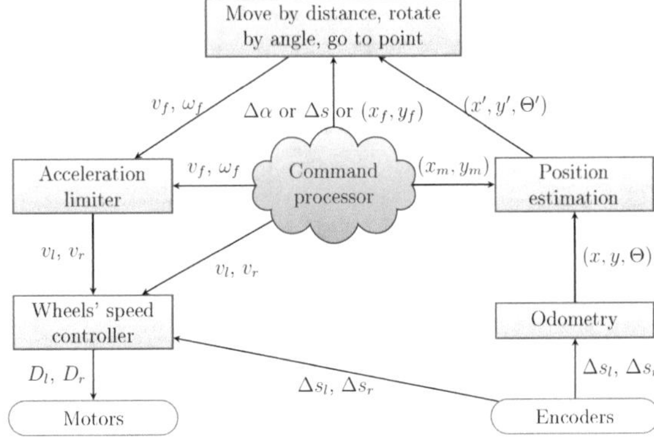

Fig. 1 A data flow between control algorithms implemented in the controller

Table 1 The layered structure of control algorithms

Highest level	Master control system
Level 3	Move by distance, rotate by angle, go to point
Level 2	Acceleration limiter
Level 1	Wheel's speed controller (PID)
Level 0 (hardware)	PWM (motors)

the layer immediately below it. The only exception is the master control system, which is not a part of controller. This system can select freely the level which it want to control by sending commands related to it. In such case, the higher layers are deactivated until they are used by the master system. For example, when last executed command is the one which sets the speed of wheels respecting the acceleration control, then the topmost algorithms are not active, and the linear and angular speed (v_f and ω_f) in the acceleration limiter are set according to the command. Since acceleration limiter is active in such case, it sets left and right wheel speed (v_l, v_r) in the wheels' speed controller. Another possibility is, for example, to set directly v_l and v_r using proper command, so the acceleration limiter will be deactivated. Meaning of the other symbols is as follows:

- D_l, D_r—a duty cycles of waveform which controls the motors
- Δs_l, Δs_r—increments of distance traveled by wheels, which is proportional to the number of pulses generated by encoders
- (x, y, θ)—estimation of current pose based on odometry
- (x_m, y_m)—estimation of current position based on vision system [6, 7]
- (x', y', θ')—estimation of current pose using both odometry and vision
- (x_f, y_f)—destination point
- Δs—distance to move
- $\Delta\alpha$—angle to rotate

It should be mentioned here, that in case of extending the controller software with new control algorithms (in the on-board system), the control level switching has to be done manually using function *ctrl_switch_level* implemented in module controllers [4]. Developing new algorithms [8–11] and computer simulation [12] are classical means to achieve progress of the technology.

2 Odometry

As we want to be able to realize algorithms related to robot's pose (like rotate by angle, go to point), it is necessary to estimate robot's position on the ground [13]. The information about it's absolute position is obtained by a separate vision positioning system and received using wireless link, however it is not sufficient. Mentioned data is received with time delay of approximately 50 ms, and with frequency not exceeding 30 Hz. It is too small for good reactive algorithm, so it is needed to estimate robot's current position using another source—the encoders. They provide current information about distance traveled by each wheel, which can be easily used to estimate successive changes of position and direction. The technique of estimating of displacement basing on data from movement actuators is called odometry.

The odometry provides quite good estimation of position for short time intervals. In longer intervals, however, it is much worse, since errors of estimations of successive displacements cumulates. In [14] many causes of both systematic and non-systematic errors were given. Generally, in odometry, systematic errors can be caused by:

- Unequal wheel diameters.
- Average of actual wheel diameters differs from nominal wheel diameter.
- Actual wheelbase differs from nominal wheelbase.
- Misalignment of wheels.
- Finite encoder resolution.
- Finite encoder sampling rate.

The first point was experimentally checked to have negligible effect on overall error. The second one was minimized by calibrating experimentally coefficients depending on wheel diameter. The wheelbase and misalignment problems do not concern current two-wheeled differential drive. The finite resolution and sampling rate probably influences results a bit. The sampling rate is 1000 Hz, and the encoders gives us a resolution of approximately $\frac{1}{28500}$ m $\approx$ 0.035 mm.

The causes of non-systematic errors given in [14] are:

- Travel over uneven floors.
- Travel over unexpected objects on the floor.
- Wheel-slippage due to: slippery floors, overacceleration, fast turning (skidding), external forces (interaction with external bodies), internal forces (castor wheels), non-point wheel contact with the floor.

Since robot is supposed to drive only on special, dedicated flat field, the first point seems to have negligible meaning here. The second one also is not very important (the field is uniform and no additional objects, except robots, should appear there). Skidding and overacceleration (including overbraking) is potentially the most important source of errors in odometry, so special control layer (acceleration control) is present [15–17]. It is described in details in Sect. 5. The second important reason of errors is interaction with external bodies. There are bands around the board and other robots may be present there, so the collision may occur. It should be avoided by higher level control algorithms, but it may happen, so it has to be taken into account, that in such cases the odometry estimation will be not correct. When no skidding or collisions happen, the main source of error is non-point wheel contact with the ground. It is especially important during turning, because the during radius depend on the distance between contact points [18]. It will be shown further, that this distance varies significantly during movement.

For differential drive, the equations describing displacement and direction are quite simple. Denoting the distance between the contact points by D and increment of travel distance of left and right wheels by Δs_l and Δs_r, the increment of robot's orientation $\Delta\theta_i$ can be computed as follows [14]:

$$\Delta\theta_i = \frac{(\Delta s_r - \Delta s_l)}{D} \tag{1}$$

and the increment of linear displacement Δs_i is the average of distance traveled by wheels:

$$\Delta s_i = \frac{(\Delta s_r - \Delta s_l)}{2} \tag{2}$$

Robot's position can be updated as follows:

$$x_{i+1} = x_i + \Delta s_i \cos\left(\theta_i + \frac{\Delta\theta_i}{2}\right) \tag{3}$$

$$y_{i+1} = y_i + \Delta s_i \sin\left(\theta_i + \frac{\Delta\theta_i}{2}\right) \tag{4}$$

and the orientation can be updated simply by adding $\Delta\theta_i$:

$$\theta_{i+1} = \theta_i + \Delta\theta_i \tag{5}$$

The values of Δs_l and Δs_r can be determined from encoders according to equations:

$$\Delta S_{l/r} = c_m N_{l/r} \tag{6}$$

where $N_{l/r}$ is number of encoder pulses, and c_m is a factor which tells how many encoder pulses corresponds to one distance unit. It was experimentally obtained, that per one meter there are approximately 28,500 pulses. In the software however, the distance unit is exactly 1 pulse, so $c_m = 1$ in this case. That minimizes the rounding errors.

The D parameter from Eq. (1) could be measured directly and converted to internal distance units, however wheels are approximately 6 mm wide (while distance between their centers is 63 mm) and it is not obvious where the contact points exactly are. Another approach—much better in this case—is to obtain it experimentally using Eq. (1). In the experiment robot was moving in circles and values of Δs_r and Δs_l were measured for certain values of $\Delta\theta_i$. In order to minimize errors, $\Delta\theta_i$ was kept large. Several measurements were done for different turning radiuses, each for 10 full rotates ($\Delta\theta_i = 10 * 2\pi$). After these measurements we received an average value of $\overline{D} = 819$ pulses $\approx$ 0.0638 m. It can be observed, that the distance between contact points varies. It may be an effect of small irregularities of the floor or tires, so we cannot compensate it. Probably, this is a main source of errors during normal movement (without collisions or skidding).

2.1 Results of Position Estimation

The trajectory estimation obtained from odometry was recorded and compared with data from vision positioning system. The second one has also significant errors, but they are not cumulative. The comparison is presented in the Fig. 2. It can be observed that general shape of trajectory was preserved, but the deviation from real position is increasing with time. This is what was expected.

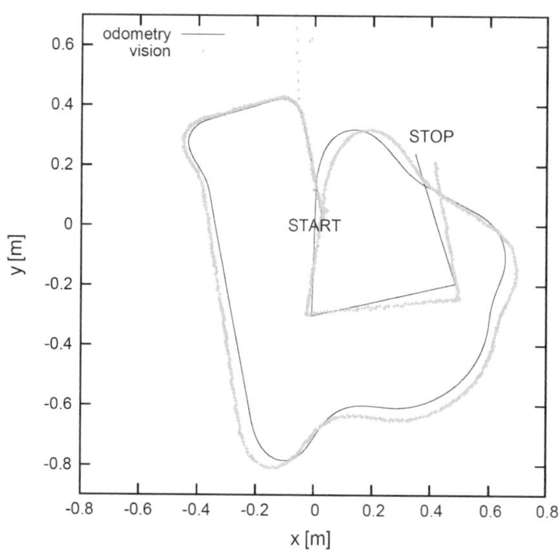

Fig. 2 A comparison of trajectory estimated by odometry and from vision positioning system

3 Combining Data from both Positioning Systems

There are two sources which provide data about current position and orientation. One is odometry, which was described in the Sect. 2. The second one is vision system based on stationary camera mounted above the field where robots are [19–21]. On the top of each robot there is a color mark which is then found in the image by the software and sent to robot using wireless link. Some tests were done in order to check the precision. The standard deviation of x and y coordinates, obtained in several measurements, were in the range 2.1–3.4 mm and standard deviation of angle was approximately 2.8° (however maximal error was almost 8°). The vision system provides 30 fps with almost full CPU usage, so per frame delay is approximately 33 ms plus time needed to deliver the data to the robot. The odometry provides better precision, much shorter delays and far greater sampling frequency, however because the maximal error of estimation increases with time, it cannot be used as the only positioning system. In the Table 2 both systems are compared. The final estimation of pose is result of combining a data from both sources.

3.1 *Implemented Solution*

Since in small time intervals the odometry estimates position very well, it is used as a primary positioning system. It is assumed, that is determines properly position in some local coordinate system (LCS), which is translated and (then) rotated with respect to the global coordinate system (GCS). Initially the translation and rotation is given by the initial robot's position and orientation in the GCS, but then these values drifts because of growing estimation error. The relation between point in GCS (denoted by $\mathbf{p}'$) and point in odometry's LCS ($\mathbf{p}$), which is translated by $\mathbf{p}_e$ and rotated by θ_e can be expressed as follows:

$$\mathbf{p}' = \mathbf{R}\mathbf{p} - \mathbf{p}_e \tag{7}$$

Table 2 Comparison of the vision positioning system and the odometry

Feature delay	Vision 40–100 ms	Odometry 1 ms
Sampling frequency	30 Hz	1000 Hz
Precision of position changes	Bad	Very good
Precision of direction changes	Very bad	Very good
Precision of absolute position	Good	Bad
Precision of absolute direction	Average	Bad
Sliding, collisions immunity	Yes	No

where:

$$\mathbf{p} = \begin{bmatrix} x \\ y \end{bmatrix}, \mathbf{p}' = \begin{bmatrix} x' \\ y' \end{bmatrix}, \mathbf{p}_e = \begin{bmatrix} x_e \\ y_e \end{bmatrix} \tag{8}$$

and R is rotation matrix:

$$\mathbf{R} = \begin{bmatrix} cos(-\theta_e) & -sin(-\theta_e) \\ sin(-\theta_e) & cos(-\theta_e) \end{bmatrix} \tag{9}$$

The idea is to use vision positioning system for estimation of $\mathbf{p}_e$ and θ_e (changing in time). The estimated current position at any moment can be obtained then using (7). The values of $\mathbf{p}_e$ and θ_e are obtained by fitting trajectory obtained from odometry into data received from vision system using weighted least squares method.

The objective function to be minimized is weighted sum of squared euclidean distances between points estimated by vision system and the ones estimated using formula (7).

$$I = \sum_{i=1}^{n} \left(w_i \sqrt{(x_i' - x_{m,i})^2 + (y_i' - y_{m,i})^2} \right) \tag{10}$$

where $x_{m,i}$ and $y_{m,i}$ denotes coordinates of robot's positions from vision system at the same time, when estimation done using formula (7) would result in x_i' and y_i'. If we apply (7) to the Eq. (10), we obtain:

$$\begin{aligned} I = & \textstyle\sum_{i=1}^{n} (w_i (x_i \cos(-\theta_e) - y_i \sin(-\theta_e) - x_e - x_{m,i})^2) \\ & + \textstyle\sum_{i=1}^{n} (w_i (x_i \sin(-\theta_e) + y_i \cos(-\theta_e) - y_e - y_{m,i})^2) \end{aligned} \tag{11}$$

At the minimum, the derivatives of I over x_e, y_e and θ_e should be 0.

$$\frac{\partial I}{x_e} = 0, \frac{\partial I}{y_e} = 0, \frac{\partial I}{\theta_e} = 0 \tag{12}$$

The derivatives over x_e, y_e can be computed rather easily:

$$\begin{aligned} \frac{\partial I}{x_e} & = \textstyle\sum_{i=1}^{n} -2w_i (x_i cos(-\theta_e) - y_i sin(-\theta_e) - x_e - x_{m,i}) \\ & = -2 \left(\textstyle\sum_{i=1}^{n} w_i (x_i cos(-\theta_e) - y_i sin(-\theta_e) - x_{m,i}) \right) \\ & \qquad + 2 \left(\textstyle\sum_{i=1}^{n} w_i \right) x_e = 0 \end{aligned} \tag{13}$$

$$x_e = \frac{S_x cos(-\theta_e) - S_y sin(-\theta_e) - S_{xm}}{S_w} \tag{14}$$

$$\frac{\partial I}{y_e} = \sum_{i=1}^{n} -2w_i(x_i sin(-\theta_e) + y_i cos(-\theta_e) - y_e - y_{m,i})$$
$$= -2\left(\sum_{i=1}^{n} w_i(x_i sin(-\theta_e) + y_i cos(-\theta_e) - y_{m,i})\right)$$
$$+2\left(\sum_{i=1}^{n} w_i\right) y_e = 0 \tag{15}$$

$$y_e = \frac{S_x sin(-\theta_e) + S_y cos(-\theta_e) - S_{ym}}{S_w} \tag{16}$$

where:

$$S_x = \sum_{i=1}^{n} w_i x_i, \quad S_y = \sum_{i=1}^{n} w_i y_i \tag{17}$$

$$S_{xm} = \sum_{i=1}^{n} w_i x_{m,i}, \quad S_{ym} = \sum_{i=1}^{n} w_i y_{m,i} \tag{18}$$

$$S_w = \sum_{i=1}^{n} w_i \tag{19}$$

The estimations should be done successively when new data from the vision system arrives. In such case the n is increased by one, and the new data is assigned to ($x_{m,n}$, $y_{m,n}$). In order to decrease complexity of computations, the exponential weighting is used, i.e. the weights are:

$$w_i = \lambda^{-(t_n - t_i)} \quad for \quad i = 1, 2, \ldots, n \tag{20}$$

where t_i is time for which x_i, y_i, $x_{m,i}$ and $y_{m,i}$ were taken and $\lambda > 1$ is some constant. That makes easy recursive computation of weights and weighted sums in consequence:

$$w_{i-1} = w_i \lambda^{(t_{i-1} - t_i)} \quad for \quad i = 1, 2, \ldots, n \tag{21}$$

It shows, that though weights depends on n, the ratios between them are constant. The weight of the last sample can be obtained from Eq. (20) for $i = n$:

$$w_n = 1 \tag{22}$$

However when the new sample is added, and the 'new' n is greater by 1 than the previous (let say n_p), the weight for the same sample will be equal now:

$$w_{n_p} = w_{n-1} = \lambda^{t_{n-1} - t_n} \tag{23}$$

Therefore when the new sample arrives, the weight previously equal to 1 changes to $\lambda^{t_{n-1} - t_n}$. It leads to conclusion (together with the fact that ratios of weights are constant), that every time when new sample arrives and n is increased, all weights are multiplied by $\lambda^{t_{n-1} - t_n}$.

$$w_i(t_n) = w_i(t_{n-1})\lambda^{t_{n-1}-t_n} \tag{24}$$

Therefore S_x from Eq. (17) can be transformed as follows:

$$\begin{aligned} S_x(t_n) &= \textstyle\sum_{i=1}^{n} w_i(t_n)x_i \\ &= \lambda^{t_{n-1}-t_n} \textstyle\sum_{i=1}^{n-1} w_i(t_{n-1})x_i + x_n \\ &= \lambda^{t_{n-1}-t_n} S_x(t_{n-1}) + x_n \end{aligned} \tag{25}$$

Analogously S_y, S_{xm}, S_{ym}, S_w can be transformed to:

$$S_y(t_n) = \lambda^{t_{n-1}-t_n} S_y(t_{n-1}) + y_n \tag{26}$$

$$S_{xm}(t_n) = \lambda^{t_{n-1}-t_n} S_{xm}(t_{n-1}) + x_{m,n} \tag{27}$$

$$S_{ym}(t_n) = \lambda^{t_{n-1}-t_n} S_{ym}(t_{n-1}) + y_{m,n} \tag{28}$$

$$S_w(t_n) = \lambda^{t_{n-1}-t_n} S_w(t_{n-1}) + 1 \tag{29}$$

From the formulas above we can conclude, that every time when new sample is added, all sums can be updated in constant time independent on n.

In order to compute the derivative of I over θ_e, the Eq. (10) has to be transformed. After applying Eqs. (14) and (16), it can be written in the following form:

$$I = \sum_{i=1}^{n} (w_i I_{c,i} + 2sin(-\theta) I_{sin,i} + 2cos(-\theta) I_{cos,i}) \tag{30}$$

where:

$$\begin{aligned} I_{c,i} &= y_i^2 + x_i^2 + x_{m,i}^2 + y_{m,i}^2 \\ &+ \left(\frac{S_{ym}}{S_w}\right)^2 + \left(\frac{S_y}{S_w}\right)^2 + \left(\frac{S_{xm}}{S_w}\right)^2 + \left(\frac{S_x}{S_w}\right)^2 \\ &-2\left(\frac{S_{ym}}{S_w} y_{m,i} + \frac{S_y}{S_w} y_i + \frac{S_{xm}}{S_w} x_{m,i} + \frac{S_x}{S_w} x_i\right) \end{aligned} \tag{31}$$

$$\begin{aligned} I_{sin,i} &= -x_i y_{m,i} + x_{m,i} y_i \\ &- \frac{S_x}{S_w}\frac{S_{ym}}{S_w} + \frac{S_{xm}}{S_w}\frac{S_y}{S_w} + \frac{S_x}{S_w} y_{m,i} \\ &- \frac{S_{xm}}{S_w} y_i - \frac{S_y}{S_w} x_{m,i} + \frac{S_{ym}}{S_w} x_i \end{aligned} \tag{32}$$

$$\begin{aligned} I_{cos,i} &= -y_i y_{m,i} - x_i x_{m,i} \\ &- \frac{S_y}{S_w}\frac{S_{ym}}{S_w} - \frac{S_x}{S_w}\frac{S_{xm}}{S_w} + \frac{S_y}{S_w} y_{m,i} \\ &+ \frac{S_{ym}}{S_w} y_i + \frac{S_x}{S_w} x_{m,i} + \frac{S_{xm}}{S_w} x_i \end{aligned} \tag{33}$$

When we substitute as follows:

$$S_c = \sum_{i=1}^{n}(w_i I_{c,i}) \tag{34}$$

$$S_{sin} = \sum_{i=1}^{n}(w_i I_{sin,i}) \tag{35}$$

$$S_{cos} = \sum_{i=1}^{n}(w_i I_{cos,i}) \tag{36}$$

Equation (30) can be written as:

$$I = S_c + 2sin(-\theta_e)S_{sin} + 2cos(-\theta_e)S_{cos} \tag{37}$$

which can be differentiated easily:

$$\frac{\partial I}{\theta_e} = -2cos(-\theta_e)S_{sin} + 2sin(-\theta_e)S_{cos} \tag{38}$$

Then, $\frac{\partial I}{\theta_e} = 0$ we obtain:

$$\begin{aligned} -2cos(-\theta_e)S_{sin} + 2sin(-\theta_e)S_{cos} = 0 \\ tan(-\theta_e) = \frac{S_{sin}}{S_{cos}} \end{aligned} \tag{39}$$

The S_{sin} and S_{cos} can be expressed as below when (32) and (33) is applied to (35) and (36):

$$S_{sin} = -S_{xym} + S_{yxm} + \frac{S_x S_{ym}}{S_w} - \frac{S_{xm} S_y}{S_w} \tag{40}$$

$$S_{cos} = -S_{yym} + S_{xxm} + \frac{S_y S_{ym}}{S_w} + \frac{S_x S_{xm}}{S_w} \tag{41}$$

where:

$$S_{xym}(t_n) = \lambda^{t_{n-1}-t_n} S_{xym}(t_{n-1}) + x_n y_{m,n} \tag{42}$$

$$S_{yxm}(t_n) = \lambda^{t_{n-1}-t_n} S_{yxm}(t_{n-1}) + y_n x_{m,n} \tag{43}$$

$$S_{xxm}(t_n) = \lambda^{t_{n-1}-t_n} S_{xxm}(t_{n-1}) + x_n x_{m,n} \tag{44}$$

$$S_{yym}(t_n) = \lambda^{t_{n-1}-t_n} S_{yym}(t_{n-1}) + y_n y_{m,n} \tag{45}$$

There are two values of θ_e in the range $(-\pi, \pi)$ that fulfills (39). One will give the maximal, and the second—the minimal value of performance index. To distinguish them, the second derivative can be computed:

$$\frac{\partial^2 I}{\theta_e^2} = -2sin(-\theta_e)S_{sin} - 2cos(-\theta_e)S_{cos} \tag{46}$$

Assuming $\cos(-\theta_e) \neq 0$, the following conversion can be made ((39) is applied)

$$\begin{aligned} \frac{\partial^2 I}{\theta_e^2} &= -2\frac{S_{cos}}{cos(-\theta_e)}\left(sin(-\theta_e)cos(-\theta_e)\frac{S_{sin}}{S_{cos}} + cos^2(-\theta_e)\right) \\ &= -2cos\frac{S_{cos}}{cos(-\theta_e)}\left(sin(-\theta_e)cos(-\theta_e)\frac{sin(-\theta_e)}{cos(-\theta_e)} + cos^2(-\theta_e)\right) \\ &= -2\frac{S_{cos}}{cos(-\theta_e)} \end{aligned} \tag{47}$$

Since at minimum the second derivative is positive, we have:

$$\frac{\partial^2 I}{\theta_e^2} = -2\frac{S_{cos}}{cos(-\theta_e)} > 0 \tag{48}$$

If $cos(-\theta_e) = 0$, another transformation, similar to (47), can be done:

$$\frac{\partial^2 I}{\theta_e^2} = -2\frac{S_{sin}}{sin(-\theta_e)} > 0 \tag{49}$$

Therefore $cos(-\theta_e)$ must have the opposite sign to S_{cos} (assuming $cos(-\theta_e) \neq 0$) and $sin(-\theta_e)$ must have the opposite sign than S_{sin} (assuming $sin(-\theta_e) \neq 0$). This leads to conclusion, that the performance index will be minimal for:

$$-\theta_e = \begin{cases} atan(\frac{S_{sin}}{S_{cos}}) & for\ S_{cos} < 0 \\ atan(\frac{S_{sin}}{S_{cos}}) + \pi & for\ S_{cos} > 0 \\ +\frac{\pi}{2} & for\ S_{cos} = 0, S_{sin} < 0 \\ -\frac{\pi}{2} & for\ S_{cos} = 0, S_{sin} > 0 \end{cases} \tag{50}$$

When both $S_{cos} = 0$ and $S_{sin} = 0$ the result is undefined. It may occur for example if all x_m and all y_m are equal (we have no trajectory to fit).

The algorithm realized when new packet from vision system arrives is simple (Algorithm 1). Then, current position can be estimated using (7).

4 Wheels' Speed Controller

The most basic thing that must be done on the robot board is setting desired revolution speed of the wheels. Whatever higher-level algorithms we want to implement,

Algorithm 1 Position estimation algorithm.

Step 1. Set $x_{m,n}$ and $y_{m,n}$ to the received coordinates and t_n to the time when frame was acquired from camera.
Step 2. Set x_n and y_n to the coordinates obtained from odometry at time t_n (the past values of coordinates are buffered, so they are available now).
Step 3. Update S_x, S_y, S_{mx}, S_{my}, S_{xym}, S_{yxm}, S_{xxm} and S_{yym} according to (25)–(29) and (42)–(45).
Step 4. Compute θ_e using (50).
Step 5. Compute x_e and y_e according to (14) and (16).

it would be much easier if they can just set the speed of wheels (and hope this speed will be reached quickly and preserved) instead of operating on motors directly. In all further part of this work instead of revolution speed, rather the linear speed of points on a perimeter of the wheel will be used, and the term 'wheel's speed' will refer to it. This is equivalent to the distance traveled in time unit by the wheel. Symbols v_l and v_r are used for the left and right wheel in the same meaning. Each motor-wheel system is controlled separately on this level. This is a classical regulation problem. There is one input for the system (PWM duty cycle), one output (encoder reading) and a reference value (desired speed). For this task the PID controller is used. PID is very popular type of controller, because it can be used effectively without knowledge of model of the system. It is widely discussed in many publications (e.g. [22–24]). Since nonzero input value is needed to hold the output (speed) at the setpoint, the integral part is necessary, so at least PI controller would be needed in this case. The derivative part is added in order to improve performance.

The most common form of the PID controller equation in the time domain is as follows:

$$u(t) = K\left(e(t) + \frac{1}{T_i}\int_0^t e(\tau)D\tau + T_d\frac{de(t)}{dt}\right) \tag{51}$$

where $u(t)$ is input of the system (output of the controller), $e(t)$ is a difference between setpoint and output at time t, and K, T_i, T_d are the constants. The constants in such a form has a physical meaning, so it is easier to choose them properly (tune the controller). For the computational purposes, however, a bit different form is better:

$$u(t) = K_p e(t) + K_i\int_0^t e(\tau)D\tau + K_d\frac{de(t)}{dt} \tag{52}$$

where:

$$K_p = K, \quad K_i = \frac{K}{T_i}, \quad K_d = KT_d \tag{53}$$

In order to implement the PID in the microcontroller, we have to discretize it. The Eq. (52) can be written in the discrete form:

$$\begin{aligned} u(nh) = K_p e(nh) + K_i h \textstyle\sum_{k=0}^{n} e(kh) \\ + K_d \tfrac{1}{h}(e(nh) - e(nh - h)) \end{aligned} \tag{54}$$

where h is a sampling time. The control signal u can take both positive and negative values. For negative values the reversed polarization is applied to motors, and the PWM duty cycle is set to the absolute value of u.

However a straightforward implementation of a controller from (54) would give poor result. Since many other algorithms depend on this controller it is important to take care of it's performance. In the [24] there are several things mentioned to take into consideration when designing the PID controller. The most important in this case are: noise filtering [25], windup, tuning, computer implementation.

4.1 Noise Filtering

The noise may have significant influence especially on the differential part [26]. In the Fig. 3 sample measured signal is presented (the 'not filtered' one). It is visible, that there is a high noise due to low resolution of measured signal. Such noise would make the differential part totally useless, since the difference between errors in successive steps is nearly independent of general trend of the signal. Moreover, the proportional part would also introduce some oscillations in case of such disturbed signal. It is necessary, therefore, to average signal in time in order to filter out the high frequency components. Two methods of averaging were tried:

- **Simple Moving Average (SMA)**. The value of filtered signal is equal to arithmetic mean of previous n samples:

$$e'(nh) = \frac{1}{n} \sum_{i=0}^{n-1} e((n - i)h) \tag{55}$$

  It can be computed quickly, but the buffer of the length proportional to n must be hold in the memory.
- **Exponential Moving Average (EMA)**. It behaves like first order lag. The current value of filtered signal depends on all previous samples, but weights decrease exponentially. It has a great advantage, that it can be implemented easily and computed quickly with no additional buffer. The recursive equation for EMA has a form:

$$y'(nh) = ae(nh) + (1 - a)e(nh - h) \tag{56}$$

  where a is a 'smoothing factor' between 0 and 1. The smaller it is, the smoother the filtered signal will be.

In the Fig. 3 both methods are compared. It can be observed, that EMA introduces much greater delay, however it can be extended (by decreasing a) without need of

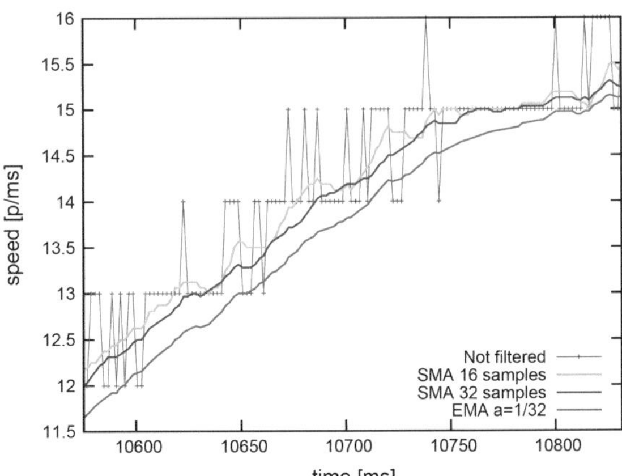

Fig. 3 An effect of several averaging filters

usage additional memory (unlike SMA) but it is possible to make it smoother. The best filtering method was chosen during PID tuning.

4.2 Windup Protection

Since the control signal range is limited (like in other physical devices), the windup phenomena must be considered. In this controller, simple approach was used—the integral is limited to such value that causes maximal control value for error and derivative equal to 0. Therefore:

$$u_{max} = K_p \cdot 0 + K_i h I_{max} + K_d \cdot 0 \tag{57}$$

$$I_{max} = \frac{u_{max}}{K_i h} \tag{58}$$

where u_{max} is a maximal absolute value of the control signal.

Therefore Eq. (54) takes the following form:

$$u(nh) = K_p e(nh) + K_i h I(n) + K_d \frac{1}{h}(e'(nh) - e'(nh - h)) \tag{59}$$

where:

$$I(n) = \begin{cases} I_{max} & \text{for } I(n-1) + e(nh) > I_{max} \\ -I_{max} & \text{for } I(n-1) + e(nh) < -I_{max} \\ I(n-1) + e(nh) & \text{otherwise} \end{cases} \tag{60}$$

with I_{max} defined by (58).

5 Acceleration Control

When wheels start rotating, robot moves because of frictional force between the wheel and the ground. This force, however, is limited to some value and if it is exceeded, the robot starts sliding [27]. As long as resultant force of interaction between ground and robot on each wheel is less than this value, we can assume that robot moves according to odometry Eqs. (1)–(5). Therefore, it is very important to ensure, that the maximal friction force is not exceeded, since odometry is critical for other algorithms [28].

The resultant force acting on each wheel of robot during motion with both linear and angular acceleration can be composed from the following: centripetal force $\overrightarrow{F_c}$ which deflects the trajectory, the accelerating force $\overrightarrow{F_{lin}}$ which gives the linear acceleration to the robot, and torque τ which gives angular acceleration to the robot. They are presented in the Fig. 4. The forces can be expressed using the following equations [29]:

$$|\overrightarrow{F_c}| = m\frac{v^2}{r} = m\omega v = m\omega^2 r \tag{61}$$

$$\overrightarrow{F_c} = m\vec{a} \tag{62}$$

$$\vec{\tau} = J\vec{\varepsilon} \tag{63}$$

where: $\vec{a}$ is linear acceleration, $\vec{\varepsilon}$ is angular acceleration, $\vec{r}$ is current turning radius, ω is current angular speed, v is current linear speed, J is robots moment of inertia. In practice, all the forces mentioned above must result from interaction force between wheels and the ground. Figure 5 shows them denoted as $\overrightarrow{F_r}$ and $\overrightarrow{F_l}$ together with their decomposition into components perpendicular ($\overrightarrow{F_{rx}}$ and $\overrightarrow{F_{lx}}$) and parallel ($\overrightarrow{F_{rx}}$ and $\overrightarrow{F_{lx}}$) to direction of motion. Comparing forces in Figs. 4 and 5 we can write the following equations (only magnitude is taken into account, since compared forces has the same direction):

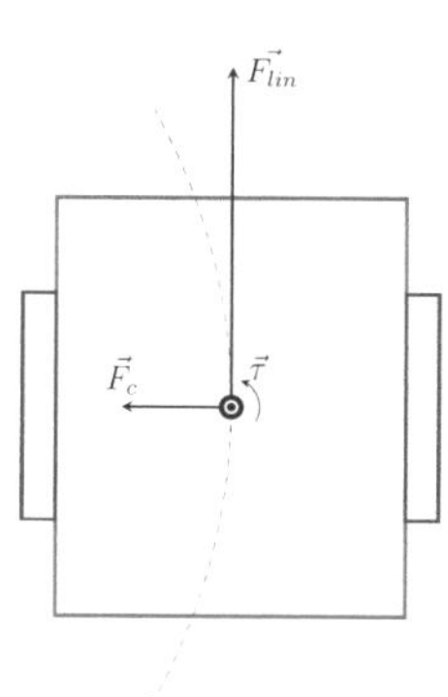

Fig. 4 The forces that acts on robot during acceleration

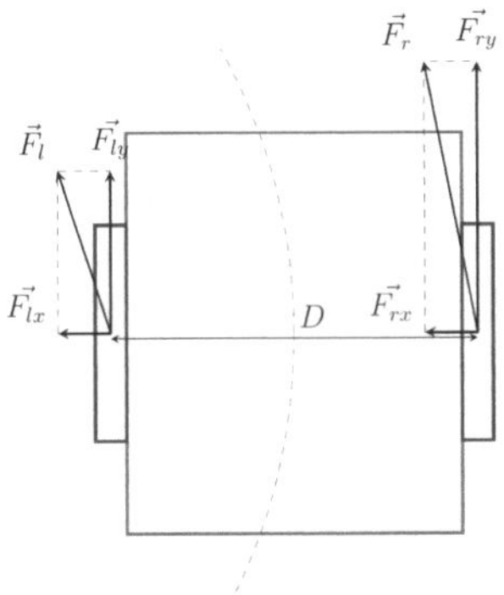

Fig. 5 The forces that acts on each wheel

$$\tau = \frac{D}{2}F_{ry} - \frac{D}{2}F_{ly} \tag{64}$$

$$F_{lin} = F_{ry} + F_{ly} \tag{65}$$

$$F_p = F_{rx} + F_{lx} \tag{66}$$

From (64) and (65) the following equations can be obtained:

$$F_{ry} = \frac{F_{lin}}{2} + \frac{\tau}{D} \tag{67}$$

$$F_{ly} = \frac{F_{lin}}{2} - \frac{\tau}{D} \tag{68}$$

Since the wheels are identical and the robot is symmetrical, we can assume that centripetal force is decomposed equally between both wheels.

$$F_{rx} = F_{lx} \tag{69}$$

therefore

$$F_{rx} = F_{lx} = \frac{1}{2}F_p \tag{70}$$

Substituting (61), (62), (63) into (67), (68) and (70) we obtain:

$$F_{ry} = \frac{m}{2}a + \frac{J}{D}\varepsilon \tag{71}$$

$$F_{ly} = \frac{m}{2}a - \frac{J}{D}\varepsilon \tag{72}$$

$$F_{rx} = F_{lx} = \frac{m}{2}v\omega \tag{73}$$

The resultant forces acting on each wheel can be expressed in a following way:

$$F_r^2 = F_{rx}^2 + F_{ry}^2 \tag{74}$$

$$F_l^2 = F_{lx}^2 + F_{ly}^2 \tag{75}$$

therefore:

$$F_r^2 = \left(\frac{m}{2}v\omega\right)^2 + \left(\frac{m}{2}a + \frac{J}{D}\varepsilon\right)^2 \tag{76}$$

$$F_l^2 = \left(\frac{m}{2}v\omega\right)^2 + \left(\frac{m}{2}a - \frac{J}{D}\varepsilon\right)^2 \tag{77}$$

As it was said earlier, both forces must not exceed the maximal value of friction force, which is given by the following equation [29]:

$$F_{max} = N\mu \tag{78}$$

where N is magnitude of normal force and μ is coefficient of static friction. In this work it is assumed, that weight is distributed equally between both wheels, and the normal force for each of them is equal to the half of weight. It is not always true, even if mass distribution in the robot is symmetrical, since the centrifugal force unweight the inner wheel at the cost of outer wheel's. However assuming that center of mass is located at low height and centrifugal acceleration is small with respect to gravitational acceleration, the inaccuracy will also be small. Therefore (78) takes the form:

$$F_{max} = \frac{1}{2}mg\mu \tag{79}$$

where m is robot's mass and g is acceleration due to gravity. Therefore if we substitute (76), (77) and (79) to:

$$F_r \leq F_{max} \tag{80}$$

$$F_l \leq F_{max} \tag{81}$$

we have:

$$\left(\frac{m}{2}v\omega\right)^2 + \left(\frac{m}{2}a + \frac{J}{D}\varepsilon\right)^2 \leq \left(\frac{1}{2}mg\mu\right)^2 \tag{82}$$

$$\left(\frac{m}{2}v\omega\right)^2 + \left(\frac{m}{2}a - \frac{J}{D}\varepsilon\right)^2 \leq \left(\frac{1}{2}mg\mu\right)^2 \tag{83}$$

which is equivalent to:

$$(v\omega)^2 + (a + k\varepsilon)^2 \leq (a_{max})^2 \tag{84}$$

$$(v\omega)^2 - (a + k\varepsilon)^2 \leq (a_{max})^2 \tag{85}$$

where:

$$a_{max} = g\mu \quad (86)$$

$$k = \frac{2J}{dm} \quad (87)$$

and finally:

$$|a| + |k\varepsilon| \leq \sqrt{a_{max}^2 - (v\omega)^2} \quad (88)$$

The a_{max} was obtained experimentally for $\omega = 0$ and $\varepsilon = 0$ by applying different values of acceleration a in order to check what is the greatest one for which no sliding is observed (in several successive trials). Then k was obtained by setting $v = 0$ and finding maximal value of angular acceleration (ε_{max}) in the similar way like previously. Then k was calculated according to transformed form of (88) for $v = 0$ and $a = 0$:

$$k = \frac{a_{max}}{\varepsilon_{max}} \quad (89)$$

The a_{max} used in the software was lowered by 50% with respect to obtained value since the coefficient of friction may vary depending on many conditions and we want to minimize the risk of any sliding.

5.1 *Acceleration Limiting*

The input of the acceleration limiting controller are current and desired linear and angular speeds (v_c, ω_c, v_f, ω_f). The output should be linear and angular acceleration values (a and ω) that satisfies the Eq. (88) and that leads finally to desired speeds. This can be, however, achieved in many ways. The problem can be formulated as finding trajectory from (ω_c, v_c) to (ω_f, v_f) under constraint (88).

Several problems have to be considered in order to make the solution universal. The first problem is that when $v\omega$ product increases, the right hand side of (88) decreases (eventually to 0) which limits the possible value of a and ω. For this reason, the following limitation is held in the implemented solution:

$$v\omega \leq \sqrt{2}a_{max} \quad (90)$$

This constraint limits area of allowed (ω, v) pairs and introduces a forbidden region for them. The question arises what to do if desired point (ω_f, v_f) lays in the forbidden region, which may put the acceleration control algorithm into confusion. The best action which can be taken depends on the overall aim of control algorithm, which is not known at this level. For this reason the decision is left to the higher level algorithm. One of three policies may be selected (see Fig. 6):

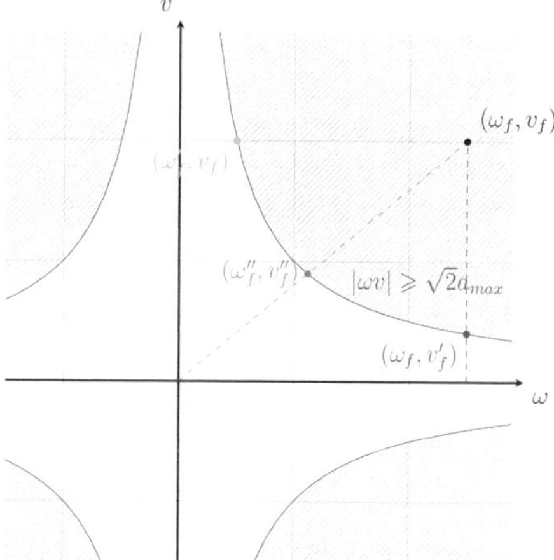

Fig. 6 Possible solutions when desired speeds are in forbidden region. The 'set v' policy is denoted by *green point*, the *blue* one denotes 'set ω' and the *orange* one is for 'set radius'

- **Set** v—v_f is unchanged, but the ω_f is modified in order to fit into allowed region.
- **Set** ω—analogously, but ω_f is unchanged an the v_f is modified.
- **Set radius**—both ω_f and v_f are modified, but their ratio (which defines turning radius) is left unchanged.

The second thing to consider is choosing the trajectory as mentioned above. The basic possibilities are achieving angular speed first and then linear ('angular first') or in the reverse order ('linear first'). Another approach is going along the shortest path in the ω, v space (which does not mean the shortest in time), let's call it 'shortest path'. A bit more complicated case is reaching proper turning radius first and then increase speed proportionally, holding this radius ('radius first'). As it will be shown later, the turning radius is dependent on the ratio $\frac{v}{\omega}$, so in this case the line connecting origin with (v_f, ω_f) point should be reached. In the Fig. 7 trajectories obtained with these methods are shown. All these solutions have, however, a serious drawback—they behaves poorly under fast changes of desired values before they are reached. Unfortunately, it can be a common case, when some higher level algorithm changes

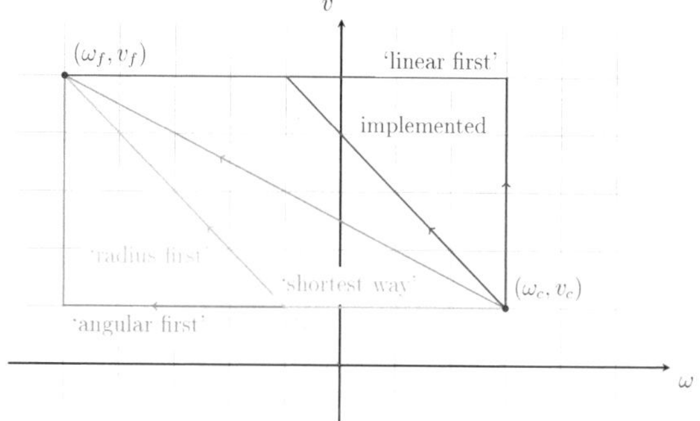

Fig. 7 Trajectories in angular-linear speed space obtained for different acceleration limiting controllers

decision often, and it should be handled properly by the controller. For example when desired angular speed jumps frequently from great positive to negative value while desired linear speed is still a bit greater than the current, all mentioned solutions, except 'linear first', will spent most of time changing the angular speed back and forth, while the linear speed is almost or totally unaffected. The implemented solution is resistant to such conditions. The main idea is simple—the ratios $\frac{|j\varepsilon|}{|j\varepsilon|+|a|}$ and $\frac{|a|}{|j\varepsilon|+|a|}$ are ensured to be not less than some assumed constants as long as both desired linear and angular speed is not reached:

$$|a| \geq \rho(|j\varepsilon| + |a|) \tag{91}$$

$$|j\varepsilon| \geq (1-\rho)(|j\varepsilon| + |a|) \tag{92}$$

After current accelerations are obtained, the setpoint of wheel's speed controller must be set to the proper value. The exact implementation of the acceleration controller checks the current and desired speeds, determines angular and linear accelerations and then computes new value of wheel's speeds. Since there is no need to compute it more frequently than it is sampled by the wheel's speed controller (1 kHz), then acceleration controller's job is done with the same frequency. The following formulas are used:

$$v(t+h) = v(t) + \Delta v \tag{93}$$

$$\omega(t+h) = \omega(t) + \Delta\omega \tag{94}$$

$$\Delta v = ha \tag{95}$$

$$\Delta\omega = h\varepsilon \tag{96}$$

where h is sampling period. In the implementation, however, rather than angular speed ω another variable is used—the differential speed between right wheel and robot's center denoted by p. The relation between p and ω can be obtained easily. The angular speed is by definition [29]:

$$\omega = \frac{d\alpha}{dt} \tag{97}$$

where $d\alpha$ is angle increment, which is equal to the length of the arc traveled by the right wheel s_r divided by the radius of this arc, which is equal in this case to half of distance between contact points D.

$$d\alpha = \frac{ds_r}{\frac{1}{2}D} \tag{98}$$

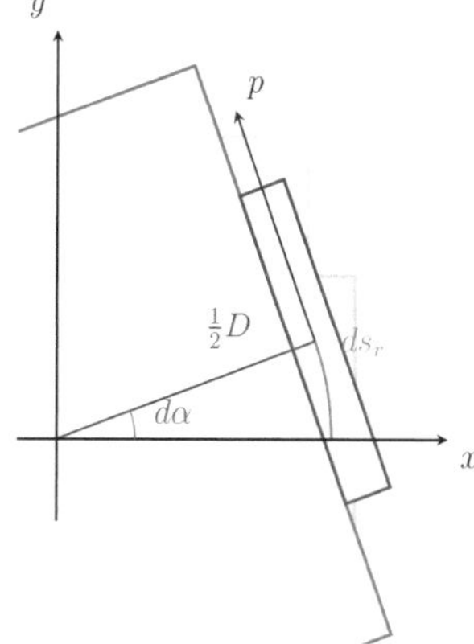

Fig. 8 The rotation of robot in time dt

On the other hand, the ds_r can be easily computed as follows (see Fig. 8):

$$ds_r = p \cdot dt \tag{99}$$

We have therefore:

$$\omega = \frac{2p}{D} \tag{100}$$

The (94) and (96) become then:

$$p(t+h) = p(t) = \Delta p \tag{101}$$

$$\Delta p = h\frac{2\varepsilon}{D} \tag{102}$$

Since rather Δv and Δp are obtained instead of a and ε, the constraint given by Eq. (88) was transformed into the following form:

$$|\Delta v| + |\frac{1}{j}\Delta p| \leq h\sqrt{a_{max}^2 - (v\omega)^2} \tag{103}$$

where $j = \frac{1}{kD}$. Then the algorithm is as follows—Algorithm 2.

The point of intersection can be obtained from the following system of equations:

$$\begin{cases} \sigma_w j\Delta v + \sigma_p \Delta p = h\sqrt{2}a_{max}j \\ \sigma_{vp} v \cdot p \cdot \frac{2}{D} = \sqrt{2}a_{max} \end{cases} \tag{104}$$

where σ_v and σ_p and σ_{vp} are defined as follows:

$$\sigma_v = \begin{cases} +1 & v_f > v_c \\ -1 & v_f < v_c \end{cases} \tag{105}$$

$$\sigma_p = \begin{cases} +1 & p_f > p_c \\ -1 & p_f < p_c \end{cases} \tag{106}$$

$$\sigma_{vp} = \begin{cases} +1 & vp > 0 \\ -1 & vp < 0 \end{cases} \tag{107}$$

These equations can be written in the following form:

$$\begin{cases} \sigma_v j \Delta v + \sigma_p p = C_1 \\ \sigma_{vp} v \cdot p = C_2 \end{cases} \tag{108}$$

where:

$$C_1 = \sqrt{2} a_{max} j - \sigma_v j v_c - \sigma_p p_c \tag{109}$$

$$C_2 = \frac{D}{2} \sqrt{2} a_{max} \tag{110}$$

That leads to the following quadratic equation:

$$\sigma_p p^2 - C_1 p + \frac{\sigma_v}{\sigma_z} j C_2 = 0 \tag{111}$$

So the value of points of intersection (p, v) can be computed as follows:

Algorithm 2 Possible solutions algorithm.

Step 1. Compute $A = h\sqrt{a_{max}^2 - (v\omega)^2}$ which denotes the maximal Δv if $\Delta p = 0$.

Step 2. Compute $\Delta_m v = \rho A$ and $\Delta_m p = j(1 - \rho)A$ which denotes (respectively) minimal linear and angular speed increase under the condition that both speeds does not reach the desired value yet.

Step 3. If both $|v_f - v(t)| \le \Delta_m v$ and $|p_f - p(t)| \le \Delta_m p$ then assume $\Delta v = v_f - v(t)$ and $\Delta p = p_f - p(t)$.

Step 4. Else, if $|v_f - v(t)| \le \Delta_m v$ then set $\Delta v = v_f - v(t)$ and $\Delta p = j(A - \Delta v)$.

Step 5. Else, if $|p_f - p(t)| \le \Delta_m p$ then set $\Delta p = p_f - p(t)$ and $\Delta v = A - \frac{\Delta p}{j}$.

Step 6. Else, assume $\Delta v = \Delta_m v$ and $\Delta p = \Delta_m p$.

Step 7. If $(v(t) + \Delta v) \cdot (p(t) + \Delta p) \cdot \frac{2}{D} < \sqrt{2} a_{max}$ (new speeds would not lie in forbidden region) then set new speeds according to (93) and (101).

Step 8. In other case set $(v(t+h), p(t+h)$ to intersection of the hyperbola given by $v \cdot p \cdot \frac{2}{D} = \sqrt{2} a_{max}$ and one of the lines given by $|\Delta v| + |\frac{1}{j} \Delta p| = h\sqrt{2} a_{max}$.

$$\begin{cases} p = \frac{1}{2\sigma_p}\left(C_1 \pm \sqrt{C_1^2 - 4\frac{\sigma_v\sigma_p}{\sigma_{vp}}jC_2}\right) \\ v = \frac{C_1 - \sigma_p p}{\sigma_v j} \end{cases} \tag{112}$$

When new values of v and p are computed, the left and right wheel speed can be obtained as follows:

$$v_r = v + p \tag{113}$$

$$v_l = v - p \tag{114}$$

and they are passed to the lower control layer (wheel's speed controller).

5.2 *Turning Radius Control*

The interface of acceleration controller allows to set v_f and p_f directly, or indirectly by giving v_f and arctangent of inverse of turning radius ($atan\frac{1}{r}$). The reason of usage arctangent rather than r itself is to be able to serve whole range of radiuses including ∞ (which mean no turning). The arctangent of $\frac{1}{r}$ was used instead of arctangent of r, since we need to distinguish 0^+ and 0^- which mean right or left rotating around own axis, while distinction between $+\infty$ and $-\infty$ is not important (there is only one way of rectilinear motion). The function that was used maps both $+\infty$ and $-\infty$ to the same value (0) while 0^+ and 0^- are mapped respectively to $+\frac{\pi}{2}$ and $-\frac{\pi}{2}$. The value of p_f can be obtained from transformed form of (100):

$$p = \frac{D}{2}\omega \tag{115}$$

This formula was obtained for the coordinate system with origin at center of robot, but not rotating together with it, so ω will be the same as in global coordinate system bound with the ground. In such case, a value of ω can be computed as follows:

$$\omega = \frac{v}{r} \tag{116}$$

We have then:

$$p = v\frac{D}{2}\frac{1}{r} \tag{117}$$

Therefore for constant turning radius, the differential speed p (as well as angular speed ω) is proportional to linear speed v. It can be expressed as follows:

$$p = vf_t(r) \tag{118}$$

where $f_t(r)$ is turning factor:

$$f_t(r) = \frac{D}{2}\frac{1}{r} \tag{119}$$

Denoting $\gamma = atan(\frac{1}{r})$ (which is in fact passed to controller) we have:

$$f_t(\gamma) = \frac{D}{2}tan(\gamma) \tag{120}$$

The linear speed can be set using command speed. Depending on which kind of turning control is preferred (turning radius or differential speed), radius or turn command can be used.

6 Rotation by Given Angle

One of the highest-level algorithms implemented in the controller is rotating by given angle. When desired orientation is reached, the robot must stop rotating. Taking into account limited angular acceleration, the task is not trivial. The idea is to compute in each step such angular speed, that will make able to stop the robot before it cross the target angle. The main constraint is limited deceleration. First, we calculate the angle by which robot will rotate if it start decelerating immediately. From [29] we know:

$$\alpha(t) = \int \omega(t)dt = \int\int \varepsilon(t)dtdt \tag{121}$$

Assuming constant angular acceleration/deceleration it can be written in the following form:

$$\Delta\alpha = \omega_0\Delta t + \frac{\varepsilon_{max}\Delta t^2}{2} \tag{122}$$

where $\Delta\alpha$ is the angle by which robot rotates in time Δt if initial angular speed is ω_0 and angular acceleration is ε in whole interval. We assume $\omega_0 > 0$ and $\Delta\alpha \geq 0$. In such case, if from now the robot decelerates as quickly as possible, we can substitute $\varepsilon = -\varepsilon_{max}$, so we have:

$$\Delta\alpha = \omega_0\Delta t - \frac{\varepsilon_{max}\Delta t^2}{2} \tag{123}$$

We want the robot to stop after whole operation, so the final angular speed is 0:

$$\omega_0 - \Delta t\varepsilon_{max} = 0 \tag{124}$$

From this, the time needed to stop can be obtained:

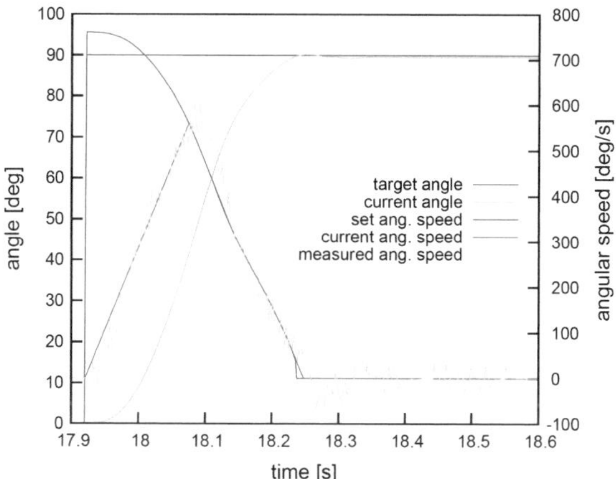

Fig. 9 Values registered during rotation by 90°

$$\Delta t = \frac{\omega_0}{\varepsilon_{max}} \tag{125}$$

So if robot start braking now, it will stop after rotation by:

$$\Delta\alpha = \frac{2\omega_0^2 - \omega_0^2}{2\varepsilon_{max}} = \frac{\omega_0^2}{2\varepsilon_{max}} \tag{126}$$

Transforming the above equation, the unknown ω_0 can be obtained:

$$\omega_0(t) = \sqrt{2\Delta\alpha(t)\varepsilon_{max}} \tag{127}$$

Therefore formula (127) allows to compute what angular speed the robot should have in order to stop exactly after rotation by $\Delta\alpha$ assuming constant deceleration ε_{max}. The implemented solution simply compute it periodically (with the same frequency as other algorithms work—1 kHz) for the remaining angle, and set it as a desired angular speed to the acceleration controller. If the angular speed is lower at the beginning, the acceleration controller causes that it changes smoothly up to ω_0. Then, the robot decelerates since ω_0 decreases with time (because of decreasing $\Delta\alpha$). The experimental results are shown in the Fig. 9. The ω_0 is called 'set ang. speed', and the 'current ang. speed' is an output from acceleration controller. A small overshot is an effect of non-perfectly tuned wheel's speed controller. There is one more thing which should be considered, however. It must be decided what value of ε_{max} will be optimal. As it was mentioned in Sect. 5, the maximal angular acceleration (and deceleration) depends on ωv product. Moreover, if desired linear speed is not the same as current, only some part of total acceleration of each wheel will result from angular acceleration of the robot (the rest will be for linear acceleration). There is, however, some guaranteed minimal value of angular acceleration/deceleration, which results from limitation of ωv (inequality (90)) and for the ratio of angular acceleration to total maximal acceleration (92). That gives the 'safe' deceleration, which is guaranteed independently of any conditions, which can be calculated as follows:

$$\varepsilon_{safe} = j(1-\rho) \cdot \sqrt{2}a_{max} \tag{128}$$

On the other hand, when robot is only rotating, and both linear acceleration and speed is equal to zero, we can assume much greater deceleration:

$$\varepsilon_{rotateonly} = j \cdot a_{max} \tag{129}$$

Both versions were implemented. The second one is intended to be used when only rotation is performed and allow much faster operation. The first one is guaranteed to work even if robot is moving with linear speed and/or accelerating linearly, but even when robot is not moving it takes longer to complete rotation.

7 Moving by Given Distance

There is also a similar algorithm to the previous, which allows to move robot forward or backward exactly by a given distance. The distance is measured on the real trajectory, so if robot is turning simultaneously the length of curve is taken into account (rather than displacement). The problem analysis and implemented solution is analogous to rotating by angle. The equation for current speed takes the form:

$$v_0(t) = \sqrt{2\Delta s(t) a_m} \tag{130}$$

where Δs is the distance left to travel.

Also in this case there are two versions of the command: one 'safe' which works properly even when robot is turning at the same time, and one which works faster, but assumes angular speed equal to zero. The experimental results obtained during operation of this algorithm are presented in the Fig. 10. It can be observed, that it behaves very similarly to the previous presented (Fig. 9) but it takes much more time to move by one meter than to rotate by 90° (2 s vs. 300 ms).

8 Reaching Given Destination Point

The last algorithm presented in this work makes the robot go to given position and stop there. The goal is to do it relatively quickly and smoothly, even if robot has some initial speed in wrong direction. Therefore the simple algorithm which just stops the robot, rotate it in the direction of destination and then move straight by proper distance would not fulfill the requirements. The algorithm should rather control current

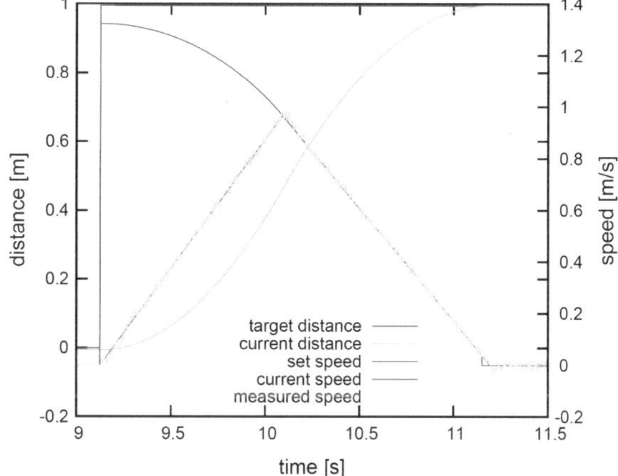

Fig. 10 Values registered during moving by 1 m forward

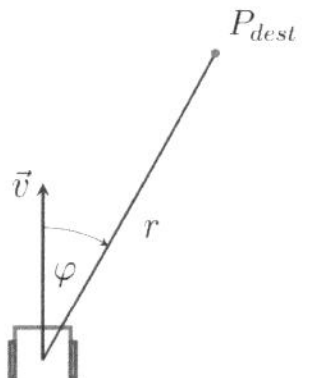

Fig. 11 The robot going to point P_{dest}

linear and angular speed in such a way, that robot turns smoothly in the direction of target point and slows down only if it is necessary. The implemented solution use simultaneously two algorithms described earlier (rotate by angle, move by distance) in their 'safe'. In the acceleration controller the 'set ω' policy is chosen (see Fig. 6), since rotation in direction of target point has a priority (there is no reason to move quickly forward if robot is not rotated properly). Moreover, when the absolute value of the angle at which robot 'see' the destination (φ in the Fig. 11) is greater than 90°, the 'move by distance' algorithm is deactivated and desired linear speed is set to 0. In such case robot decelerates until 90°) threshold is crossed. Then both algorithms become active again. The values passed to them (distance and angle) are computed periodically with the 1kHz frequency. The distance by which robot should move is set always to the distance between robot and the destination (r in the Fig. 11). Unless $\varphi = 0$, this is not real length of trajectory that will be traveled before reaching the point, but taking into account that φ must become 0 finally, this estimation is convergent to proper value. It should be noticed, that the value passed to 'move by' it never greater than real distance that will be traveled, which ensures that robot will have enough time for braking. The only negative effect is that the time of whole operation may not be as short at it could be if the distance was estimated more precisely.

In the Fig. 12 the exemplary trajectory is presented. This shows that algorithm works properly, also in the case when initial speed is non zero. This is also evidence that 'move by' and 'rotate by' algorithms can be used effectively (Fig. 13).

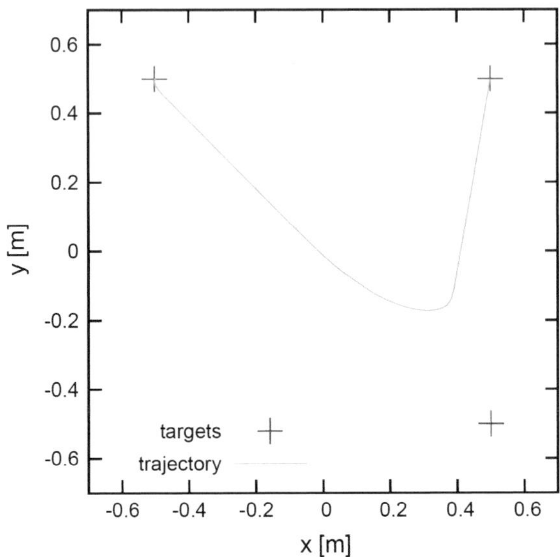

Fig. 12 Trajectory registered for going to point algorithm. Motion from the *upper left* to the *bottom position* was interrupted by new target (*upper right*)

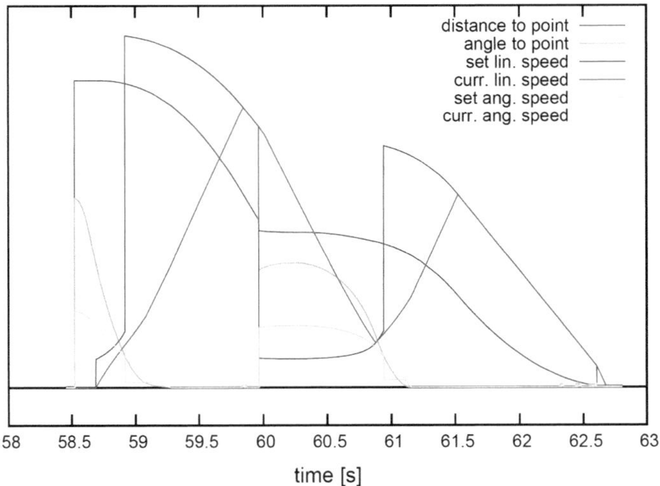

Fig. 13 Values registered during motion shown in the Fig. 12

9 Conclusions

The final version of the controller mets all requirements that were stated. It provides a reliable platform for development of distributed control system. Most commonly used actions are already implemented, and proved to work properly, fully taking the advantage of the robot's capabilities. Moreover, there is a reserve of computational power that allows implementing some other, more specialized algorithms on board. Speaking more precisely, the following goals were reached. Wheel speed controller sets properly desired speed. Acceleration limiter fulfill it's role, allowing to achieve great speed without skidding.

Desired values of linear and angular speed can be changed freely and the controller behaves properly in any conditions. Robot can rotate by given angle with precision better than 1°. Robot can move forward backward by given distance and stop with precision better than 2 mm. Robot can reach any destination point and stop in distance less than 5 mm from it. It works properly for any initial conditions in reasonable time. The MCU based on the ARM core that was applied provides enough computational power for implementation of nontrivial algorithms with high sampling rates [30]. Generally, everything works properly, however the following things could be improved. The PID is may be tuned better. Especially overshot could be reduced and oscillations may be eliminated. It should be kept however, that it is more important to obtain smooth ramp response without oscillations than improving step response. Maybe some other controller for wheel's speed could be used [31]. This is a wide field for research. If the ramp response was improved considerably, the maximal acceleration that not cause sliding would increase, so the robot may reach greater speed in the same time. The commands that rotate or move the robot may be extended in order to respect other parameters like desired time of reaching target or maximal speed (linear or angular).

The algorithm that make robot going to destination point may check if going backward will not allow to reach destination faster. Some algorithm that allow robot track some path (planned by the master system) could be implemented in the controller.

Part of presented solution was implemented in the other mobile robots [32, 33].

References

1. K-Team. (2015). *Website of the K-team mobile robotics*. http://www.k-team.com.
2. Barrientos, A., Vidal, J., Quesada, E., Oliver, J., Macotela, F., & Dominguez, M. (2013). Design and construction of mini-robot for gas LP detection using a mobile device. *Latin America Transactions, IEEE (Revista IEEE America Latina), 11*, 1295–1300.
3. Barrios-Aranibar, D., & Alsina, P. J. (2007). Imitation learning, an application in a micro robot soccer game. In *Mobile robots, the evolutionary approach*, Studies in computational intelligence, ch. 10 (pp. 201–219). Berlin: Springer.
4. Jaskot, K., & Łakota, T. (2016). Experimental mobile robot—Hardware. In *Innovative simulation systems, studies in systems, decision and control*, ch. 15 (Vol. 33, pp. 277–289). Springer.
5. Jaskot, K., & Łakota, T. (2016). Experimental mobile robot-software. In *Innovative simulation systems, studies in systems, decision and control*, ch. 15 (Vol. 33, pp. 303–316). Springer.
6. Skrzypczyński, P. (2005). Uncertainty models of vision sensors in mobile robot positioning. *Int. J. Appl. Math. Comput. Sci., 15*, 73–88.
7. Babiarz, A., Bieda, R., & Jaskot, K. (2013). A distributed control group of mobile robots in a limited area with a vision system. In *Vision based systems for UAV applications*, Studies in computational intelligence, ch. 10 (pp. 157–175). Springer.

8. Switonski, A., Josinski, H., Jedrasiak, K., Polanski, A., & Wojciechowski, K. (2010). Classification of poses and movement phases. *Lecture Notes in Computer Science*.
9. Ryt, A., Sobel, D., Kwiatkowski, J., Domzal, M., Jedrasiak, K., & Nawrat, A. (2015). Real-time laser point tracking. In *International Conference on Computer Vision and Graphics*, pp. 542–551.
10. Sobel, D., Jedrasiak, K., Daniec, K., Wrona, J., Jurgas, P., & Nawrat, A. (2014). Camera calibration for tracked vehicles augmented reality applications. In *Innovative control systems for tracked vehicle platforms* (pp. 147–162).
11. Nawrat, A., & Jedrasiak, K. (2008). Fast colour recognition algorithm for robotics. *Problemy Eksploatacji* (pp. 69–76).
12. Daniec, K., Iwaneczko, P., Jedrasiak, K., & Nawrat, A. (2013). Prototyping the autonomous flight algorithms using the prepar3D® simulator. In *Vision based systems for UAV applications* (pp. 219–232).
13. Laumond, J.-P. (1998). *Robot motion planning and control (Lecture notes in control and information sciences)*. Springer.
14. Borenstein, J., Everett, H. R., & Feng, L. (1996). *Where am I?: Sensors and methods for mobile robot positioning*. Technical report, University of Michigan.
15. Ward, C. C., Iagnemma, K. (2007). Model-based wheel slip detection for outdoor mobile robots. In *Proceedings of the IEEE International Conference on Robotics and Automation, 2007, Roma, Italy*, pp. 2724–2729.
16. Gonzalez, R., Fiacchini, M., Alamo, T., Guzman, J., & Rodriguez, F. (2009). Adaptive control for a mobile robot under slip conditions using an LMI-based approach. In *Proceedings of the European Control Conference, 2009, Budapest, Hungary*, pp. 1251–1256.
17. Yu, T., & Nilanjan, S. (2014). Control of a mobile robot subject to wheel slip. *Journal of Intelligent & Robotic Systems, 74*, 915–929.
18. Topalov, A. V. (2011). *Recent advances in mobile robotics*. InTech.
19. Amitava, C., Anjan, R., & Nirmal, S. N. (2013). Mobile robot navigation. In *Vision based autonomous robot navigation*, Studies in computational intelligence, ch. 1 (pp. 1–20). Berlin: Springer.
20. Babiarz, A., Bieda, R., & Jaskot, K. (2013). Vision system for group of mobile robots. In: *Vision based systems for UAV applications*, Studies in computational intelligence, ch. 9 (pp. 129–156). Springer.
21. Bereska, D., Daniec, K., Jedrasiak, K., & Nawrat, A. (2013). Gyro-stabilized platform for multispectral image acquisition. *Vision based systems for UAV applications* (pp. 115–121).
22. Åström, K. J., & Hägglund, T. (2004). Revisiting the Ziegler-Nichols step response method for PID control. *Journal of Process Control, 14*, 635–650.
23. Gessing, R. (2004). *Control fundamentals*. Silesian University of Technology.
24. Åström, K. J., & Hägglund, T. (2006). *Advanced PID control*. ISA—The Instrumentation: Systems, and Automation Society.
25. Segovia, V. R., Hägglund, T., & Åström, K. J. (2014). Measurement noise filtering for PID controllers. *Journal of Process Control, 24*, 299–313.
26. Segovia, V. R., Hägglund, T., & Åström, K. J. (2014). Measurement noise filtering for common PID tuning rules. *Control Engineering Practice, 32*, 43–63.
27. Khan, H., Iqbal, J., Baizid, K., & Zielińska, T. (2015). Longitudinal and lateral slip control of autonomous wheeled mobile robot for trajectory tracking. *Frontiers of Information Technology & Electronic Engineering, 16*, 166–172.
28. Takeuchi, M., Ikeda, T., & Minami, M. (2002). Modeling of a mobile robot including slipping of carrying objects. In *Proceedings of the 41st SICE Annual Conference, 2002, Osaka, Japan*, pp. 2412–2417.
29. Halliday, D., Resnick, R., & Walker, J. (2014). *Fundamentals of physics extended* (10th ed.). Wiley.
30. Ostafew, C. J., Schoellig, A. P., Barfoot, T. D., & Collier, J. (2014). Speed daemon: Experience-based mobile robot speed scheduling. In *Canadian Conference on Computer and Robot Vision, 2014, Montreal, Canada*, pp. 56–62.

31. Ruiz, U., Marroquin, J. L., & Murrieta-Cid, R. (2014). Tracking an omnidirectional evader with a differential drive robot at a bounded variable distance. *International Journal of Applied Mathematics and Computer Science*, *24*, 371–385.
32. Jaskot, K., & Knapik, K. (2014). Building the environment map using the group of mobile robots. *Przegląd Elektrotechniczny*, *12*, 30–39.
33. Babioch, K., & Jaskot, K. (2015). Inspection robot. *Przegląd Elektrotechniczny*, *1*, 55–64.

Hierarchical Game Approach to Solve Conflicts in Multiagent Systems

Witold Brandys, Adam Gałuszka, Karol Jędrasiak, Joanna Radziszewska and Krzysztof Daniec

1 Introduction

In an environment where there are many agents operating independently, each of them have incomplete knowledge about the environment, trying to achieve their own goals, conflicts of interest arise. The agreement between the agents can be achieved through the exchange of information and negotiation. In this case, each of the active agents may seek to: present convincing arguments to carry out its purpose, to offer in exchange relevant information, or offer to perform some action which another agent is not able to perform [1]. In order to achieve these effects it is imperative to exchange of information between agents. In the literature solutions often lead to a hierarchical approach, in which selected agents act as so-called. leaders, and others—the so-called. followers (e.g. [2–6]).

However, there are situations in which communication is impossible. The reason for this may be, for example, by finding in the environment that prevents contact, the high cost of maintaining communication or failure of the communication system. There are situations in which there is no possibility of contact, or agent cannot wait for its restoration. In this case, the system has to apply an algorithm that will allow agents to make decisions independently. A special case of such a situation is one in which the communication is one-way. This means that one of the agents do not have any information about the activities of other agents, and they may react having access to the information conveyed to them. A similar situation occurs when one of the agents must take action without waiting for the remaining agents.

W. Brandys · A. Gałuszka · K. Jędrasiak · K. Daniec (✉)
Institute of Automatic Control, Silesian University of Technology,
Akademicka 16, 44-100, Gliwice, Poland
e-mail: krzysztof.daniec@polsl.pl

J. Radziszewska
VR Technology, Gliwice, Poland
e-mail: j.radziszewska@vrtechnology.pl

A. Nawrat et al. (eds.), *Advanced Technologies in Practical Applications for National Security*, Studies in Systems, Decision and Control 106,
https://doi.org/10.1007/978-3-319-64674-9_8

The proposed in the article method of finding a solution uses the properties of reversibility of certain planning systems in artificial intelligence and a methodology to seek a solution under conditions of ambiguous information about the situation of the initial problem. Because the solution found in the general case is incomplete (person specifies the actions of agents), showing you how to specify the elements of the hierarchical plan using game theory (Stackelberg non-cooperative equilibrium).

The problems of this type can be modeled as a STRIPS system (blocks world environment) with one initial and target a number of alternative states.

If STRIPS planning problem is reversible, it is possible to use the mechanism of planning in the presence of incomplete information to solve the inverse problem and to find a solution to the problem of the original [7].

Illustration of a situation with a number of agents may be an example in which the role of agents perform work with the grippers capable of moving objects in the environment blocks world.

The subject of the work is to find solutions to the conflicts between the agents and to investigate the possibility of applying game theory. Due to the existing hierarchy to clarify the plan Stackelberg non-cooperative equilibrium was used.

2 Problem Definition—Assumptions

In order to model the situation being examined, the following assumptions [7]:

- Robots have autonomy of action—act as agents.
- The initial state contains a finite number of blocks on the table for an unlimited amount of space.
- Two (or more general case) robots (agents) are trying to transform the initial state, each in a separate way (each agent wants to achieve its own goal).
- Purpose of each robot is composed of sub-objectives.
- Sub-goal Everyone has their own preferences (sub-objectives are more or less important for agents) expressed as numerical values.
- Agents have different ability to influence the external environment. Each robot must not be able to perform all operations leading to the realization of its sub-objectives.
- Agents can not cooperate (this assumption is justified, e.g. In the case of the environment in which communication is prohibited or devices for communication should fail).
- One of the agents must perform an action without waiting for the others. Such a situation may occur when one of the robots working in real-time system and the key for him is to complete the task within a certain time. The purpose of the second agent is a rational response to the decision of the first agent.
- Profit is defined as the sum of completed sub-objectives.

The solution will be understood as finding a plan (action sequence) leading to the maximization of the profits of each of the agents.

2.1 STRIPS System

The STRIPS (STanford Research Institute Problem Solver) uses modeling environment robot classic environment called World of Blocks (Blocks World) [8]. Blocks World domain can be described by the system STRIPS by four letters (C, O; I; G) [9]:

- C—a finite set of basic formulas/atoms (facts), called conditions;
- O—finite set of operators (called. “Operators”), also called shares;
- I—a finite set of predicates indicating the initial state (called. “Initial state”);
- G—a finite set of predicates indicating the status of the destination (called. “Goal state”).

For example situation of the world of blocks set of facts can be a list:

(on x y)—block x is located on block y,
(on-table x)—block x is located on table,
(clear x)—on block x there is no other block,
(arm-empty)—arm of robot is empty,
(holding x)—arm of robot is holding block x.

Operators (O) in the STRIPS representation have:

- a name which can contain arguments—for example PICK-UP x, where x is the argument, and the PICK-UP name;
- a list of preconditions (pre), which is a list of facts that need to be true to be able to apply the action;
- a list of deletions (del), which is a list of facts that ceases to be true after applying the operator;
- a list of references (add), which is a list of facts that become true after applying the operator.

Construction of the operator (PICK-UP x), which means lifting the pad from the table, as follows [10]:

List of preconditions (pre): (on-table x) (x clear), (arm-empty).
List deletions (del) (on-table x) (x clear), (arm-empty).
List of references (add): (holding x).
The operators:
(PUT-DOWN x)—the location of the pad on the table
List of preconditions (pre) (holding x)
List deletions (del) (holding x)
List of references (add): (on-table x) (x clear), (arm-empty).
(STACK x y)—position of the block on the stack (another block)
List of preconditions (pre) (holding x) (y clear).
List deletions (del) (holding x) (y clear).
List of references (add): (arm-empty), (x on y), (x clear).
(UNSTACK x y)—raising the block from the stack (from another block)

List of preconditions (pre): (arm-empty), (clear x) (x on y).
List deletions (del): (arm-empty), (clear x) (x on y).
List of references (add): (holding x) (y clear).

The solution type STRIPS is to find a set of operators transforming the initial state to target state of the world with the task.

2.2 STRIPS System for Planning Problem for the Environment with Many Agents as a Problem Inverse to the World of Block with Ambiguous Initial Situation

The problem, in which there are many possible initial states (we are unable to determine which of them really exists) and a target state, is called the problem of planning with incomplete (incomplete) information [11]. In the absence of complete information about the external environment, you can look for proof plan. In this case, we analyze the effects of an operator in every possible situation so as to achieve the target state, regardless of the possible initial state. Incomplete immune plan is one that does not specify a set of variables for the operator. For example STACK1 B will mean the position of the block B by the robot 1 on some other—unspecified (Fig. 1).

As a problem opposite to it is the situation in which one initial state and the target number of states [12]. This corresponds to the problem of planning in the environment many agents (robots), where each of them wants to get its own purpose (Fig. 2). The source of the conflict is to identify mutually conflicting sub-goals

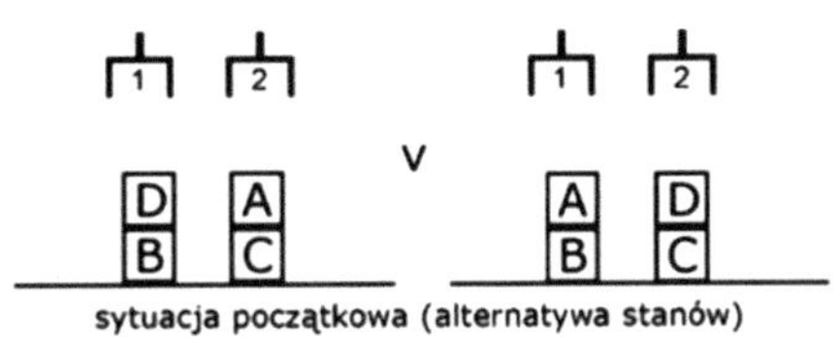

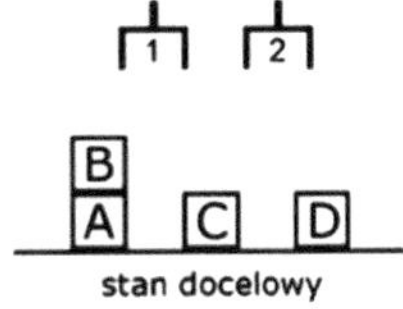

1. (UNSTACK1 A) (UNSTACK2 D)
2. (PUT-DOWN1 A) (PUT-DOWN2 D)
3. (PICK-UP1 B)
4. (STACK1 B)

Fig. 1 Ambiguous initial state and conformant plan

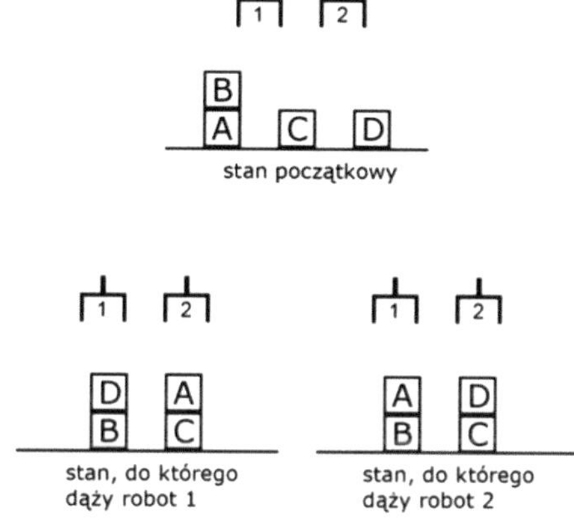

Fig. 2 Inverse problem with incomplete conformant plan

1.(UNSTACK1 B)
2.(PUT-DOWN1 A)
3.(PICK-UP1 A) (PICK-UP2 D)
4.(STACK1 A) (STACK2 D)

of individual agents. Agent 1 (robot 1) tends to set the block D on the block B, while the two agent (robot 2) tends to set the block D on the block C. Likewise, block A. If we assume further, that the manipulation of the blocks A and B may carry the agent 1, and the blocks C and D agent 2, the robots must cooperate with each other in spite of contradictory final objectives. Given in Fig. 2 incomplete plan resistant leads to alternative states. If you are unable to find a plan for scheduling problem in the presence of incomplete information, it is possible to obtain on the basis of solutions for planning in the environment many agents [7]. In practice, this involves the conversion of operators on the reverse (UNSTACK on STACK, PICK-UP on the PUT-DOWN STACK on UNSTACK and PUT-DOWN PICK-UP) and the reversal of the order of operators.

Plans for solving problems with Figs. 1 and 2 are imprecise and therefore ambiguous. This is due to the fact that the sub-objectives agents (robots) are contradictory. The conflict can not be avoided, if you are looking for a plan taking into account the different initial situations. To solve it we use techniques known from game theory.

3 The Solution of the Conflict with Use of Non-cooperative Equilibrium

On the basis of vague plan and the non-cooperative equilibrium will be shown how to obtain a clear plan, a compromise for the conflict situation. The basis for constructing the game, which is an illustration of the conflict between robots are the priorities for achieving specific sub-objectives. It is assumed that these priorities may represent profits.

3.1 Non-cooperative Equilibrium in Pure Strategies

For the problem of the game between the two players and the matrix representation, we assume that the player D1 selects lines and the player D2 selects columns of the matrix. Both players try to minimize quality indicator, defined respectively arrays MA = {aij} and MB = {bij}. The result pair selection strategy (i, j) is therefore a pair of results (aij, bij).

Assume the following assumption: 1 agent is a robot working in a real-time system, for which overriding priority is to complete the task within a certain time.

If we assume that the agent 2 can follow the agent's actions 1—solve the conflict situation in the sense of balance von Stackelberg. Role players are then unbalanced—one of the players, leader, has the ability to override its strategy in relation to the other player—Follower. The task Follower is a rational response to the leader's decisions.

The set of rational reaction (optimal response) Follower (player D2):

$$\mathrm{R(i)} = \{\mathrm{k} : \forall \mathrm{j}\ \mathrm{bik} \leq \mathrm{bij}\}$$

Von Stackelberg strategies for leader i0 (S*—leader cost)

$$\mathrm{maxj} \in \mathrm{R(i0)}\ \mathrm{ai0\ j} = \mathrm{mini\ maxj} \in \mathrm{R(i)}\ \mathrm{aij} = \mathrm{S} * (\mathrm{A}).$$

Element j ϵ R(i0) is the answer follower strategy i0 leader. Pair (i0, j0) is a solution of equilibrium Stackelberg.

The same solution we use when the agent 2 is ahead of agent 1. It will change only their role. Leader is the agent 2, and the follower is agent 1. A remedy to the situation in the sense of von Stackelberg, including obtaining a complete plan is shown later in this paper, together with the example.

If the agent performs two actions at the same time as the agent 1, the conflict situation solved in the sense of a Nash equilibrium (e.g. [13]), as shown in [10].

3.2 The Method of Solving the Problem

Algorithm of generating planning problem in the World of Blocks up to obtain a complete plan can be summarized as follows:

First, we look for a plan solving the problem of ambiguous initial situation. Found plan includes "incomplete" STACK operators as shown in Fig. 2. Thus, the tile can be placed by a robot (agent) in any other block, realizing at the same time, their sub-goal, or sub-goal of another robot (agent). This leads to a conflict situation. If you specify different priorities sub-objectives, it will be possible to obtain a state leading to the final, which will be a compromise for all agents (robots) are in conflict. It should be noted that a compromise usually arises in the achievement of only some sub-goals for individual agents. To achieve such a compromise non-cooperative equilibrium is proposed.

Example. The problem in PDDL is defined as follows:

```
(define (domain problem1) (:requirements :strips :equality
:uncertainty)
(:predicates (on ?x ?y) (on-table ?x) (clear ?x) (arm1-empty)
(arm2-empty)
(holding1 ?x) (holding2 ?x) )
(:action pick-up
:parameters (?ob1)
:precondition (and (clear ?ob1) (on-table ?ob1) (arm-empty))
:effect (and (not (on-table ?ob1)) (not (clear ?ob1)) (not (arm-
empty)) (holding ?ob1))
(:action put-down
:parameters (?ob)
:precondition (holding ?ob)
:effect (and (not (holding ?ob) (clear ?ob) (arm-empty) (on-
table ?ob)))
 (:action stack
:parameters (?sob ?sunderob)
:precondition (and (holding ?sob) (clear ?sunderob))
:effect (and (not (holding ?sob)) (not (clear ?sunderob)) (clear
?sob) (arm-empty) (on ?sob ?sunderob)))
(:action unstack
:parameters (?x)
:precondition (and (clear ?x) (arm-empty))
:effect (and (holding ?x) (not (clear ?x)) (not (arm-empty))
(when (on ?x C) (and (clear C) (not (on ?x C)) ) ) (when (on ?x
D) (and (clear D) (not (on ?x D)).
```

The assumption that the robot 1 moves the blocks A, B, C, and the robot 2 moves the blocks D, E, F will modify the action stack on STACK1 and stack2 and appropriate action unstack, pick-up and put-down on the corresponding shares taking into account the manipulation only selected blocks.

Description planning problem—the initial situation (init) and the end (goal) is defined as follows (description in PDDL):

```
(define (problem s1) (:domain problem1) (:objects A B C D E F)
(:init (clear A) (clear E) (arm1-empty) (arm2-empty) (on-table B)
(on-table F)
(or (and (on C B) (on A C) (on D F) (on E D)
 (not (on A D)) (not (on E C)) (not (on D B)) (not (on C F)))
 (and (on A D) (on D B) (on E C) (on C F)
 (not (on A C)) (not (on C B)) (not (on E D)) (not (on D F)))))
(:goal (and (on A E) (on B D) (on C F) (on-table E) (on-table D)
(on-table F))
```

Resistant plan solving the problem therefore takes the following form:

```
step7 - ((((stack2 E)))
step6 - (((pick-up2 E)) ((stack1 A)))
step5 - (((stack2 D)) ((unstack1 A)))
step4 - (((pick-up2 D)) ((stack1 C)))
step3 - (((put-down2 D)) ((unstack1 C)))
step2 - (((pick-up2 D)) ((put-down1 B)))
step1 - (((unstack1 B)))).
```

This plan is imprecise, e.g. operator STACK1 A does not describe where to put the block A. Performing a reverse plan resistant by both robots leads to conflict. Since the robot 1 performs its action before the robot 2, the solution can be achieved by the use of von Stackelberg's non-cooperative equilibrium. Let the matrix MA = {aij} describes profit of robot 1, and the matrix MB = {bij} profit of robot 2. Profit is defined as the sum of preferences generated by sub-goals:

$$a_{ij} = l_{iR1} + l_{iR2}, b_{ij} = m_{iR1} + m_{iR2}, \quad (1)$$

where:

l_{iR1} robot 1 profit, when reached subgoal by executing *i*-th action,
l_{jR2} robot 1 profit, when reached subgoal by executing *j*-th action by robot 2,
m_{iR1} robot 2 profit, when reached subgoal by executing *i*-th action by robot 1,
m_{jR2} robot 2 profit, when reached subgoal by executing *j*-th action (Tables 1 and 2).

Table rows represent possible realizations operators STACK by the robot 1, resulting from the plan found. Accordingly column tables show the possible actions a robot 2. Thus, for example. If the robot 1 decides to perform the action {stack1 C}

Table 1 Matrix MA—agent's 1 profits

Agent1	Agent2			
	Stack D	Stack D	Stack E	Stack E
	B	F	C	D
Stack C B	3	3 + 2	**(3)**	3 + 4
Stack C F	0	2	0	4
Stack A C	1	1 + 2	1	1 + 4
Stack A D	0	2	0	4

Table 2 Matrix MB—agent's 2 profits

Agent1	Agent2			
	Stack D	Stack D	Stack E	Stack E
	B	F	C	D
Stack C B	1	0	**(2)**	0
Stack C F	1 + 3	3	2 + 3	3
Stack A C	1	0	2	0
Stack A D	1 + 4	4	2 + 4	4

due to its plan for implementing the sub-goal (on C B) to gain profit liR1 = 3 (the first row of the matrix A). Profit can be increased by 2, if the robot 2 performs an action in realizing the Sub-objective (on D F)—Column 2, or increased by 4 if the robot 2 performs an action in realizing the Sub-objective (on E D)—column 4.

Assuming hierarchy of robotic activities proposed to reach a compromise with the use of Stackelberg balance. This strategy will lead to the maximization of the sum of the priorities of all robots in points balance.

For the matrix of profits shown above, assuming that the robot 1 is the leader and the follower robot 2 solution to the conflict situation is the row 1 and column 3. The values profits robots are shown in bold in parentheses.

The use of such a strategy leads to clarify the plan and the final situation in which each of the agents achieves its objectives in terms of equilibrium von Stackelberg.

Precise plan that solves the problem thus takes the form:

```
step7 - ((((stack2 E)))
step6 - (((pick-up2 E C)) ((stack1 A D)))
step5 - (((stack2 D F)) ((unstack1 A)))
step4 - (((pick-up2 D)) ((stack1 C B)))
step3 - (((put-down2 D)) ((unstack1 C)))
step2 - (((pick-up2 D)) ((put-down1 B)))
step1 - (((unstack1 B)))),
```

while the target state resulting from the balance of von Stackelberg is

```
(and (on-table B) (on C B) (on E C) (clear C) (on-table F) (on D
F) (on A D) (clear A) )
```

4 Conclusions

The proposed method in the article allows you to search for solutions to the problems of planning in an multi-agent environment of hierarchical structure in which there is no communication between agents. This method uses the properties of reversibility of certain planning systems in artificial intelligence and a methodology to seek a solution under conditions of ambiguous information about the situation of the initial problem. Because the solution found in the general case is incomplete (person found specifies the actions of agents), showing you how to specify the elements of the hierarchical plan using game theory (Stackelberg non-cooperative equilibrium). The presented in the article solutions can be applied in real world with using specialized hardware [14–17] and software [18, 19, 20, 21, 22, 23, 24, 25, 26, 27, 28].

References

1. Kraus, K., Sycara, K. P., & Evenchik, A., Reaching agreements through argumentation: a logical model and implementation. *Artificial Intelligence, 104*, 1–69.
2. Peng, T., & Li, S. (2014). Formation Control of Multiple Wheeled Mobile Robots via Leader-Follower Approach. In *Proceedings of 26th IEEE Chinese Control and Decision Conference (CCDC)* (pp. 4215–4220).
3. Ćosić, A., Šušić, M., Graovac, S., & Katić, D. (2013). An algorithm for formation control of mobile robots. *Serbian Journal of Electrical Engineering, 10*(1), 59–72.
4. Madhevan, B., & Sreekumar, M. (2013). Tracking algorithm using leader follower approach for multi robots. In *International Conference on Design and Manufacturing (IConDM2013), Procedia Engineering* (Vol. 64, pp. 1426–1435).
5. Edelkamp, S., & Hoffmann, J. (2004). PDDL2.2. *The language for the classical part of the 4th International Planning Competition*. Technical report, 195, Albert-Ludwigs-Universitat Freiburg, Institut fur Informatik.
6. LaValle, S. M. (2006). *Planning algorithms*. Cambridge University Press.
7. Gałuszka, A., & Świerniak, A. (2002, July). Planning in multi-agent environment as inverted STRIPS planning in the presence of uncertainty. In *Recent advances in computers, computing and communications* (pp. 58–63). WSEAS Press.
8. Nilson, N. J. (1980). *Principles of artificial intelligence*. Palo Alto, CA: Toga Publishing Company.
9. Bylander, T. (1994). The computational complexity of propositional STRIPS planning. *Artificial Intelligence, 69,* 165–204.
10. Gałuszka, A., & Latawiec, M. (2006). *Rozwiązywanie konfliktu pomiędzy robotami z wykorzystaniem elementów teorii gie*r. Zeszyty Naukowe Politechniki Śląskiej, s. Organizacja i Zarządzanie, z. 38, s. 35–41.
11. Weld, D. S., Anderson, C. R., & Smith, D. E. (1998). Extending Graphplan to Handle Uncertainty and Sensing Actions. In *Proceedings of the 15th National Conference on AI* (pp. 897–904).
12. Koehler, J., & Hoffmann, J. (2000). On reasonable and forced goal orderings and their use in an agenda-driven planning algorithm. *Journal of Artificial Intelligence Research, 12*(2000), 339–386.
13. Mc Kinsey, J. C. (1952). *Introduction to the theory of games*. New York: Mc Graw Hill.
14. Jędrasiak, K., Daniec, K., & Nawrat, A. (2013). The low cost micro inertial measurement unit. In *2013 IEEE 8th Conference on Industrial Electronics and Applications (ICIEA)*. IEEE.
15. Bereska, D., Daniec, K., Jędrasiak, K., & Nawrat, A. (2013). Gyro-stabilized platform for multispectral image acquisition. In *Vision based systems for UAV Applications* (pp 115–121). Springer International Publishing,

16. Jedrasiak, K., Nawrat, A., Daniec, K., Koteras, R., Mikulski, M., & Grzejszczak, T. (2012, September). A prototype device for concealed weapon detection using IR and CMOS cameras fast image fusion. In *International Conference on Computer Vision and Graphics* (pp. 423–432). Heidelberg: Springer.
17. Skrzypczyk, K., Gałuszka, A., Ilewicz, W., & Antas, T. (2015). Synthesis and evaluation of the smart electric powered wheelchair route stabilization concept–a simulation study. *Archives of Control Sciences, 25*(2), 263–273.
18. Jedrasiak, K., Mariusz A., & Nawrat. A. (2014). SETh: the method for long-term object tracking. In *International Conference on Computer Vision and Graphics*. Springer International Publishing.
19. Nawrat, A., & Jedrasiak, K. (2009). SETh system spatio-temporal object tracking using combined color and motion features. In S. Chen (Ed.), *WSEAS International Conference. Proceedings. Mathematics and Computers in Science and Engineering*. No. 9. World Scientific and Engineering Academy and Society.
20. Josinski, H., Switonski, A., Jedrasiak, K., & Kostrzewa, D. (2012). Human identification based on gait motion capture data. In *Proceedings of the 2012 International MultiConference of Engineers and Computer Scientists, IMECS* (Vol. 12).
21. Ryt, A., Sobel, D., Kwiatkowski, J., Domzal, M., Jedrasiak, K., Nawrat, A. (2014). Real-time laser point tracking. In *International Conference on Computer Vision and Graphics*. (pp. 542–551). Springer International Publishing.
22. Nawrat, A., & Jedrasiak, K. (2008). Fast colour recognition algorithm for robotics. Problemy Eksploatacji (pp. 69–76).
23. Sobel, D., Jedrasiak, K., Daniec, K., Wrona, J., Jurgas, P., & Nawrat, A. (2014) Camera calibration for tracked vehicles augmented reality applications. In: *Innovative control sytems for tracked vehicle platforms* (pp. 147–162). Springer International Publishing.
24. Bieda, R., Jaskot, K., Jędrasiak, K., & Nawrat, A. (2013). Recognition and location of objects in the visual field of a UAV vision system. In *Vision based systems for UAV applications* (pp. 27–45). Springer International Publishing.
25. Babiarz, A., Bieda, R., & Jaskot, K. (2013). Vision system for group of mobile robots. In *Vision based systems for UAV applications* (pp. 139–156). Springer International Publishing.
26. Daniec, K., Iwaneczko, P., Jedrasiak, K., & Nawrat, A. (2013). Prototyping the autonomous flight algorithms using the prepar3D® simulator. In *Vision based systems for UAV applications*. (pp. 219–232). Springer International Publishing.
27. Galuszka, A., Skrzypczyk, K., & Ilewicz, W. (2014). On transformation of conditional action planning to linear programming. In *Methods and models in automation and robotics (MMAR)* (pp. 764–769).
28. Switonski, A., Josinski, H., Jedrasiak, K., Polanski, A., & Wojciechowski, K. (2010). *Classification of poses and movement phases, ICCVG 2010.*Lecture Notes in Computer Science: Springer.

Suppressing Disturbances in the UAV's Control System Based on the Modified BLT Method

Zygmunt Kuś and Aleksander Nawrat

1 Introduction

The UAV is an example of the controlled plants for which the particular control channels cannot be treated as not influencing each other. It results in the fact that we have to treat such an object as a multivariable control system. We can find in the body of literature the various methods of the synthesis of the multivariable control systems. As the examples, we can notice the following methods: Ziegler-Nichols method [1], BLT method [2], Dominant Pole Placement Tuning method [3], LQ method [4], Davison method [5] or decoupling method [6]. The following paper proposes the further development of the modified BLT method presented in [7]. The authors will consider the disturbances which appear in the input signal of the controlled plant. The modification of BLT method presented in [7] was based on the assumption that the parameters of the PI regulator for the processed main channel are tuned with the use of any condition (method) which guarantees an appropriate control quality for this main control channel.

2 The Linear Multivariable Model for the UAV Taking into Account Disturbances

We will consider the model of controlled multivariable plant presented in [8]. The transfer functions defining the main and coupling channels will be assumed on the basis of literature on helicopters' behaviour [7–10] similarly to [8]. We assume that

Z. Kuś (✉) · A. Nawrat
Institute of Automatic Control, Silesian University of Technology,
Akademicka 16 St., Gliwice, Poland
e-mail: zkus@interia.pl

A. Nawrat
e-mail: aleksander.nawrat@polsl.pl

A. Nawrat et al. (eds.), *Advanced Technologies in Practical Applications for National Security*, Studies in Systems, Decision and Control 106,
https://doi.org/10.1007/978-3-319-64674-9_9

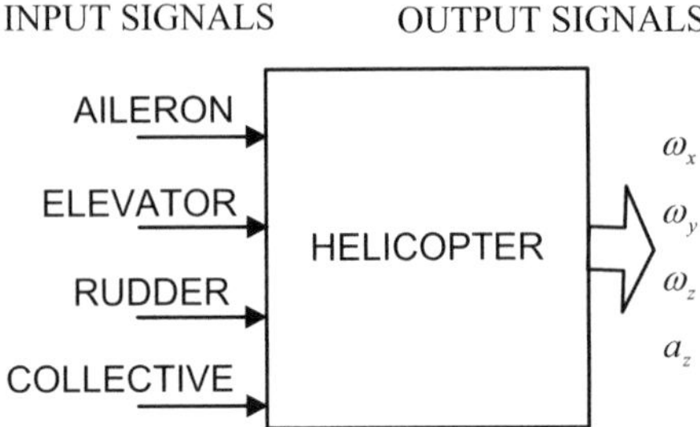

Fig. 1 Input and output signals for the helicopter model

there are four input signals (the inputs of the multivariable model) and four controlled signals (the outputs of the multivariable model). The appropriate transfer functions for each control channel will be chosen as it is presented in Fig. 1. We will consider UAV's (helicopter) model as a system [12] with four inputs and four outputs on the lowest control level.

According to [8] the authors define three coordinate axes which are connected with the helicopter: longitudinal X, lateral Y and vertical Z. There are defined three angular velocities around X, Y, Z axes $-\omega_x$, ω_y, ω_z.

Signal notation will be assumed as follows:

Input signals:

AILERON $= U_{AI}$, ELEVATOR $= U_{EL}$, RUDDER $= U_{RU}$, COLLECTIVE $= U_{CO}$.

Output signals:

Angular velocities: $-\omega_x$, ω_y, ω_z,
Helicopter linear acceleration in vertical axis: a_z.

The helicopter is a multidimensional and non-linear plant. Therefore the synthesis of the first control level uses the identification of the control channels based on a step response [9]. The analysis of the step response and behaviour of the helicopters caused that we assume the transfer functions, the elements of $K_o(s)$, for the main channels presented in (1).

$$\begin{aligned} K_{11}(s) &= \frac{1}{0.07*s^2+0.2*s+12} \quad K_{22}(s) = \frac{1+0.8*s}{0.2+4*s+0.6*s^2+0.04*s^3} \\ K_{33}(s) &= \frac{1}{0.4+0.8*s} \quad\quad K_{44}(s) = \frac{0.1}{1+0.1*s} \\ K_{ii}(s) &= \frac{0.1}{0.01s+1} \text{ for all } i \neq j \text{ where i, j} = 1, 2, 3, 4. \end{aligned} \tag{1}$$

The scheme of the multivariable control system with two excitations (disturbance and reference signal) is presented in Fig. 2.

In Fig. 2, we assume the notation of the input/output signals for the particular control channels according to (2).

$$\begin{aligned} w_1 = U_{AI}; \quad w_2 = U_{EL}; \quad w_3 = U_{RU}; \quad w_4 = U_{CO}; \\ y_1 = \omega_x; \quad y_2 = \omega_y; \quad y_3 = \omega_z; \quad y_4 = a_z; \end{aligned} \tag{2}$$

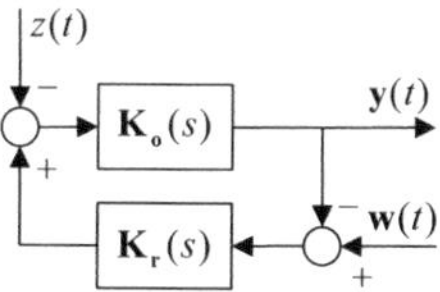

Fig. 2 Multivariable control system. ***Kr***(*s*)—diagonal multivariable controller, ***Ko***(*s*)—multivariable plant, *z*(*t*)—disturbance signal, ***w***(*t*)—reference signal, ***y***(*t*)—output signal

3 Modified BLT Method for the Control System with Disturbances

The disturbances which appear in the control system may establish the strict requirements for the controllers. The key element of the proposed solution will be the fact that we consider the properties of the disturbances during PI controllers tuning. The modified BLT method allows to use various approaches to PI controllers tuning which guarantees the stable system in the main channels and the required quality of control system operation. Figure 2 presents the control system with the disturbance. We assume that this disturbance has sinusoidal form. There is a step of the BLT method in which PI controllers for main channels are tuned. In this step, we will base on the method presented in [11] which provides the analytical condition. For all the frequencies ω for which this condition is satisfied the error amplitude in the close-loop system is decreased more than $1/\Delta$ times in relation to the open-loop error amplitude.

The tuning method is conducted in the manner discussed below.

We will consider the control system presented in Fig. 1. We assume that the open-loop and closed-loop systems are stable. This assumption results from the basic properties of BLT method. The disturbance $z(t)$, as it is presented in Fig. 1, influences the input of the controlled plant. Generally, the disturbance may have the form of a step signal or sinusoidal signal. The integral part of PI controller provides an error in the steady state equal zero for the constant value of the step disturbance. Thus we will consider the response of the control system for the sinusoidal disturbance $z(t) = A\,sin(\omega t)$ in the steady state. According to [11], we assume that $\hat{Z}$, $\hat{E}_o$, and $\hat{E}_z$ denote the symbolic amplitudes of disturbance $z(t)$, error $e_o(t)$ in the open-loop system and error $e_z(t)$ in the closed-loop system, respectively. We assume that the only excitation in the control system is disturbance $z(t)$. However, due to linearity of the system the obtained results will be useful for the system with nonzero value of the reference signal $w(t)$. We assume that $K(s) = K_o(s)K_r(s)$ is the product of the transfer functions of the plant and controller. As it was presented in [11], in order to obtain the ratio of amplitude $|\hat{E}_z|$ to amplitude $|\hat{E}_o|$ smaller than Δ (it is a given small number), it is necessary and sufficient that $|1/(1 + K(j\omega))| < \Delta$. It may be presented in an approximate version $|K(j\omega))| < 1/\Delta$. In the same time, this condition may be presented with the application of Bode plot of the magnitude $L(\omega) = 20log|K(j\omega)|$. In this case, the condition can be formulated as $L(\omega) > 20log(1/\Delta)$. The authors propose to use this condition in order to synthesise the PI controllers. It will be treated as a part of the modified BLT method applied for the UAV's control system.

The modified BLT method was presented in [7]. The first step in this algorithm was to tune PI controllers for the main channels. We will choose controllers' parameters in such a way that the condition $L(\omega) > 20log(1/\Delta)$ will be fulfilled.

In the Sect. 4, there is the example of the control system with controllers tuned as above.

4 An Example of the Suppression of Disturbances in the UAV's Control System

The examples presented in this section assumes that there appears sinusoidal disturbance in the control system. In order to illustrate the suppression of disturbances in the UAV's control system based on the modified BLT method, we can assume that the controlled plant is the lowest control level of the small helicopter described in [9] and Sect. 2. The BLT method assumes that the main channels are described by means of stable transfer functions. We can state that this assumption is fulfilled by the analysis of the Nyquist characteristics (Fig. 3) and root locus images (Fig. 4) for the transfer functions of the objects in main channels.

We assume that $L(\omega) = 20log|K_r(j\omega)Ko(j\omega)|$ and PI controller has transfer function $Kr(s) = k_r(1 + 1/sT_c)$. Looking at the Bode characteristic (magnitude) of the PI

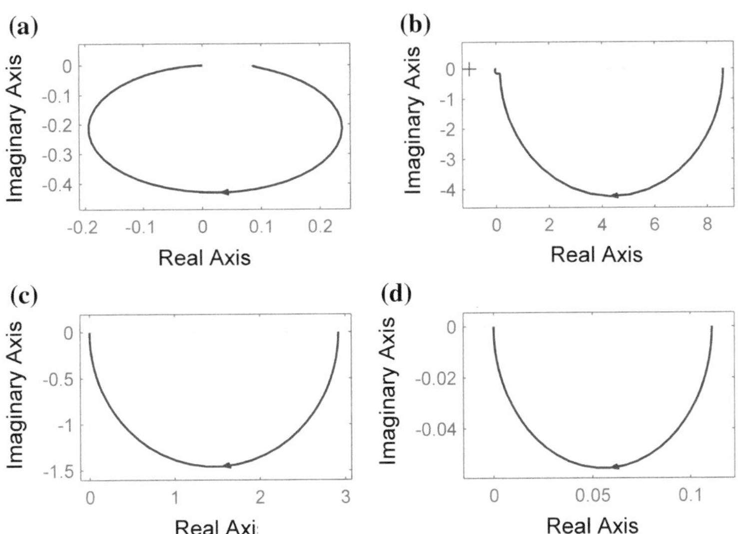

Fig. 3 The Nyquist characteristics for the transfer functions of the objects in main channels. **a** $K_{11}(s)$, **b** $K_{22}(s)$, **c** $K_{33}(s)$, **d** $K_{44}(s)$

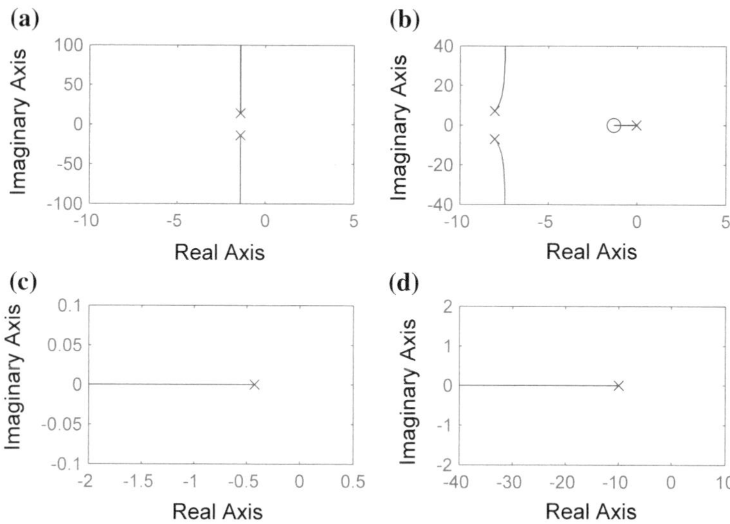

Fig. 4 The root locus graphs for the transfer functions of the objects in main channels. **a** $K_{11}(s)$, **b** $K_{22}(s)$, **c** $K_{33}(s)$, **d** $K_{44}(s)$

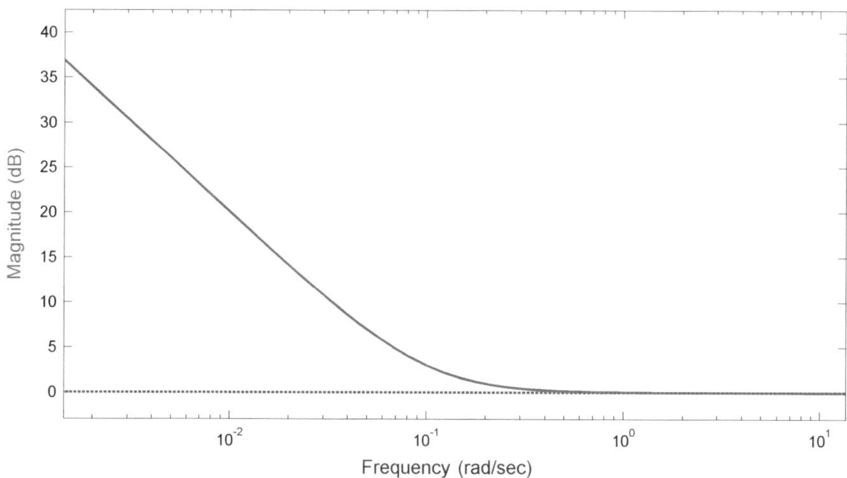

Fig. 5 Bode characteristic (magnitude) of the PI controller with $k_r = 1$ and $T_c = 10$

controller (Fig. 5), we may state that the condition $L(\omega) > 20log(1/\Delta)$ will be fulfilled more easily for the wide range of frequencies in the case when k_r is large and T_c is small. The BLT method provides PI controller's parameters which fulfil this requirement. Our task is to compute initial PI parameters. What is more we have to find, for the parameters of PI controller calculated by BLT, the range of disturbance frequencies, for which the disturbance will be $(1/\Delta)$ times suppressed. This

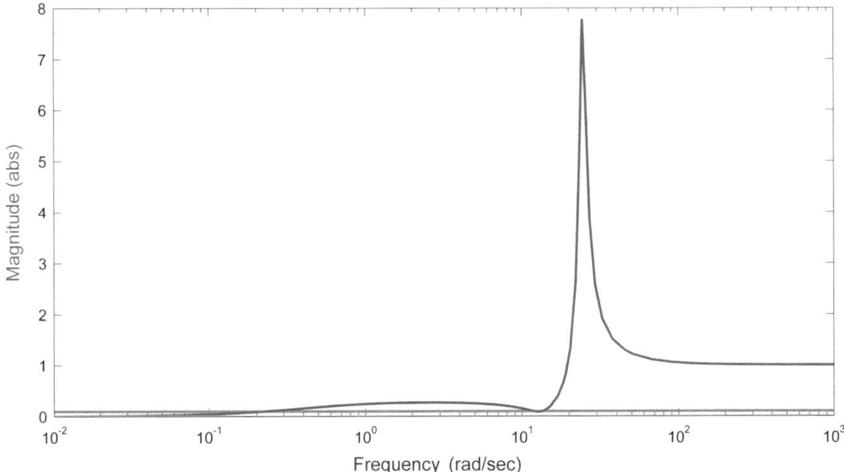

Fig. 6 Bode characteristic (magnitude) of module $|1/(1 + K(j\omega))|$ where $K(j\omega) = K_{11}(j\omega)PI_{11}(j\omega)$

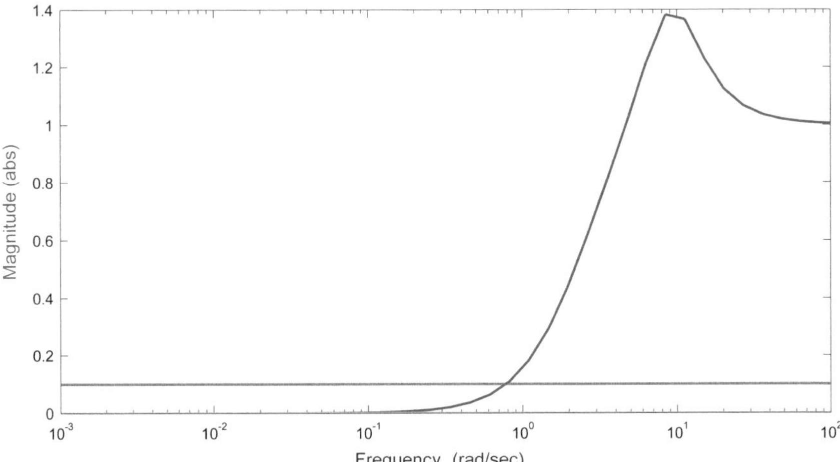

Fig. 7 Bode characteristic (magnitude) of module $|1/(1 + K(j\omega))|$ where $K(j\omega) = K_{22}(j\omega)PI_{22}(j\omega)$

range may be found in the Figs. 6, 7, 8 and 9 as the range for which the condition $|1/(1 + K(j\omega))| < \Delta$ is fulfilled. We assume $\Delta = 1/10$.

For the model presented in Sect. 2 (according to [8]), we conducted the synthesis of the multivariable regulator. PI regulators which were obtained on the basis of BLT method are presented in (3).

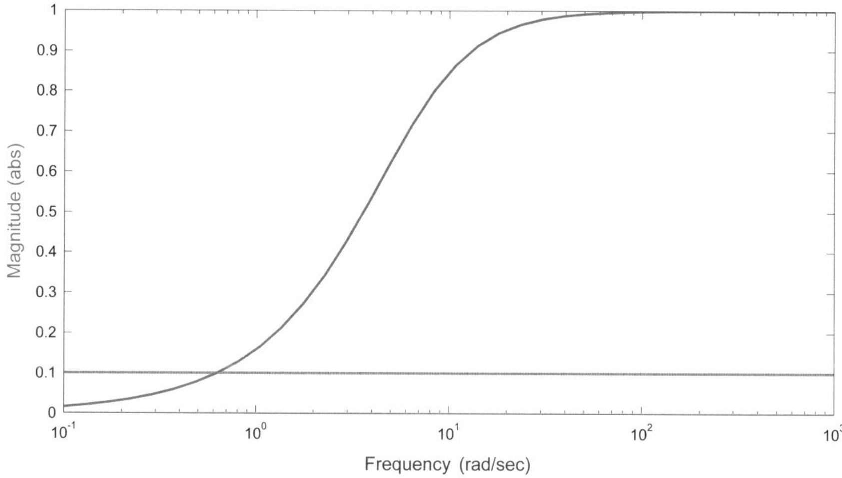

Fig. 8 Bode characteristic (magnitude) of module $|1/(1+K(j\omega))|$ where $K(j\omega) = K_{33}(j\omega)PI_{33}(j\omega)$

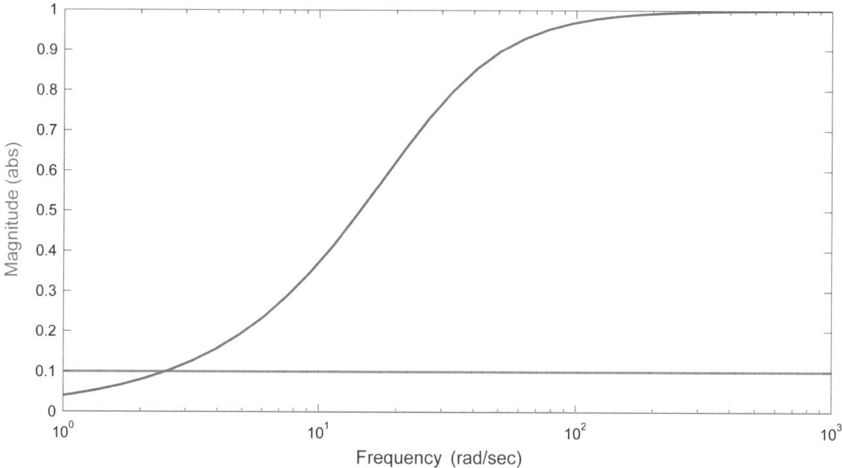

Fig. 9 Bode characteristic (magnitude) of module $|1/(1+K(j\omega))|$ where $K(j\omega) = K_{44}(j\omega)PI_{44}(j\omega)$

$$\begin{array}{rr} PI_{11}(s) = 30\frac{1.2s+1}{1.2s} & PI_{22}(s) = 2.22\frac{0.1s+1}{0.1s} \\ PI_{33}(s) = 5\frac{2s+1}{0.1s} & PI_{44}(s) = 25\frac{0.1s+1}{0.1s} \end{array} \tag{3}$$

We assume that there are sinusoidal disturbances $z(t) = A\,sin(\omega t)$ in the control system. The input of the controlled plant is the sum of these disturbances and the output of the controller. We assume that the disturbances with different parameters (amplitude A, frequency ω) are connected to the input of each plant's channel.

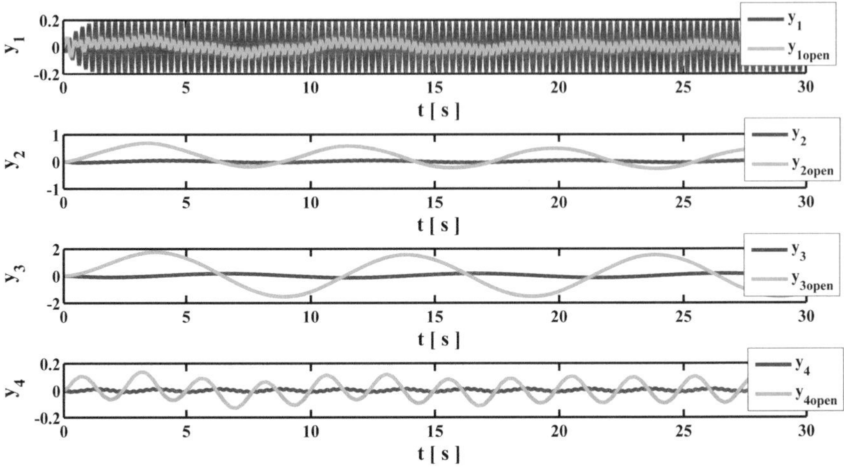

Fig. 10 Transients $y(t)$ and $y_{open}(t)$ for all control channels

Experiment 1

The frequency of the disturbance in one channel will be chosen from the outside of the good suppression range whereas other frequencies will be chosen from the inside of this range. For better clarity of the figures we will assume reference signal $w(t) = 0$. The Fig. 10 presents transients $y(t)$ (closed loop control system) and $y_{open}(t)$ (open loop control system) for all control channels. According to Fig. 10, we may conclude that one disturbance $z(t)$ with the frequencies out of good suppression range may result in the weak suppression of the disturbance in this channel whereas the disturbances in other channels are suppressed correctly.

Experiment 2

The frequency of the disturbance in all channels will be chosen from the outside of the good suppression range. The Fig. 11 presents transients $y(t)$ (closed loop control system) and $y_{open}(t)$ (open loop control system) for all control channels. According to Fig. 11, we may conclude that all disturbances $z(t)$ with the frequencies out of good suppression range result in the weak suppression of the disturbance in all channels.

Experiment 3

The frequency of the disturbance in all channels will be chosen from the inside of the good suppression range. The Fig. 12 presents transients $y(t)$ (closed loop control system) and $y_{open}(t)$ (open loop control system) for all control channels. According to Fig. 12, we may conclude that all disturbances $z(t)$ with the frequencies in good suppression range should result in the good suppression of the disturbance in all channels.

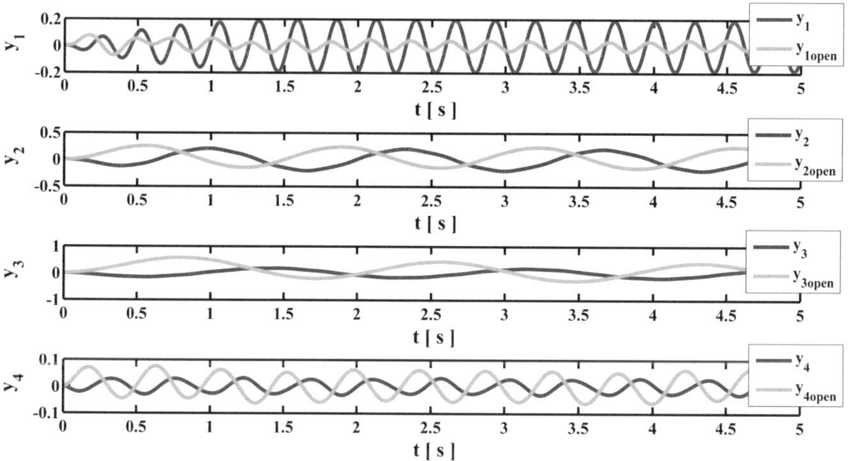

Fig. 11 Transients $y(t)$ and $y_{open}(t)$ for all control channels

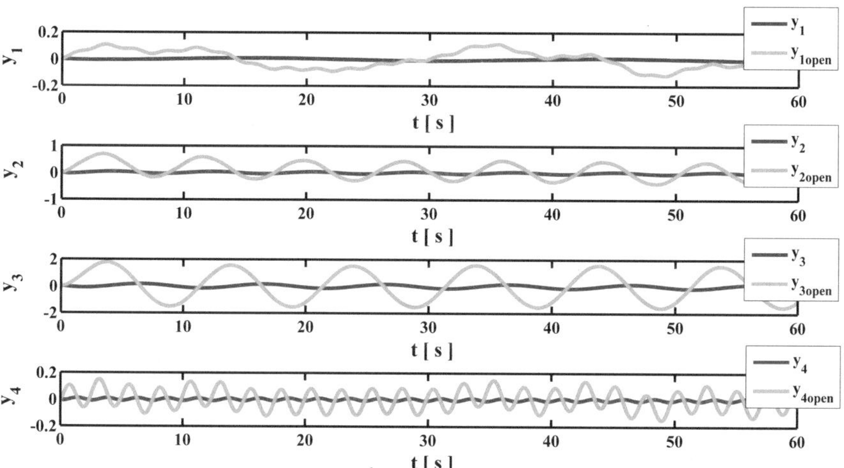

Fig. 12 Transients $y(t)$ and $y_{open}(t)$ for all control channels

5 Conclusions

To conclude, the main goal of the paper concerned the UAV's control system resistant to disturbances. We focused on the lowest control level which was examined as a linearised multivariable system. As an example of the UAV, we used a small helicopter. The disturbances influencing the helicopter behaviour were modelled as additive sinusoidal disturbances at the controlled plant input. The authors proposed using the Bode (magnitude) characteristics for main control channels to assess the range of good disturbance suppression. For the synthesis of the control system, the modified BLT method was used in order to obtain the multivariable control system on the basis of single variable systems. We applied this method to control the linear velocity of ascending and angular velocities of the UAV's rotation around the symmetry axes. Finally, the presented examples illustrated the correctness of the operation of the modified BLT method in disturbance suppression task during the UAV flight [12].

References

1. Wang, Q. G., Lee, T. H., & Zhang, Y. (1998). Multiloop version of the modified ziegler-nichols method for two input two output processes. *Industrial and Engineering Chemistry Research.*
2. Luyben, W. L. (1986). Simple method for tuning SISO controllers in multivariable systems. *Industrial and Engineering Chemistry Process Design and Development.*
3. Zhang, Y., Wang, Q. G., & Astrom, K. J. (2002). Dominant pole placement for multi-loop control systems. *Automatica, 38*(7).
4. Stein, G., & Athans, M. (1987). The LQG/LTR procedure for multivariable feedback control design. *IEEE Transactions on: Automatic Control.*
5. Davison, E. J. (1976). The robust control of a servomechanism problem for linear time-invariant multivariable systems. *IEEE Transactions on: Automatic Control.*
6. Falb, P., & Wolovich, W. (1967). Decoupling in the design and synthesis of multivariable control systems. *IEEE Transactions on Automatic Control.*
7. KuIJ, Z. (2016). The modified BLT method for multivariable control systems. In Nawrat, A., Jedrasiak, K. (Eds.) *Innovative simulation systems.* Springer, ISBN 9783319211176 9783319211183.
8. KuIJ, Z., & Nawrat, A. (2016). The application of the modified BLT method for the synthesis of UAV's control system. In Nawrat, A., & Karol, J. (Eds.) *Innovative simulation systems.* Springer, ISBN 9783319211176 9783319211183.
9. Kuś, Z., & Fraś, S. (2013). Helicopter control algorithms from the set orientation to the set geographical location. In *Advanced technologies for intelligent systems of national border security.* Studies in computational intelligence (p. 440). Springer.
10. Manerowski, J. (1999). *Identyfikacja modeli dynamiki ruchu sterowanych obiektów latających.* Warszawa: WN ASKON.
11. Gessing, R. (2004). *Control fundamentals.* Gliwice: Silesian University of Technology.
12. Jedrasiak, K., Nawrat, A., & Wydmanska, K. (2013). SETh-link the distributed management system for unmanned mobile vehicles. In *Advanced technologies for intelligent systems of national border security* (pp. 247–256). Berlin Heidelberg: Springer.

The Airspeed Automatic Control Algorithm for Small Aircraft

Sławomir Samolej, Marek Orkisz and Tomasz Rogalski

1 Introduction

Currently, there are several types of automatic flight control systems installed on both manned and unmanned flying vehicles. It is obvious, the control system is an element of the airplane's equipment directly affecting the flight safety. The control system can safely and efficiently guide the airplane through the desired trajectory or proceed any mission autonomously only on the condition control and measurement algorithms work inherently. Just, the right structures of both control laws and measurement algorithms determine flight control quality (Fig. 1). The above-mentioned flight control system quality factors are the most important ones, however some others can be taken into consideration as in [1–4] papers.

Generally, the aircraft control system ought to maintain desired values of selected flight parameters efficiently. Modern flight control systems are requested to provide not only the basic functionality, such as desired navigation and attitude flight parameters maintenance, but also some advanced mission control functionalities. For instance, 4-D trajectory management system, which is a part of the mission management system is also often requested.

Both attitude and airspeed control functions are sometimes called low level ones. They are used by higher level (altitude, curse, track) control algorithms which are in turn supervised by such systems as mission control and management, trajectory control, and recovery [5, 6].

S. Samolej · M. Orkisz · T. Rogalski (✉)
Rzeszow University of Technology, Rzeszow, Poland
e-mail: orakl@prz.edu.pl

S. Samolej
e-mail: ssamolej@prz.edu.pl

M. Orkisz
e-mail: mareko@prz.edu.pl

A. Nawrat et al. (eds.), *Advanced Technologies in Practical Applications for National Security*, Studies in Systems, Decision and Control 106,
https://doi.org/10.1007/978-3-319-64674-9_10

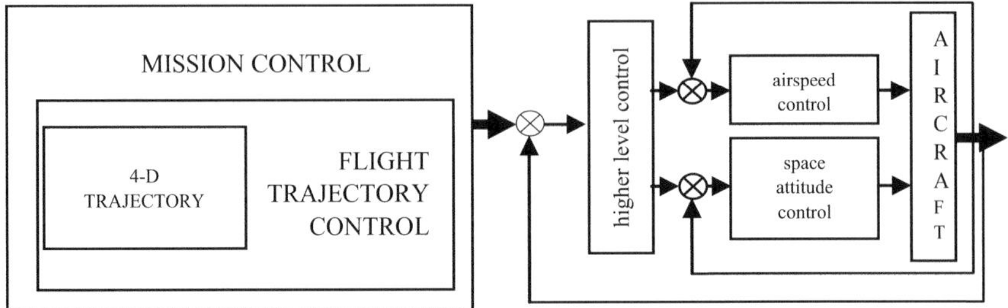

Fig. 1 Typical aircraft control system's structure in general

It must be outlined some classical approaches to the aircraft control and stability issue, taking into consideration aircraft's specific features, use set of control channels which are not affected by each other. There is also a control channel containing airspeed control algorithms among them. Those approaches can be applied for regulator synthesis process for linear models (Fig. 1) pretty successfully. Unfortunately, real aircraft are characterized by some substantial interactions between flight parameters affecting each other. These phenomena restrict the usage of classical (linear) theories significantly. An impact of pitch angle on airspeed may not be neglected [1], for instance. Consequently, authors have been caring out research under some aircraft airspeed control algorithms unlike classical (linear) ones. Some algorithms have been developed, implemented into a real control system and tested in flight.

This paper presents a new airspeed control algorithm developed for small single engine piston airplane (SEPL) with constant-pitch propeller in tractor configuration. It was successfully applied for following airborne systems: the air target imitator [3] and the flying platform of remotely piloted air system LOT [2, 7]. This paper also discusses some selected results recorded during real and simulated flights conducted under the new algorithm supervision.. They present the level of airspeed control process's quality achieved.

2 The Plant

A subject of research was a group of SEPL class airplanes flying up to flight level—FL 95. Their operational airspeed was in the range from 70 to 250 (km/h) and operational weight was approximately from about 10 kgs up to 500 kgs. The real air target imitator and the MP-02A "Czajka" aircraft (flying platform for LOT)—were used for flight tests. Obviously, test aircraft were equipped with all necessary dedicated flight data sensors and recorders [3, 8, 9].

The applied mathematical and functional models of aircraft were simplified according to well-known engineering methods. The main of them were as follows:

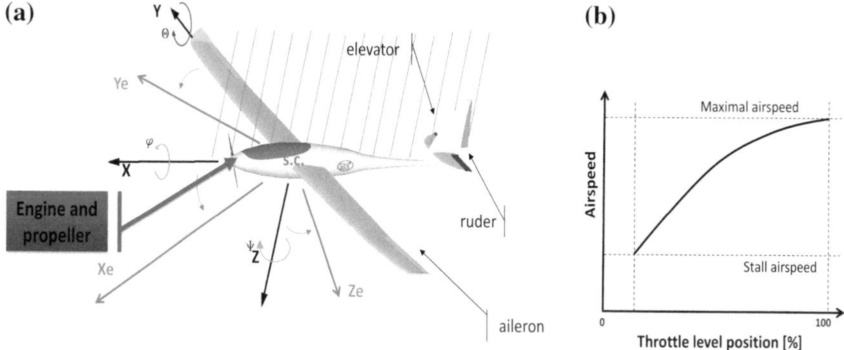

Fig. 2 The plant. **a** Aircraft in classical configuration. **b** Engine-propeller system is characterized by sample typical static characteristic like this one in [10]

- Plane's mass did not vary during flight, or mass changes were irrelevant and did not impact on the control process.
- The flight altitude (between 0 m above mean sea level up to FL95) did not affect engine's performance in a significant way.
- The plane was in a classical configuration [1].
- The classical system of control surfaces was used to control aircraft attitude (Fig. 2a).
- The airspeed was controlled by thrust directly, but could be also controlled by pitch angle's variations indirectly.
- Only aircraft's longitudinal dynamics was analyzed. Lateral motion did not affect airspeed significantly and was neglected.
- The airplane was powered by carburetor piston engine. Thrust was controlled by throttle position only [11].
- The constant-pith propeller in tractor configuration with static characteristic like shown on Fig. 2b was used.

3 The Control Algorithm

Several research concerning airplane airspeed control system had been carried out by Rzeszow University of Technology team over the period 2000–2006. They were a significant part of a bigger projects focused on the fly-by-wire flight control system developed for small general aviation airplanes [12]. In those years several series of flight tests of various control algorithms were also conducted. The results received within above-mentioned research were a starting point for activity described in this paper.

Some static and dynamic characteristic of the engine-propeller-airframe system and experiences received in the previous research suggest that the desired quality of control process can not be achieved, if some classical linear method is applied solely. Both nonlinearities existing in the system as well as external factors affecting it impose the new control algorithm investigation.

It is commonly known, pilots flying small airplanes have been able to control airspeed by engine's thrust settings since some time. Moreover, medium-skilled pilots could achieve it pretty easily, according to their opinions. The author's experience in a small airplanes as well RPAS's manual control was a signpost for activity they conducted. It was clear that the main goal of research should have been to developed algorithm which would be able to imitate pilot's activity in general. The algorithm's task would be to merge the pilot's approach to airspeed control process as well as airplane's features incoming both from practical and theoretical investigations. The heuristic expert's knowledge should have also been applied to the control law synthesis process. The significant role in proposed approach to airspeed control process played the pilot's knowledge about their actions usually conducted with engine's controls (throttle lever particularly) to maintain desired airspeed.

Taking into consideration abovementioned premises following general requirements defining limitations of the usage of control laws and quality of the control process were established:

- the control system is going to cover full range of airplane's operational airspeeds (just above stall airspeed to never exceed airspeed—typical operational range),
- the control precision 15 (km/h); ±7.5 (km/h) is required (defined by experts),
- the overshoot magnitude depending on flight state is no more than 20 (km/h), but only if its magnitude is accepted by pilots and operators,
- the regulation time of the system is similar to regulation time which medium skilled pilots can establish (to be assessed by experts),
- the time delay is less 0.5[d] (defined by experts),
- there should not be any oscillations in the system,
- the only throttle lever position is used to control airspeed,
- the operational pitch angle is in the range of [−15°; 20°] and bank angle less than 20°,
- the flight altitude may not depredate the control quality,
- any engine's technical limitations are not taken into consideration,
- the control signal's value is in the range of [0; 1].

If the algorithm is to emulate pilot's activities and actions pilot conducts to maintain speed, the main rules governing this process need to be identified at the beginning. Highly skilled pilots and UAV operators were asked to describe the way they used throttle lever to maintain desired flight speed.

The first experience the pilots reported was that they learn aircraft's responses to their inputs. Some kind of aircraft model was automatically built in pilot's mind during the preliminary flights of the new aircraft. It enabled pilot to predict aircraft's response and adapt applied control to plane's dynamic and flight conditions accordingly.

The second experience the pilots noticed was that there existed only a few main (clear and easy to be understood for human) rules to govern the control process. They can be listed in a following form:

- Rule No. 1—There are two steps of the control process, in fact. Firstly, pilots set throttle inaccurately. They do it only on the basis of experience they have got and airplane model they have built in their minds. Within the second step they do small precise corrections to achieve the desired airspeed and to maintain requested control quality.
- Rule No. 2—A number of throttle corrections must be reduced to very minimum.
- Rule No. 3—Some inaccuracy, depending on flight state and flight conditions is acceptable.

The developed control algorithm applies the listed rules as follows. According to the rule No. 1, the formula (1) defines desired position of the throttle lever (δT) as a sum of two components: draft/preliminary position (δT_0) and correction component (δt_0) to tune precisely the control signal having regard to actual airspeed deviation (Fig. 3).

$$\delta T = \delta T_0 + \delta t \tag{1}$$

When the control law's structure is defined, next step is calculation of components from formula (1). If a new airspeed is desired and if it requires new throttle setting, the algorithm pilots apply to set preliminary thrust (δT_0) uses knowledge and experiences they got before. Pilots and operators know, more or less, how to move the throttle lever to set proper thrust. The preliminary throttle position is created on the basis of some characteristics defining relation between throttle lever position and flight airspeed (Fig. 2b). These characteristics can have several forms and both theoretical analysis and practical experiments can produce them. The primary position of the throttle (δT_0) linked with desired airspeed (U_d) can be defined by table of discrete values for instance. The table is prepared as follows. The operational airspeed domain should be split into several subdomains.

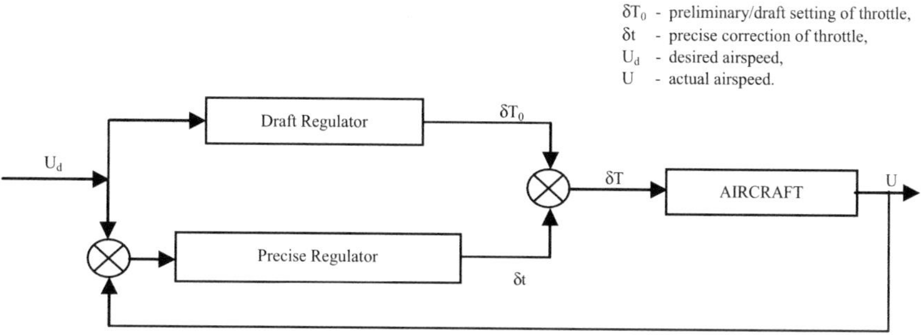

Fig. 3 The structure of the developed control law

The number of subdomains and their range are affected by different factors e.g. operational conditions, engine's dynamic characteristics, airframe's flight characteristics and so on. It is defined on the basis of expert's knowledge mainly. If the structure of subdomains is created, there should be reference airspeeds ($U_{d,n}$—n is a number of the subdomain) assigned to each of them (compare Fig. 4).

Despite of the structure of subdomains and their reference airspeeds are formed by expert, there are a few rules recommended to be held:

- The minimal reference airspeed should be 120% of a stall airspeed. This value outcomes from commonly used definition of minimal operational airspeed for small planes.
- The maximal theoretical flight airspeed at horizontal flight is a maximum reference airspeed.
- The reference airspeed is a median of the subdomain (excluding the last one).

The sample set of defined subdomains and reference airspeeds assigned to them evaluated for an airplane during flight tests, are put into Table 1.

The final precise corrections pilots usually do are affected by airspeed deviations and plane's time response to primary thrust settings. The number of pilot's correction actions is reduced to absolute minimum. Corrections are generated only when pilots (on the basis of flight parameters) predict that the deviation is not going to be reduced definitely.

Corrections of the control signal are determined by both value of airspeed deviation e and its rate $\dot{e}$ (Fig. 6) as follows: If airspeed deviation goes to zero value the control signal's value remains constant. If airspeed deviation increases the control signal needs to be tuned. The PID^2 regulator theory (Fig. 5), successfully applied for automatic control at pitch and roll angle channels [1, 13], has been adapted to do this.

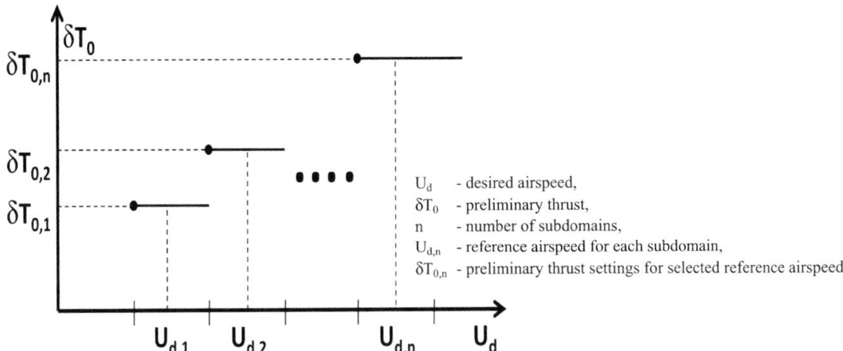

Fig. 4 Preliminary thrust as a function of selected reference airspeeds

Table 1 Operational airspeed subdomains and reference airspeeds

N	Reference airspeed (km/h)	Airspeed range (km/h)	Comments
1	125	To 150	Stall airspeed around 80 (km/h)
2	175	From 150 to 200	–
3	225	Above 200	Maximum airspeed around 260 (km/h)

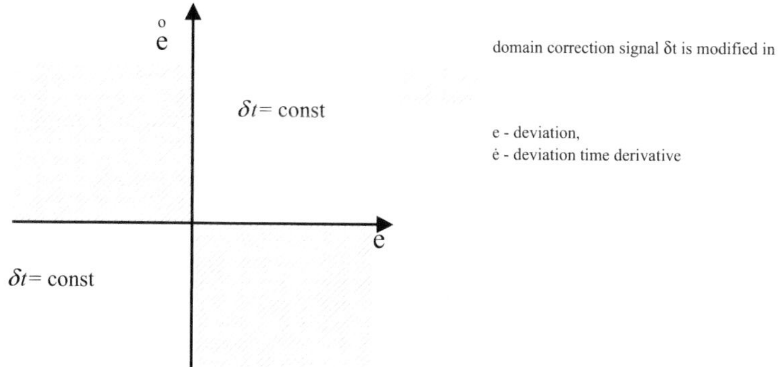

Fig. 5 The rule defining how the PID² regulator works

The magnitude of control signal δt can be found from (2), (3):

$$\delta t = \int k_e sgn(e)\; e\; \dot{e}\; dt \tag{2}$$

$$k_e = \begin{cases} gain & if \quad sgn(e) = sgn(\dot{e}) \\ 0 & remaining\ cases \end{cases} \tag{3}$$

If rules (2) and (3) are applied, the regulator can identify system's tendency to reduce or to increase the deviation. In fact, the control system tunes the magnitude of the control signal only if it detects the airspeed deviation is going to be increased.

There are several ways the precise regulator following the above mentioned rules can be created in. The two possible development paths are presented in this paragraph.

The first of them uses expert's knowledge. It is similar to the draft regulator tuning methodology. The formula defining gain's magnitude ke (3) can be obtained by interpolation of specific values found experimentally for reference airspeeds at each defined subdomains (Fig. 6). This method was used for regulator tuning during flight tests of the experimental aircraft [2].

The second approach assumes the precise (augmenting) regulator is active near some selected point of work only. Therefore, if the system is linearized at a set of working points referred to reference airspeeds, some methods intended for linear

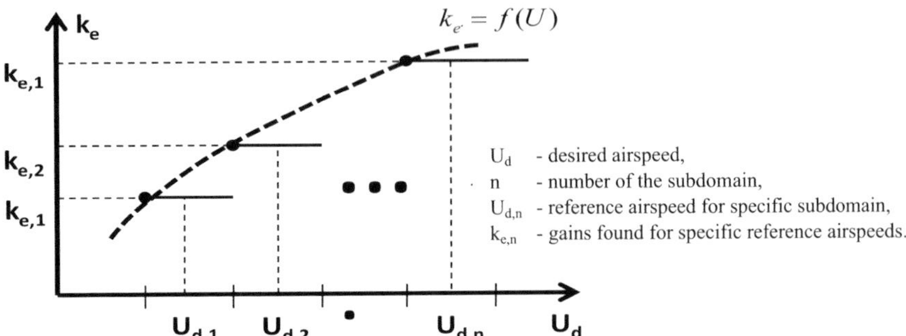

Fig. 6 The gain of precise regulator as a function of reference airspeed

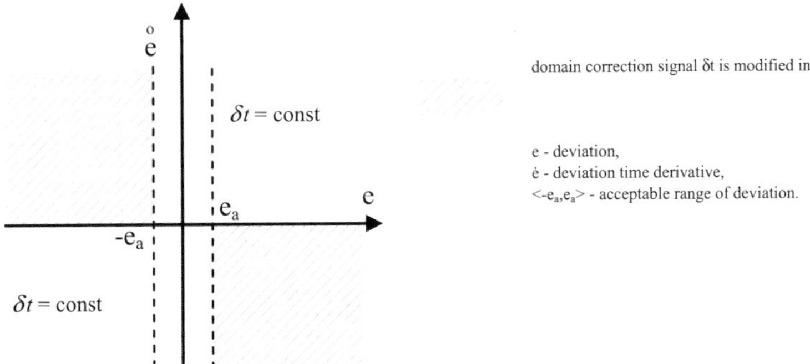

Fig. 7 The rule the PID^2 regulator works according to, if acceptable magnitude of deviation taken into consideration

systems can by applied for the regulator synthesis process, in these cases. Achieved results are interpolated on entire range of operational airspeeds next. The process of synthesis of the precise regulator is not the subject of this work so its details are neglected in this paper.

The second approach assumes the precise (augmenting) regulator is active near some selected point of work only. Therefore, if the system is linearized at a set of working points referred to reference airspeeds, some methods intended for linear systems can by applied for the regulator synthesis process, in these cases. Achieved results are interpolated on entire range of operational airspeeds next. The process of synthesis of the precise regulator is not the subject of this work so its details are neglected in this paper.

The rule No. 3 pilots apply to control airspeed says: “Some inaccuracy, depending on flight state and flight conditions is acceptable”. Consequently, the magnitudes of airspeed deviation e as e ϵ <-e_a,e_a> (Fig. 7) don’t provoke any pilot’s action. Finally Eq. (3) yields Eq. (4).

$$k_e = \begin{cases} gain & je\acute{s}li \quad sgn(e) = sgn(\dot{e}) i |e| < a_e \\ 0 & remaining\ cases \end{cases} \tag{4}$$

The range of the acceptable airspeed deviation is defined by the expert.

4 The Implementation Case

The algorithms presented in this paper were implemented into the real control system being a part of a small aircraft autopilot [2]. To improve the systems scalability and safety its software have been developed according to ARINC 653 specification suggestions [14–16]. The subsequent controllers such as mission controller, altitude/track/course controller and airspeed with space attitude controllers have been reallocated into separate spatially and temporally isolated software partitions (Fig. 8) prepared for VxWorks operating systems.

Within the partitions the control algorithms are executed as separate real-time tasks. They exchange data using safe inter-partition communication channels preventing from deadlock and starvation phenomena. A separate network driver makes it possible to exchange data among the "external" modules of control system, such as pilot, aircraft's actuators and sensors as well as remote controller. The ARINC 653-like control system software structure systematizes its development and improves its safety.

The introduced control algorithms were practically (within in-flight tests) evaluated under following conditions. The operational airspeeds range of the plane for both configuration and completion during flight tests was from 100 to 260 (km/h). Three reference airspeeds (see Table 1) were selected.

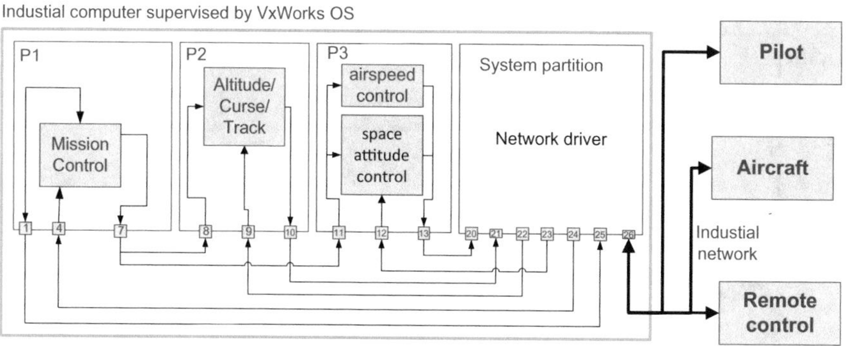

Fig. 8 Hardware/software structure of algorithm implementation

An acceptable airspeed deviation value ±7, 5 (km/h) was set. Reference airspeeds were linked with draft throttle magnitudes by experiments. The precise regulator was tuned by expert experimentally also.

A test flight was conducted at following weather conditions: wind speed 3 ÷ 5 (m/s), vertical wind's component −1 ÷ 1 (m/s). Figures 9, 10 and 11b present sample flight data when airspeed was controlled with the usage of the introduced control law.

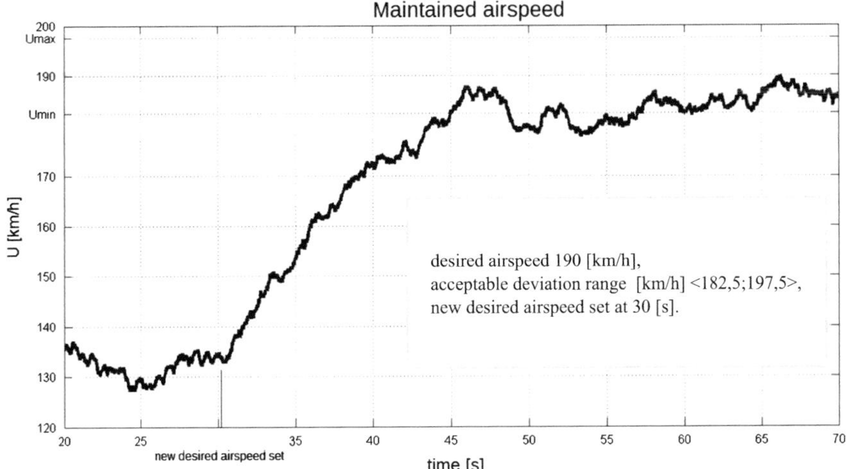

Fig. 9 Sample airspeed maintaining process

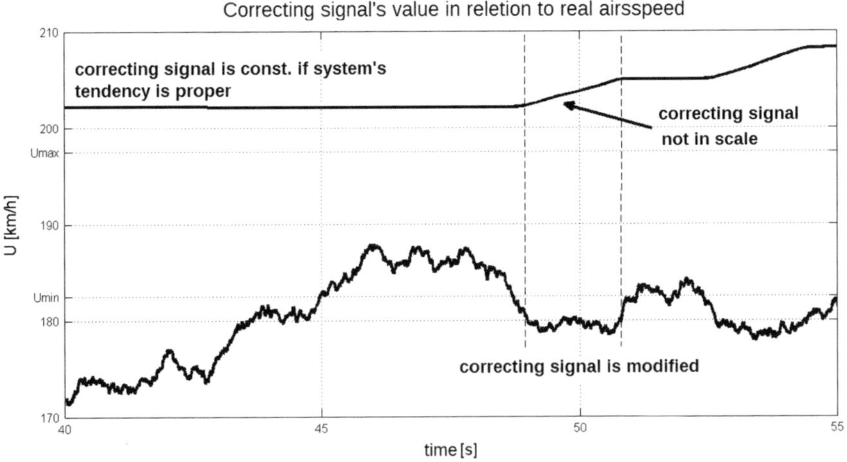

Fig. 10 The sample relation between airspeed and correction signal δt

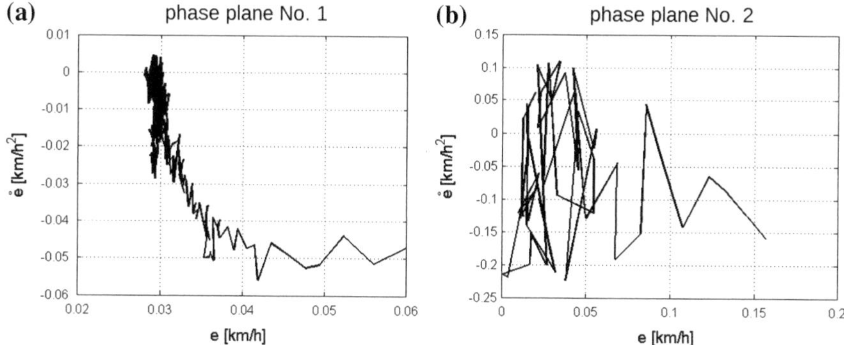

Fig. 11 Example phase planes for airspeed change from 130 to 190 (km/h); **a** simulation results with atmospheric interferences taken into consideration, **b** real flight data collected during the in-flight test depicted in Figs. 9 and 10

Graphs in Figs. 9 and 10 prove that the desired airspeed regulation process fulfills defined requirements regarding to time delay (below 0.5 s), time of regulation (proper in expert's opinion), overshoot (doesn't exist), control's quality (deviation inside of desired range), system's oscillations (no observed oscillations) defined for this control channel. The control signal is modified only when it is absolutely necessary—only if system goes to increase the deviation over acceptable range.

During both simulated and real test flights the system was stable. The limit cycles were identified (Fig. 11) and each system's instability was eliminated.

5 Summary

Simulated and real test flights allow to allege the presented method can be used for the small aircraft autopilot's airspeed control channel synthesis process. The principle features of the control process such us regulation time, overshot, oscillations, control precision, deviation meet requirements defined for this control channel by some classical approaches in general and by experts for this particular case.

The work presents the approach to the airspeed regulation process using two regulators: the draft one and the precise (augmenting) one working simultaneously (Fig. 3). The goal of the paper is to focus on the control channel concept omitting ways they are usually prepared in. Synthesis methods of these regulators were not the subject of the paper.

The future author's research should focus on search more formal methods proper for the process of control algorithm adjustment. This in turn should reduce the share of expert's knowledge. To the authors knowledge this approach to the airspeed control process can be directly implemented into autopilot's structure.

References

1. Bociek, S., & Gruszecki, J. (1999). *Układy sterowania automatycznego samolotem*. Rzeszów: Oficyna Wydawnicza Politechniki Rzeszowskiej.
2. Gruszecki, J. (2014). System obserwacji terenu—problemy praktyczne. *III Konferencja Międzynarodowa nt. Specjalizacje w oddziałach ratunkowych, medycznych i systemów wsparcia*. Arłamów.
3. Jaromi, G., Rogalski T., Rzucidło P., & Wałek Ł. (2007). Integracja układu sterowania z mini BSL. *V Konferencja Awioniki*. Rzeszów.
4. Krawczyk, M., Graffstein, J., & Maryniak, J. (2000). Mathematical model of UAV in numerical simulation of the recovery maneuvers during perturbed flight. *Journal of Theoretical and Applied Mechanics, 38*(1), s.121–130.
5. Kopecki, G., & Rzucidło, P. (2014). Integrated modular measurement system for in-flight tests. *Polskie Towarzystwo Diagnostyki Technicznej, Diagnostyka* (Vol. 15, No. 1, pp. 53–60). Warszawa.
6. Kopecki, G., Tomczyk, A., & Rzucidło, P. (2013). Algorithms of measurement system for a micro UAV. In *Solid state phenomena* (Vol. 198, pp. 165–170). Zurich: Trans Tech Publications Inc.
7. Basmadji, F. L., Gruszecki, J., Rzucidlo, P., & Kordos, D. (2012). Development of ground control station for a terrain observer—hardware in the loop simulations. In *AIAA modeling and simulation technologies conference*, Minneapolis, US, 13–16, AIAA-2012-4629.
8. Gosiewski, Z., & Kulesza, Z. [red]: Mechatronic systems and materials IV.
9. Pieniążek, J. (2014). *Kształtowanie współpracy człowieka z lotniczymi systemami sterowania*. Oficyna Wydawnicza Politechniki Rzeszowskiej, ISBN 978-83-7199-912-7. Rzeszów.
10. Rogalski, T., & Horyń, D. (2015) MP-02 *Czajka airspeed characteristic, internal report avionics and control department*. Rzeszow: Rzeszow University of Technology.
11. Rogalski, T., Tomczyk, A. (2001). Eksperymentalny układ sterowania pośredniego dla samolotu ogólnego przeznaczenia. Zeszyty Naukowe Politechniki Rzeszowskiej Nr 186, seria “Mechanika” z. 56, Awionika, tom I, s. 111–118. Rzeszów.
12. Pieniążek, J., Cieciński, P., Wałek, Ł., & Nowak, D. (2015). *Integrated measurement system for UAV*. IEEE: Metrology for Aerospace.
13. Dołega, B., & Rogalski, T. (2009). Control system for medium-sized flying target. *Aviation* (Vol. 13, pp. 11–16). Vilnius: Technika. doi:10.3846/1648-7788.
14. Rogalski, T., Samolej, S., & Tomczyk, A. (2011). *ARINC 653 based time-critical application for european SCARLETT project, AIAA guidance, navigation, and control conference, 08–11 August*. Oregon, USA: Portland.
15. Samolej, S. (2011). ARINC specification 653 based real-time software engineering, e-informatica. *Software Engineering Journal, 5*(1), 39–49.
16. Samolej, S., & Rogalski, T. Experimental real-time Arinc 653 based pitch angle control application, from requirements do software: Research and practice, scientific papers of the polish information processing society scientific council.

UAV Swarm Management Using Prepar3D

Mariusz Domżał, Karol Jędrasiak, Paweł Iwaneczko, Krzysztof Jaskot and Aleksander Nawrat

1 Introduction

The aim of the article was to prepare the management system for swarm of UAVs in the simulated Prepar3D environment. Goal of the system was to coordinate their actions in order to optimize speed of searching the area defined by the user. At the same time the system had to provide users with access to current information about the location of UAVs (Fig. 1) [1–4].

From the user's perspective, the main advantages of such a system over a single machine would be: speed, under the assumption of a sufficiently large number of vehicles, no need to restock the UAV reducing the number of steps that must be taken by the user, and elimination of the problem of transport spare batteries or fuel. A system of this type could be used for instance:

- searching for missing persons, due to fact that time plays an important role in this case, more machines can speed up the search process [5–7],
- 3D mapping of the terrain [8],
- infrastructure inspection, e.g. in the event of natural disasters system could provide information about the operability of roads for emergency services [9–11].

For the purpose of the project C ++ language was used along with standard libraries supplemented by SimConnect SDK [12], which allows communication with the simulation environment Prepar3D. The code was developed in Visual Studio 2012 Ultimate and compiled by the compiler supplied with the IDE and Visual Studio 2010.

M. Domżał · K. Jędrasiak (✉) · P. Iwaneczko · K. Jaskot · A. Nawrat
Institute of Automatic Control, Silesian University of Technology, Akademicka 16, 44-100 Gliwice, Poland
e-mail: karol.jedrasiak@polsl.pl

A. Nawrat et al. (eds.), *Advanced Technologies in Practical Applications for National Security*, Studies in Systems, Decision and Control 106,
https://doi.org/10.1007/978-3-319-64674-9_11

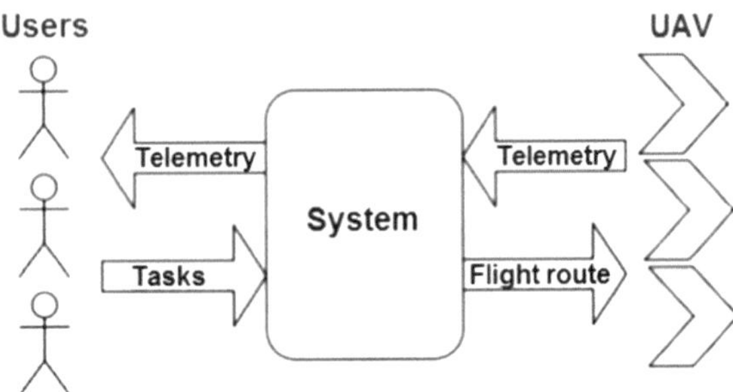

Fig. 1 Graphical representation of system use

Prepar3D a simulation environment implemented by Lockheed Martin which provides a high degree of realism, and is capable of simulating flying, floating and ground-based objects, which can be used in various types of tests, e.g. involving the study of the control system in the form of executable code on a simulated object, called SIL simulations or actual control system of a simulated object—HIL simulation [13, 14]. In this work we have used the SIL type simulation. In addition, it is worth to mention the possibility of simulating damage caused by the influence of overload and multi-user mode, which allows to test, e.g. algorithms of swarm management.

2 System Architecture

The system is composed of 3 main parts: the agents (or simulators), server and GUI applications that communicate with each other via TCP/IP (or more precisely via library WinSock2). Agent and the server have been provided with a simple configuration files, you can select the server address and port.

2.1 Agent

Agent is a software that connects directly to the server and simulation environment Prepar3D in order to take control of the UAV. It is responsible for the connection with the simulation environment and implementation of control algorithm proposed in [15]. The Agent software is divided into 2 parts:

- The inner loop—responsible for testing and control of the current orientation and speed of the UAV. Based on 3 PID controllers that regulate angles of pitch, roll and speed (yaw control does not make sense due to the fact that it used to flying wing UAV), regulators in this case were tuned by engineering method.
- The outer loop—responsible for processing of the current location, altitude, and set route in the form of points values for the inner loop. The height is controlled by the PI controller, whose output becomes the reference slope, the route points

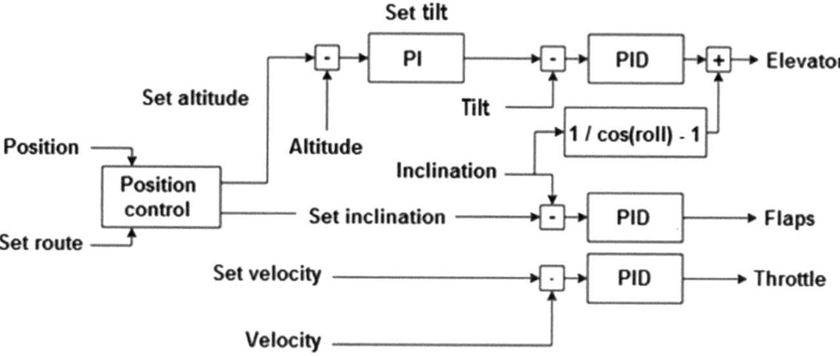

Fig. 2 Control system of a single UAV

are processed in a different way. A destination is determined at a certain distance from the UAV, which becomes the aim of the object. Then the angle between the speed of the UAV and the vector connecting the object with the destination point is determined. On the basis the tilt value is computed. If the agent does not have any destination it enters the waiting mode.

Agent receives route as ECEF points, which are converted into LLA coordinates, which held the calculations associated with the outer loop. Diagram of the control system was presented in Fig. 2. The last module of the program is a TCP client which connects to the server and allow 2-way communication.

2.2 *Server*

The program provides a connection between multiple applications with GUI and many agents and implements the logic needed to complete the mission. The server treats all tasks as searching for a limited pre-defined set of points in the area. For this reason it does not accept the tasks consisting of 1 or 2 points. A job of 0 points can stop the active job, and send the next job. It overwrites the previous one. Server can be divided into 5 main components: 2 types of servers, 2 types of protocols and logic, in which can be distinguished object that stores data on a map. The logic is also coupled to the object by visualization the current state of the map using the OpenGL (an example in the Fig. 4). Mutual relations between modules are shown in Fig. 3. A characteristic difference between used servers is that the server is communicating with a GUI to broadcast information to all users, and the server UAV sends information only to the selected object.

The first step in the implementation of the task is to determine the interesting points from the position of covering a given area. For this purpose the extreme values of the received points in the system LLA (lowest and highest values of latitude and longitude) are found. Then the molded rectangular area consisting of

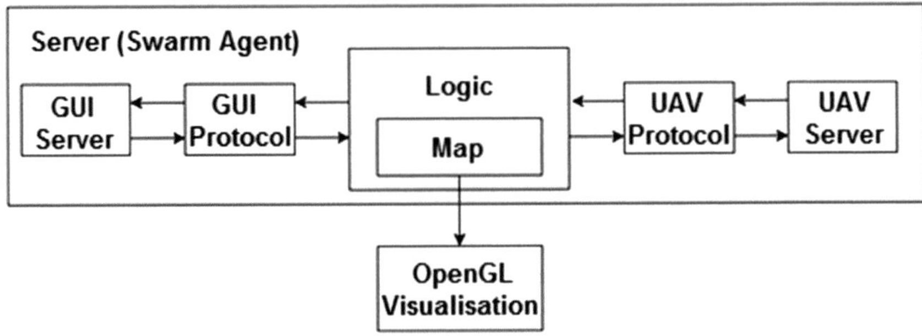

Fig. 3 The flow of information between elements of the server

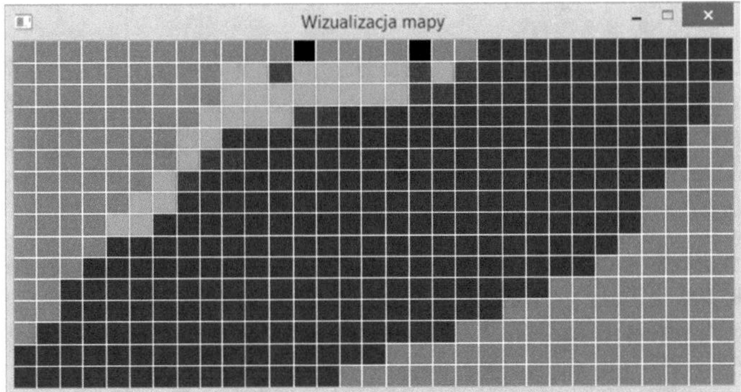

Fig. 4 An example of map visualization used by the server. *Gray* and *black* cells are unimportant for the mission, *black* cells are occupied by objects. *Red* fields represent important cells and occupied by UAVs at the moment, the *blue* cells are relevant to the mission but not yet explored, *green* fields are already examined

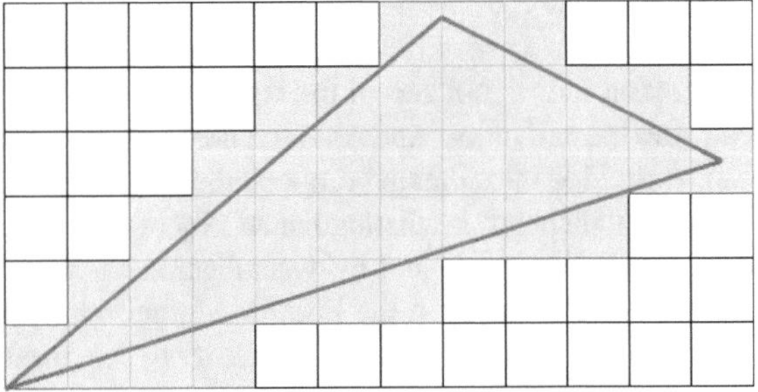

Fig. 5 The choice of important cells from the point of view of the mission (*green*) on the basis of the area of the mission (limited *red* lines)

square cells having a side with a predetermined length is determined. In the next step, selected cells are those that have a common part with the given area (shown in Fig. 5). Separation of 2 convex polygons [16], which imposes the need to inflict areas was implemented. After separation, the server starts cyclically, for instance every 1 s update the map based on the received telemetry data.

A modified algorithm is used as proposed in [16] based on the PSO. The method selects one of the neighboring (North, South, East, West) cells as a target, and during each iteration of the algorithm is trying to bring closer the object to the destined point with the distance smaller than the tolerable deviation. Preventing diagonal traffic eliminates the problem associated with taking images at an angle (Fig. 6). The behavior of the UAV can be divided into four situations:

- In the neighboring fields there is unfathomable and unallocated cell—in this case the object will be directed first to such a field,
- In the vicinity of the vehicle cells are already checked, but the area is not yet fully explored—in this case the determination of the direction of flight is carried out by PSO algorithm, which can be described by the equation:

$$V(t+1) = wv(t) + c_1 rand()(p(t) - x(t)) + c_2 rand()\left(p_g(t) - x(t)\right), \quad (1)$$

where V normalized velocity of UAV, p the closest unchecked cell, pg It is responsible for dispersion of vehicle groups (Fig. 7), x- location of UAV, w, $c1$ and $c2$ constant parameters. In case of control only single machine the algorithm can be simplified to:

$$V(t+1) = wv(t) + c_1(p(t) - x(t)). \quad (2)$$

Determined in this way the speed is converted to the flight direction by calculating the angle between it and the direction of the east. For angle in interval –45° to 45° direction is easy, 45°–135° north, 135°–225° west and for 225°–315° it is south. If the chosen direction is for some reason unavailable (the cell is occupied or selected as a target for another UAV) once again algorithm is performed (maximally 10 time) in single iteration.

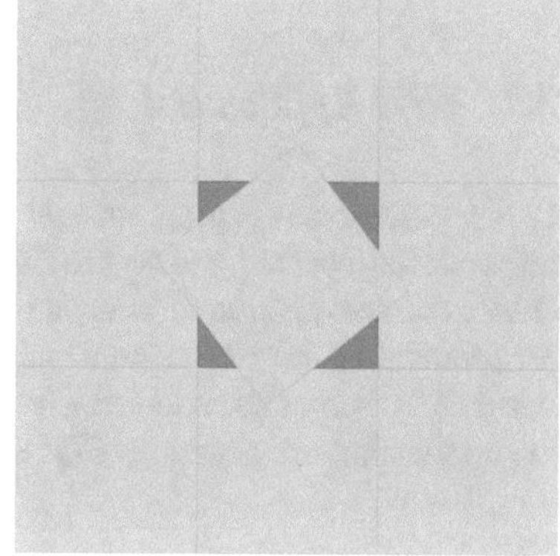

Fig. 6 Illustration of the problem with taking pictures at an angle. Area highlighted in *green* is covered with a photo. Areas marked in *gray* area not covered. Photo from an angle can result in unexplored areas

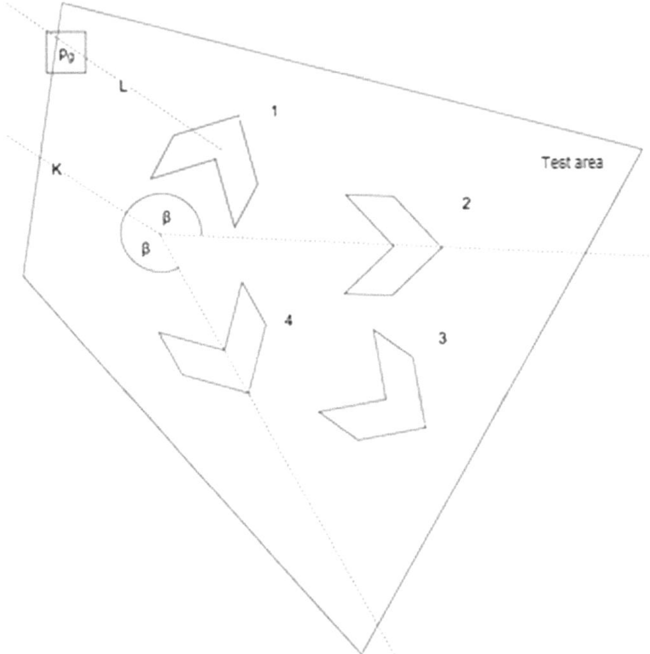

Fig. 7 Postion-determining scheme pg for object number 1 (*red*). Directions in which the objects flying 2 and 4 are the closest to the direction in which the UAV 1 is moving (respectively clockwise and counterclockwise). Ray K is determined in such a way that β parameter was largest. Ray L is parallel to Ray K and has its origin in location of UAV number 1, pg is the most distance cell from UAV 1, belonging to the area i and intersecting with a ray L

- It is located outside the area—in this case the server will send a plane to the nearest unoccupied and unassigned cell.
- The entire area has been explored—the server will stop sending information to the UAV, but continues to monitor their condition.

2.3 GUI Application

GUI application is a program that provides graphical environment to provide the information received and to provide additional information. It allows to track the UAVs and the creation of jobs on the interactive map by applying to I location tags (or removal of already present) and send them to the server. It shall also inform whether the server has adopted a new mission or rejected it. Example screenshots of the application are shown in Fig. 8.

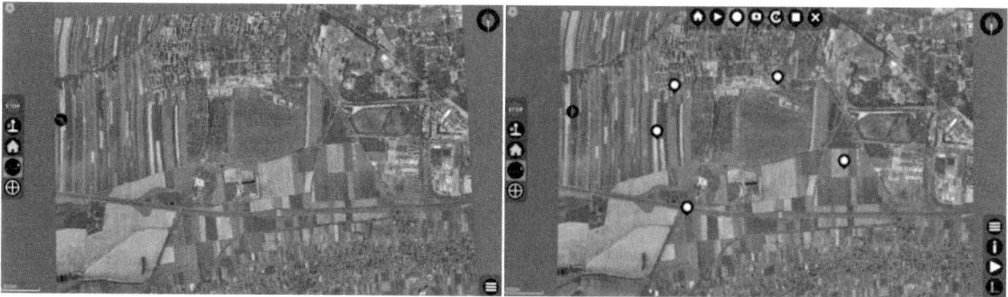

Fig. 8 Examples of screen shots of the application GUI

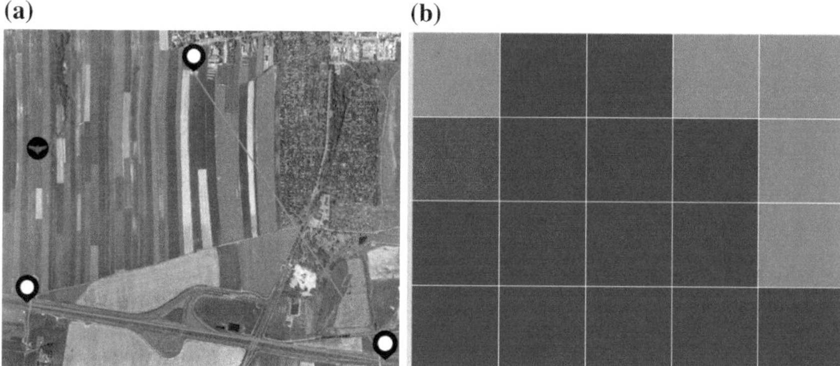

Fig. 9 Initial mission state: **a** visible in application GUI, **b** visible in the state of the server visualization (color analogous to Fig. 4)

3 Tests and Results

Testing was based on the use of a fixed number of UAVs and manipulation of parameters of Eqs. 1 and 2. The algorithm depends only on the ratio of the parameters w, $c1$, $c2$, therefore $w = 1$ was selected for all tests.

3.1 Single Object

Parameter $c2$ does not affect the operation of the algorithm for a single machine, so changed was the only the parameter $c1$. The area depicted in Fig. 9a was used for tests. For all values of the tested parameter $c1$ the algorithm managed to examine the whole area and get the same final map visualization (Fig. 11). You may also notice some differences in the case of the route UAV. Deformations of the routes shown in Fig. 10 are the result of too low value of the parameter $c1$. The algorithm

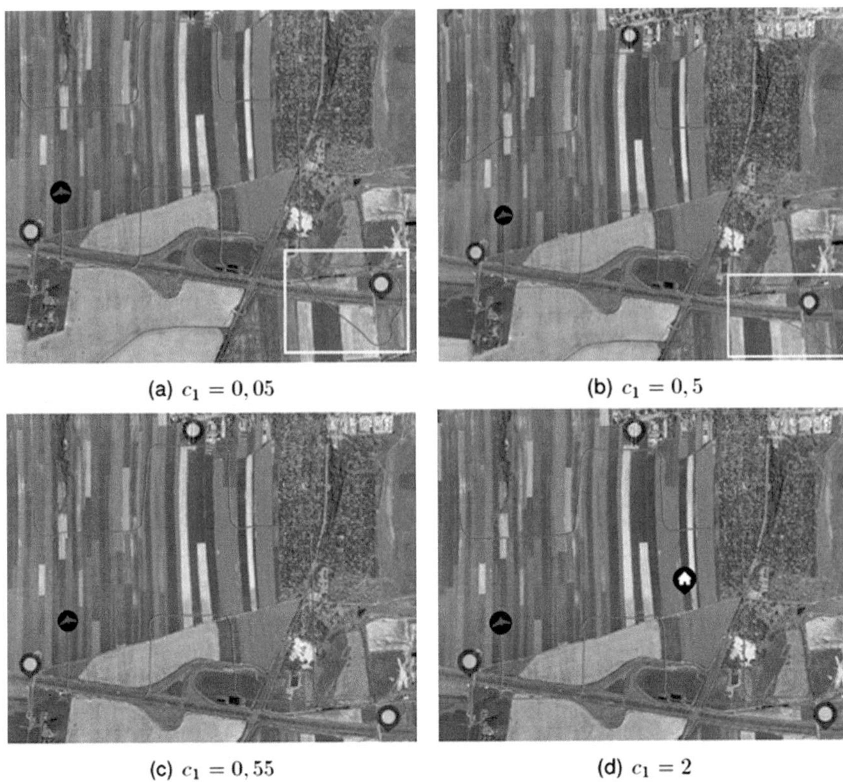

Fig. 10 The impact of parameter $c1$ on the route, *yellow* box indicates the area where there was a route deformation

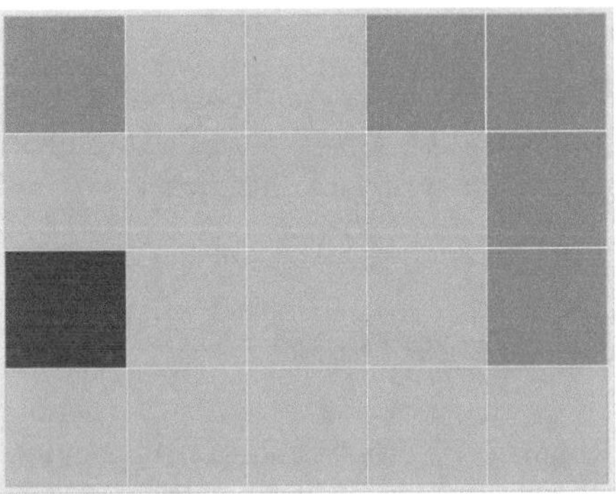

Fig. 11 The final state of the server map

is not able to find its goal until the UAV at least partly turn into its direction. This can cause taking pictures at the wrong angle (Fig. 10), values above 0.5 eliminate this problem.

3.2 Multiple Objects

To perform the following tests we used: 4 machines, 1 agent program, 3 agent simulators that simulate objects moving at the speed of 20 m/s. On each screenshot of the application simulated UAV in environment Prepar3D is highlighted in red. In addition, due to the use of an algorithm based on the randomization each pair of coefficients $c1$, $c2$ was three times tested. The slope line of routes from simulators is due to the fact that the server is preparing another route point in time when UAV's set point is at a certain distance (Fig. 12).

The results shown in Fig. 13 indicates a problem in the form of frequent changes of direction in the case of one of the UAV (purple indication) in the initial phase of exploration and deformation of route (blue and yellow), which could lead to less information about the studied area. The first of the problems stemmed from the proximity of other machines that blocked the neighbouring cells (algorithm began to look for alternative exit), while in the second case there are two reasons for alternative:

- The first is to turn out outside the test area (marked in yellow), it indicates that the effect of the parameters was too small.
- The second reason was the occupation of cells that could be a target for UAVs, so for some time the vehicles remained in the waiting mode to avoid a collision, then again back on the road.

Figure 14 shows the influence of parameter $c2$ on computed routes. Selected area marked in green, where one of the UAV was brought up in the opposite direction to the other three. White lines on one of the pictures highlighted the route of the vehicle in a simulated environment Prepar3D. You will notice that the route was quite chaotic and for most of the time UAV was in the fields already visited. This illustrates the effect of pg parameter from the Eq. 1. There are also purple, yellow and blue markings with the same meanings as before. In the second illustration length of the blue area is striking, which was due to the following of a slower object.

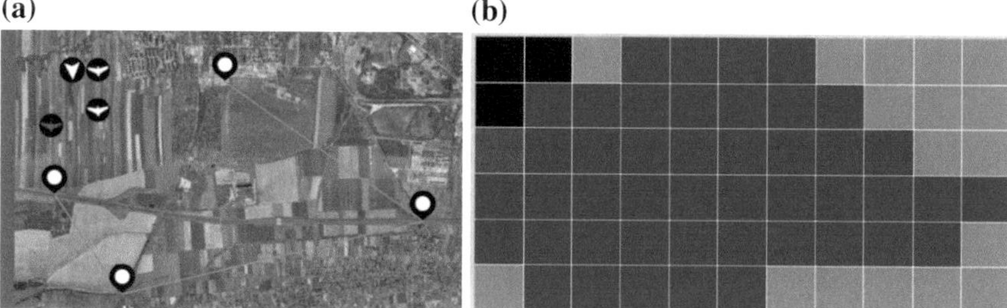

Fig. 12 Initial mission state: **a** visible in application GUI, **b** visible in the state of the server visualization (color analogous to Fig. 4)

Fig. 13 Routes for $c1 = 0.75$ and $c2 = 0.25$, meaning of the indications is explained in Sect. 3.2 of the article

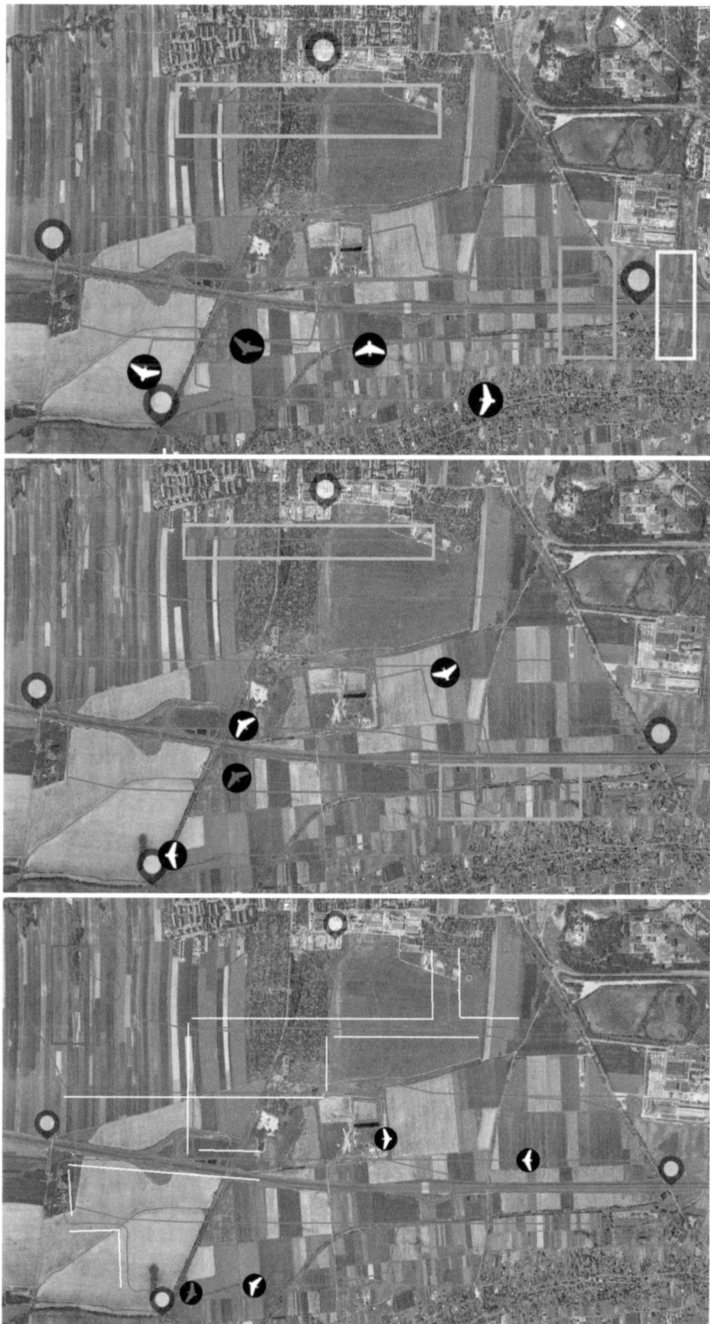

Fig. 14 Routes for $c1 = 0.75$ and $c2 = 5$, meaning of the indications is explained in Sect. 3.2 of the article

Flight path shown in Fig. 15 are characterized by similar problems as those shown in Fig. 13. However, we managed to avoid a situation where the UAV left area of the map. The parameter values were selected to maintain the proportions of the first test ($c1 = 0.75$ and $c2 = 0.25$). As you can see an increase in their value allowed to remove one problem, however, revealed another: areas covered by the white rectangles indicate that the plan has failed, as in blue areas, however in the case of simulation, and not an agent.

The last values of parameters $c1$ and $c2$ tested were respectively 2 and 2.5. The results obtained for these parameters are shown in Fig. 16 and have a limited route distortion caused by collision avoidance marked in white and blue. Therefore, further results will be obtained using the parameter values.

3.3 The Impact of the Number of Objects and Their Initial Positions

During the test the impact of the number of UAVs as well as their initial location was researched. Two cases were assumed: vehicles scattered or gathered in a group (Tables 1 and 2).

The results obtained for the region presented in Fig. 17.

Attempts have been made for objects number equal to 1 and 6. The results were obtained for distributed initial position (Fig. 18) and other parameters value like $c1 = 2$ and $c2 = 1$ for 6 vehicles. Increasing the number of objects significantly reduces the average time of exploration. The results in Table 3 to 3 and 4 vehicles, however, show that there is no guarantee that more UAVs will provide faster search area. Reducing the coefficient of dispersion and different starting positions shown in Table 4 also affect the time. In the first case, it allowed it to gain a better average time, but it increased in the time gap. In the third case, which managed to get an even better average time and a smaller time gap. Due to this change in Fig. 19 there are compared routes for different initial positions.

The results obtained for the test region (Fig. 20) under the assumption that the sectioning of the system was the same as in the previous test.

Tables 3 and 4 shows the test results. Additionally comparison of routes for distributed and grouped starting positions is presented in Fig. 21. The initial state for distributed positions is presented in Fig. 20. The results obtained for 4 or more vehicles do not differ significantly from each other in the average time. Once again, the time gap does not have any direct relationship with the number of units. However, this time the use of other values also resulted in less dehiscence (and, as previously thought average time). Initial dispersion achieved by far the best average time and small time gap, which results in the conclusion that the algorithm can have a problem with tightly coupled groups of vehicles.

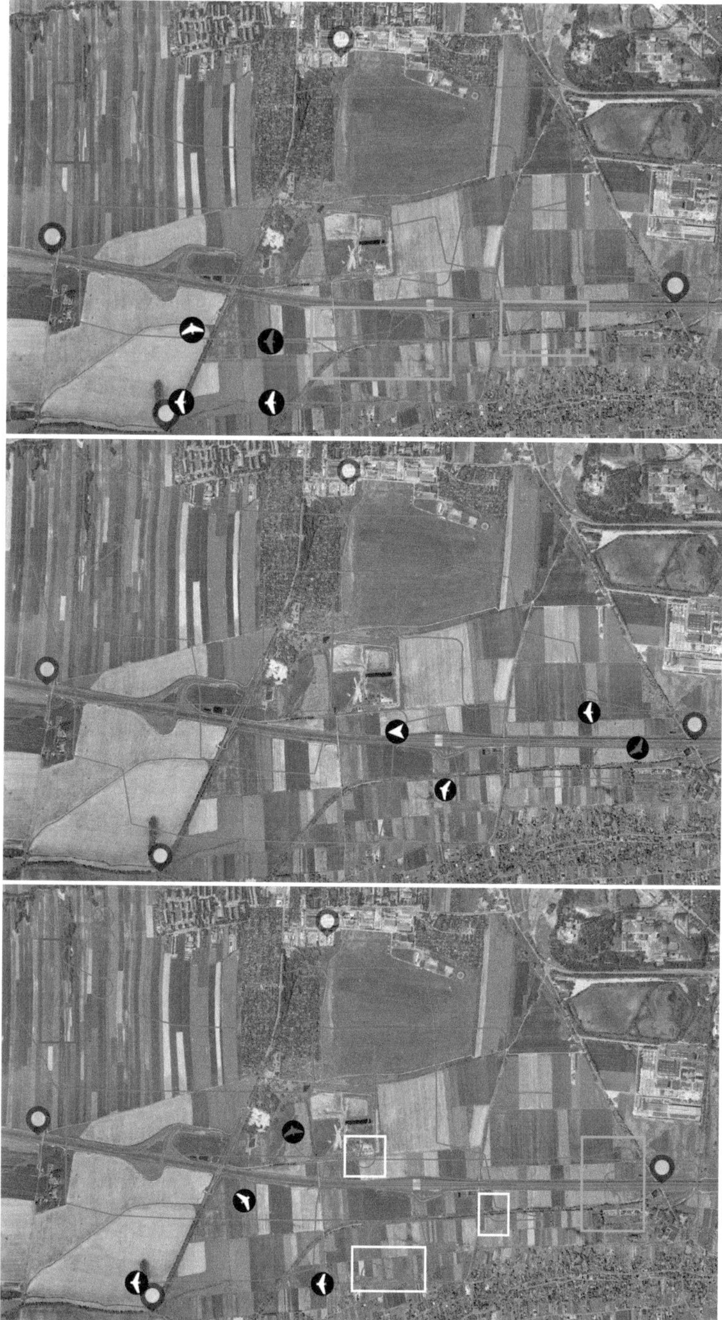

Fig. 15 Routes for $c1 = 2.5$ and $c2 = 0.8$, meaning of the indications is explained in Sect. 3.2 of the article

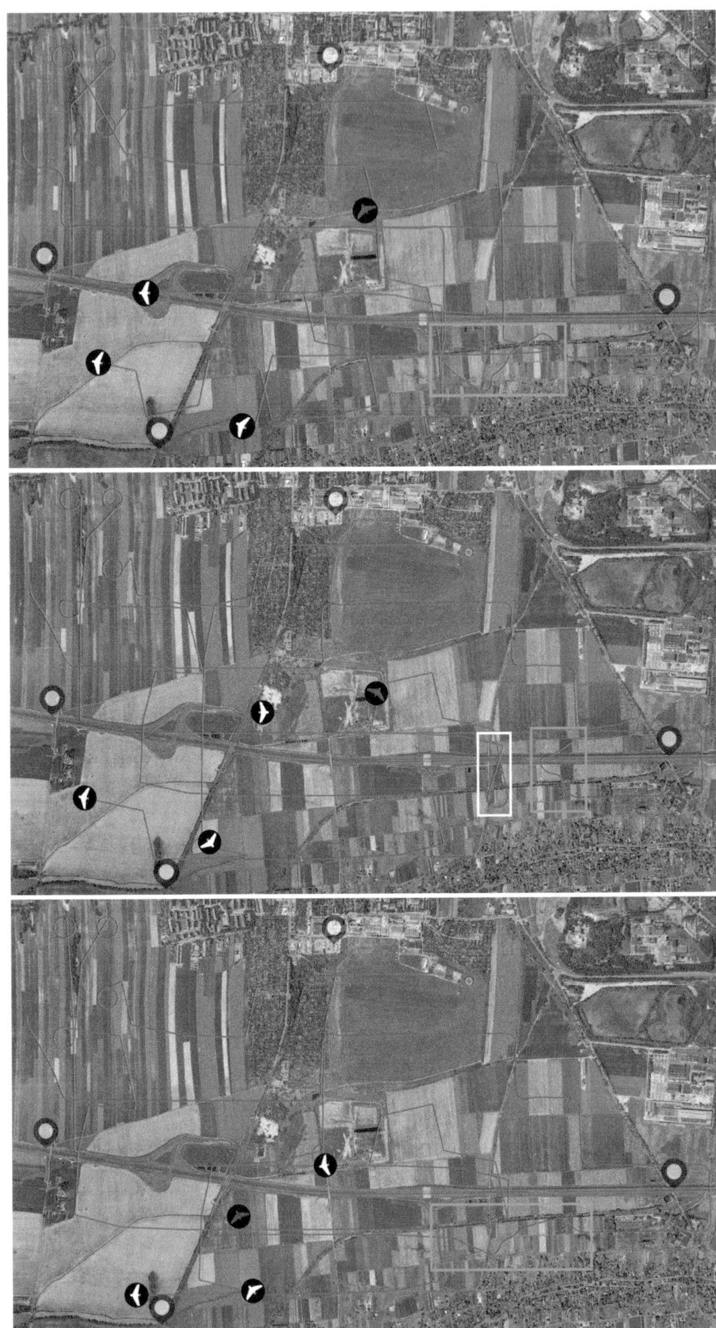

Fig. 16 Routes for $c1 = 2$ and $c2 = 2.5$, meaning of the indications is explained in Sect. 3.2 of the article

Table 1 Summary of the results obtained for $c1 = 2$ and $c2 = 2.5$

Number of passes	Examination time (s)			Avg. time (s)	Time gap (s)	Time gap/Avg. time
	Flight 1	Flight 2	Flight 3			
1	1458	–	–	1458	–	–
2	993	1021	845	953	176	0.1847
3	741	762	817	773.3	76	0.0983
4	669	610	782	687	172	0.2504
5	561	604	582	582.3	43	0.0738
6	464	544	547	518.3	83	0.1601

Table 2 The first row shows the results obtained for the parameters $c1 = 2$, $c2 = 1$, while the second is based on a previously used parameters during distributed starting positions

Number of passes	Examination time (s)			Avg. time (s)	Time gap (s)	Time gap/Avg. time
	Flight 1	Flight 2	Flight 3			
6	531	405	512	482.7	126	0.261
6	477	447	445	456.3	32	0.0701

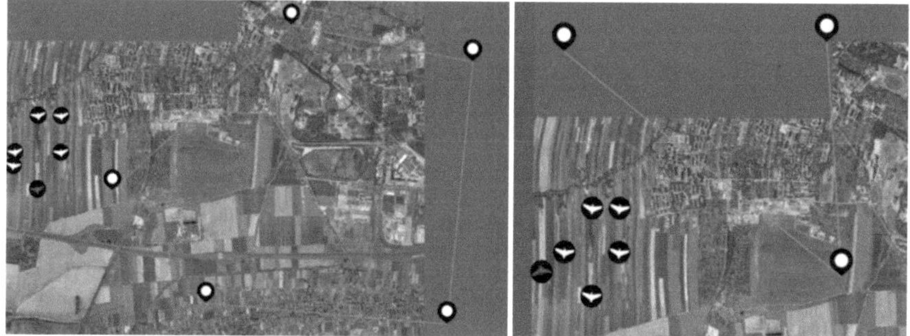

Fig. 17 Areas used during the tests

3.4 Research on the Influence of Area's Shape

The aim of the tests in this situation was to verify the correctness of generating internal server map based on transmitted points. The photos are presented in pairs: the first (left) is the actual shape of the map sent to the server using GUI application, while the second (right) is the server map visualization, which has been visualized without grid because of the small size of the cells. (Figures 22, 23, 24, and 25)

The server can cope with convex figures. In the case of concave figures there are additional cells that are classified as important for the mission, although there should not be any. Their number depends on the measure of concave angle. For

Fig. 18 UAV's initial locations

Table 3 Summary of results obtained for $c1 = 2$ and $c2 = 2.5$

Number of passes	Examination time (s)			Avg. time (s)	Time gap (s)	Time gap/Avg. time
	Flight 1	Flight 2	Flight 3			
1	574	–	–	574	–	–
2	396	378	394	389.3	18	0.0462
3	361	282	331	324.7	76	0.2341
4	296	268	236	266.7	60	0.225
5	251	256	255	254	5	0.0197
6	243	229	273	248.3	44	0.1772

Table 4 The first row shows the results obtained for the parameters $c1 = 2$, $c2 = 1$, while the second is based on a previously used parameters at distributed starting positions

Number of passes	Examination time (s)			Avg. time (s)	Time gap (s)	Time gap/Avg. time
	Flight 1	Flight 2	Flight 3			
6	225	235	236	232	11	0.0466
6	142	135	136	138	8	0.058

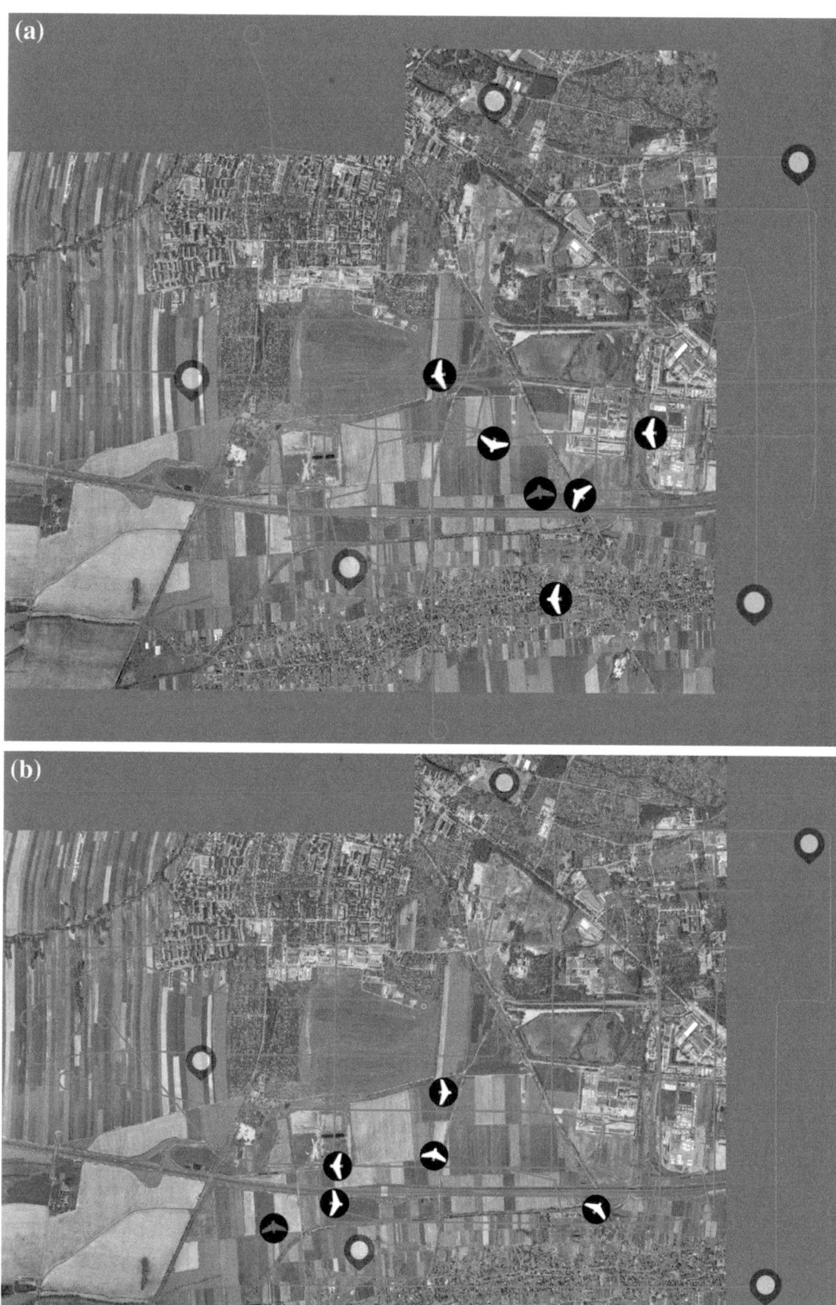

Fig. 19 Comparison of routes for different initial positions, **a** distributed, **b** grouped

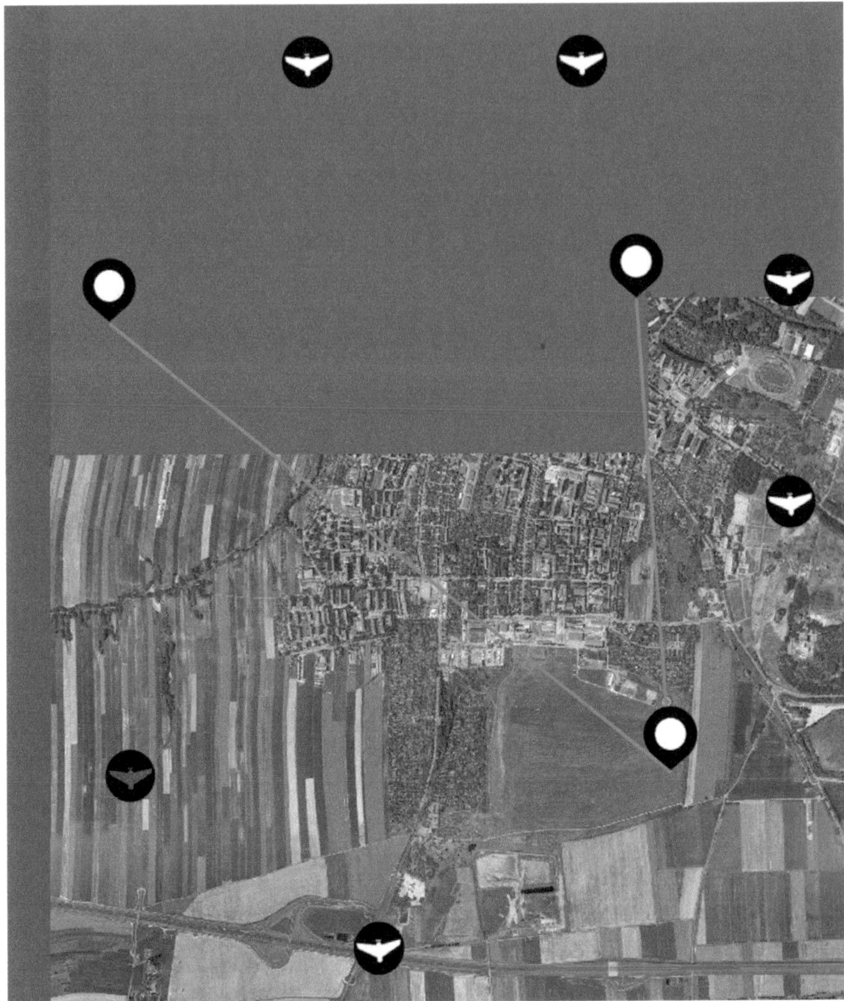

Fig. 20 Distributed initial positions

instance in Fig. 4.16, there is an angle with value close to 270° there are multiple erroneous cells. In Figs. 4.17 and 4.15 there are concave angles closer to180° and the number of erroneous cells is much smaller. Assuming that in any polygon there are concave angles n $\alpha_1, i = 1, 2, \ldots, n$ the server converts them into $360° - \alpha$.

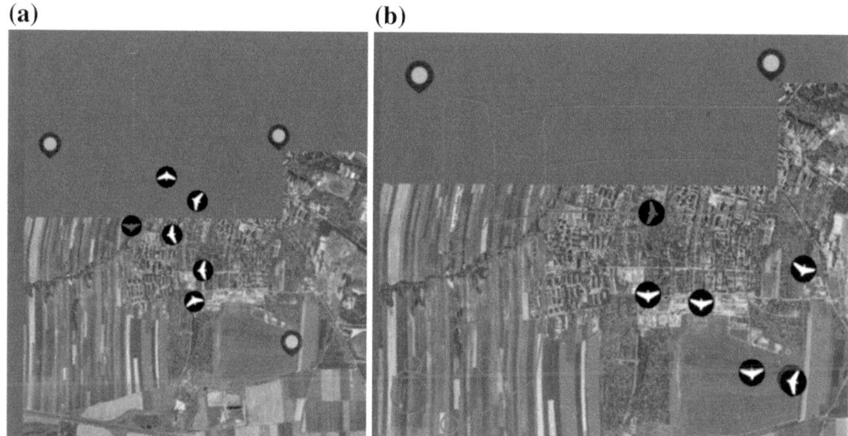

Fig. 21 Comparison of routes for different initial positions, **a** distributed, **b** grouped

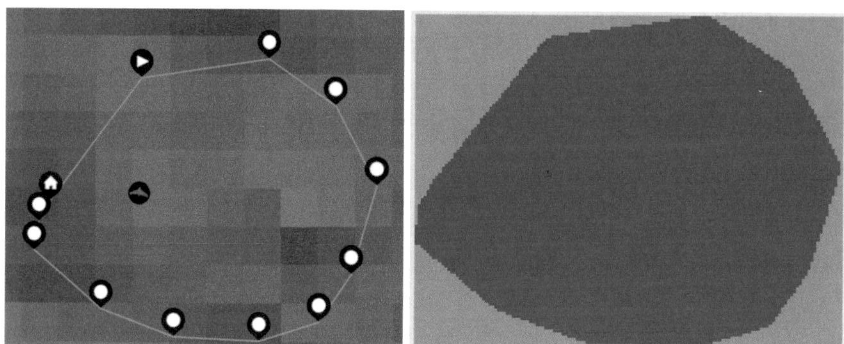

Fig. 22 13-angle convex poligon mapping

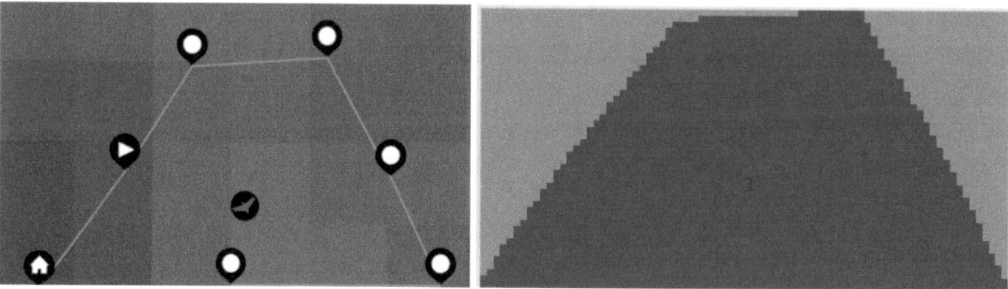

Fig. 23 7-angle trapezoid-like (not a convex figure) mapping

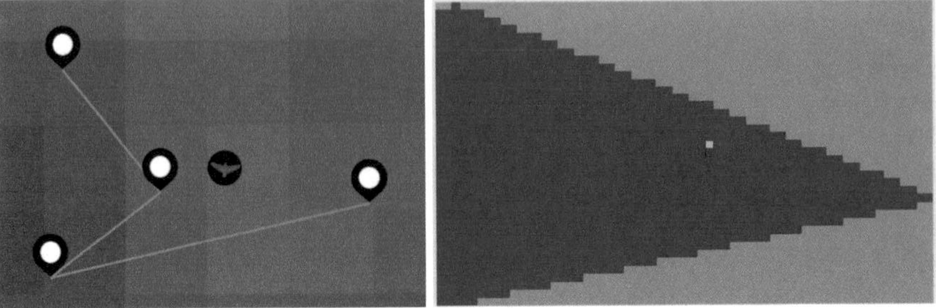

Fig. 24 Concave quadrilateral mapping

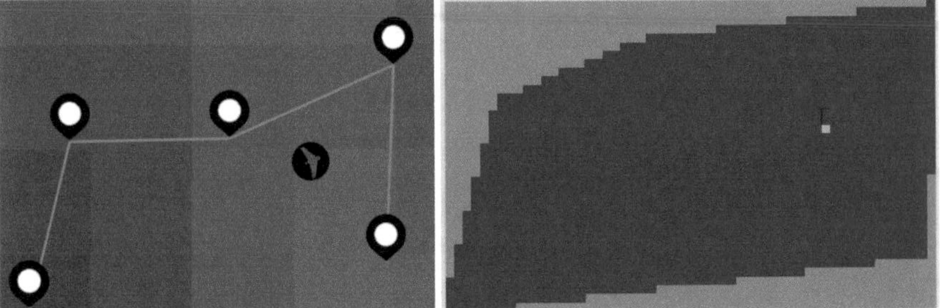

Fig. 25 Concave pentagon mapping

3.5 Research on the Influence of Multiple UAVs Participation

The scenario of this test was as follows: during a mission for 1 vehicle (it took 424 s for single vehicle) 2 additional vehicles were added (after 150 s since mission start), with goal of supporting the first vehicle. The objective has been achieved and the search time was reduced from 424 s to 387 s. Initial situation is shown in Fig. 26, supporting vehicles are marked yellow. Routes acquired after the test are presented in Fig. 27.

3.6 Impact of the Loss of Communication with UAV

In this case, the test began with two vehicles. After some time, vehicle #1 has been disconnected from the system. The effects of such an action is shown in Fig. 28. Visualization of server's map shows the location of a missing machine (upper right). The system should not direct there to any of vehicles. It is in accordance with the route of the second UAV—it avoided the cells already visited by the first

Fig. 26 Initial positions of UAVs, marked in *yellow* joined later

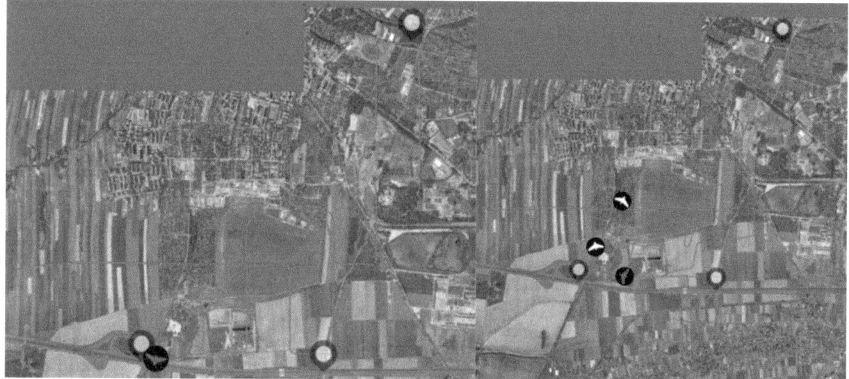

Fig. 27 On the *left* is the route of single UAV, on the *right* there are routes of multiple UAVs which joined the task at a later time

Fig. 28 On the *left* there is UAV's route, on the *right* the state of the server's internal map

vehicle and, in particular, cells in which #1 had disappeared. This behavior of the system is negative from the point of view of the exploration of a given area, but it increases the security of the controlled UAVs. They can thus avoid mistakes of their predecessors.

4 Conclusions

Completed software solution developed during the work allow creation of the swarm of simulated objects with goal of searching the area. In conjunction with the application GUI it allows convenient management of the swarm. Increasing the number of vehicles in the swarm increases the number of turns on the route, however it allows for faster testing of the areas. As one of the advantages of the presented approach it can be added the fact that for all studied situations, including parameter sets and the number of vehicles, in all cases swarm managed to examine the entire area. The selection of the appropriate parameters provides a better route for the coverage of area with photographs and allow to do it faster. Despite multiple intersecting routes there have been no cases impending collision. The system also allows to attach during the run additional vehicles and copes with the loss of connection with the units. A very positive impact on the results of a scattering of objects before starting the search, therefore future version of this system could use some algorithm that ensures dispersion of the swarm before exploration. Another possible modification is to use an alternative method of dispersing vehicles in the course of the algorithm—the current version can in 2 subsequent iterations fruit in points lying on opposite sides of the map. An example of an alternative may be based, among others, on electrostatic interactions, unvisited cells would have the opposite charge to the vehicle and other vehicles the same, which would prefer to areas with a large number of unexplored fields away from other vehicles and alter slowly.

The problem is the lack of concave areas. One way of solving this problem might be to divide into sub-convex, which can be handled by the current algorithm, e.g. by means of triangulation. One of the possibilities for the development of this system is to provide multiple, posed by different users missions. In conjunction with the ability to estimate the time (especially the upper limit) duration of the mission, depending on the units used UAVs at the time of receipt, it could allow the management of swarm in such a way that executed in parallel missions ended within the time specified by the user. The system would assign vehicles for specific missions. A system of this type could be used in reconnaissance missions or searching for people missing in uninhabited regions [17, 18], or for example extensive research of infrastructure [19]. The system could also generate a 3D map of the area.

References

1. Wang, Y., Liang, A., Guan, H. (2011). Frontier-based multi-robot map exploration using particle swarm optimization. In *2011 IEEE Symposium on Swarm Intelligence*.
2. Jedrasiak, K., Nawrat, A., & Wydmanska, K. (2013). SETh-link the distributed management system for unmanned mobile vehicles. In *Advanced technologies for intelligent systems of national border security* (pp. 247–256). Berlin, Heidelberg: Springer.
3. Daniec, K., Iwaneczko, P., Jedrasiak, K., & Nawrat, A. (2013). Prototyping the autonomous flight algorithms using the prepar3D® simulator, vision based systems for UAV applications (pp. 219–232). Springer International Publishing.
4. Galuszka, A., Skrzypczyk, K., & Ilewicz, W. (2014). On transformation of conditional action planning to linear programming. In *Methods and models in automation and robotics (MMAR)* (pp. 764–769).
5. Jedrasiak, K., Andrzejczak, M., Nawrat, A. (2014). SETh: The method for long-term object tracking. In *Computer vision graphics*, 8671 (Vol. 316, pp. 302–315). Lecture Notes in Computer Science.
6. Ryt, A., Sobel, D., Kwiatkowski, J., Domzal, M., Jedrasiak, K., & Nawrat, A. (2014). Real-time laser point tracking. In *International conference on computer vision and graphics* (pp. 542–551). Springer International Publishing.
7. Nawrat, A., & Jedrasiak, K. (2008). Fast colour recognition algorithm for robotics. *Problemy Eksploatacji*, 69–76.
8. Sobel, D., Jedrasiak, K., Daniec, K., Wrona, J., Jurgas, P., & Nawrat, A. (2014). Camera calibration for tracked vehicles augmented reality applications. In *Innovative control systems for tracked vehicle platforms* (pp. 147–162). Springer International Publishing.
9. Bieda, R., Jaskot, K., Jędrasiak, K., & Nawrat, A. (2013). Recognition and location of objects in the visual field of a UAV vision system. In *Vision based systems for UAV applications* (pp. 27–45). Springer International Publishing.
10. Babiarz, A., Bieda, R., & Jaskot, K. (2013). Vision system for group of mobile robots. In *Vision based systems for UAV applications* (pp. 139–156). Springer International Publishing.
11. Kus, Z., & Nawrat, A. (2013). Object tracking in a picture during rapid camera movements. In *Vision based systems for UAV applications* (pp. 77–91). Springer International Publishing.
12. Martin, L. (2015). SimConnect SDK. http://www.prepar3d.com/SDK/SDK.php. internet source, visited: 2015-12-01.
13. Bereska, D., Daniec, K., Fras, S., Jedrasiak, K., Malinowski, M., & Nawrat, A. (2013). System for multi-axial mechanical stabilization of digital camera. In *Vision based systems for UAV applications* (pp. 177–189). Springer International Publishing.
14. Jedrasiak, K., Nawrat, A., Daniec, K., Koteras, R., Mikulski, M., & Grzejszczak, T. (2012, September). A prototype device for concealed weapon detection using IR and CMOS cameras fast image fusion. In *International conference on computer vision and graphics* (pp. 423–432). Berlin, Heidelberg: Springer.

15. Elkaim, G. H., Lie, F. A. P., & Gebre-Egziabher, D. (2015). Principles of guidance, navigation, and control of UAVs (pp. 347–380). Springer Netherlands.
16. Bittle, W. (2015). Twierdzenie o prostej separującej. http://www.dyn4j.org/2010/01/sat/. online, internet source, visited: 2015-12-15.
17. Josinski, H., Switonski, A., Jedrasiak, K., & Kostrzewa, D. (2012). Human identification based on gait motion capture data. In *Proceedings of the 2012 international multiconference of engineers and computer scientists, IMECS* (Vol. 12).
18. Switonski, A., Josinski, H., Jedrasiak, K., Polanski, A., & Wojciechowski, K. (2010). *Classification of poses and movement phases, ICCVG 2010*. Lecture Notes in Computer Science: Springer.
19. Skrzypczyk, K., Gałuszka, A., Ilewicz, W., & Antas, T. (2015). Synthesis and evaluation of the smart electric powered wheelchair route stabilization concept–A simulation study. *Archives of Control Sciences, 25*(2), 263–273.

Part III
Computer Models and Simulations

One of the most important and challenging aspects of creating Practical Applications for National Security is to create realistic computer models and simulations. This chapter contains scientific description of three issues: geometric and photometric calibration of the image projection system, probabilistic model of radio propagation delay in indoor environment and kinematic-dynamic analysis of human upper arm. Proposed method of manual calibration of image projection system enables calibration of any number of Digital Light Processing (DLP) projectors, on-screen consisting of many planar surfaces, that leads to result of developing the method a software arose facilitating conducting the geometrical and colorful calibration. Presented software was subjected to qualitative and quantitative tests, what enables to distinguish the distorting factors in the initial visualization. Probabilistic model of describing radio propagation delay in indoor environment is based of delta functions sequence to describe retransmissions between a transmitter and a receiver. Model has been verified by measurement results obtained by using the experimental system, which includes commonly occurring disturbances in buildings, such as walls. The presented model of upper arm motion has been derived using the Euler–Lagrange equation. Mathematical models use direct kinematics to develop compute position and orientation of arm based on the given human joint position. Dynamics of the arm refers to the interaction between forces in the system and change of state of the system. Based on the dynamic equation of motion of the arm non-linear and linear model of human upper arm has been defined to achieve realistic simulation.

Advanced Ballistic Model and Its Experimental Evaluation for Professional Simulation Systems

Karol Jędrasiak, Jarosław Cymerski, Przemysław Recha, Damian Bereska and Aleksander Nawrat

1 Introduction

High mobility training of officers, soldiers, guards and security personnel as representatives of the state security system responsible for ensuring public safety is a complex and multifaceted process, especially in terms of use of firearms. This process should be conducted at each level of tasks, which include the level of intervention, tactic and strategic while taking into account practical experience gained during domestic and abroad operations [1]. In addition, it should be emphasized that the aim of the firearms training process is a theoretical and practical preparation of service representatives to perform their tasks [2]. The above issues also tends to carry out considerations on the use of tools to support the process of training of tactics and the application of laws relating to the use of the firearms in situations threatening the life or health.

For the purposes of this study devoted to the use of models of ballistic systems simulation authors will refer only to the most important areas from the point of view of the realism of the training process conducted on the basis of simulation systems designed to support the training of officers and soldiers.

The first area is the "environment" and its specific circumstances in which the above designated representatives of the state security system entities carry out tasks which should have representation in professional systems simulation. For the purposes of this article the term "environment" should be understood as the area and the conditions prevailing in it, taking into account the conditions of the

K. Jędrasiak (✉) · P. Recha · D. Bereska · A. Nawrat
Institute of Automatic Control, Silesian University of Technology,
Akademicka 16, 44-100 Gliwice, Poland
e-mail: karol.jedrasiak@polsl.pl

J. Cymerski
Government Protection Bureau, Warsaw, Poland
e-mail: j.cymerski@bor.gov.pl

A. Nawrat et al. (eds.), *Advanced Technologies in Practical Applications for National Security*, Studies in Systems, Decision and Control 106,
https://doi.org/10.1007/978-3-319-64674-9_12

infrastructure specific to the site. In the literature there is division of the operational environment into urban area (city) and open area (extra-urban). In urban areas during the operation, the "black tactics" referred to as a Close Quarters Battle (CQB) [3] and the Military Operation in Urban Terrain (MOUT) [4], are used. However, in the suburban areas, the "green tactics" [5, 6] are used. At the same time it should be noted that defining "environment" should take into account the conditions resulting from natural phenomena that may occur in the indicated areas.

Another area is the legal conditions delegating permissions to use of firearms, along with the obligations arising from the consequences of its use. Both of these areas "environmental" and legal, seem to be important from the point of view of the necessity to introduce the realism of the situation in the shooting training due to the consequences of both the physical and legal, that may accompany the use of firearms.

In the Republic of Poland legal basis for the use of firearms is the Act of 24 May 2013 on means of direct force and firearms (Dz. U from 2013, 628 with later changes). The act on means of direct force and firearms defines a closed list of entities authorized to use of firearms, which are the officers of: Internal Security Agency, Intelligence Agency, Government Protection Bureau, Customs Service, the Central Anti-Corruption Bureau, the Police, Prison Service, the Border Guard, Railway Security Guard, Guard Park, officers and soldiers: Military Counterintelligence Service, the Military Intelligence Service, the soldiers of the Military Police or military law enforcement agencies, security guards: the National guard hunting, the National Fisheries guard, local guards (urban), Forest guard, Marshal guard, inspectors and employees of tax inspection, inspectors of the Road Transport inspection, security staff entitled to use firearms under the provisions of the Act of 22 August 1997 on the protection of persons and property (Dz. U 2014. pos. 1099).

In the abovementioned Act of 24 May 2013 on means of direct force and firearms (Dz. U 2013, 628 with later changes). The legislature precisely defined the terms "use of firearms" and "application of firearms." "use of firearms" was defined as taking a shot in the direction of a person using penetrating ammunition, while the "application of firearms" will be taking a shot with the use of penetrating ammunition in the direction of the animal, object or in another direction not posing a threat to people. The above presents wide range of entities, whose representatives have the right to use of firearms. It also presents the spectrum of the various effects of impact of penetrating ammunition [7–10], and thus translates into a multidimensional training process. In addition, the legal content which is an integral part of the process of training in the use of firearms requires stating that due to its multithreading it should be included in the process of building professional simulation systems.

These environmental and legal regulations regarding the use of firearms require that officers, soldiers, guards and workers' protection obligations related to acquiring and developing skills of using a firearm during the different circumstances surrounding the execution of tasks. Acquiring and developing theoretical and practical skills provides training process conducted on the basis of professional methodology using modern training tools. Undoubtedly, such tools include training facilities holding in its infrastructure combat shooting ranges equipped with innovative simulation systems reflecting the realism of the shooting situation and taking into account the consequences of the use of firearms. Achieving realism of the shooting situation is possible, among other things as a result of the implementation of developed ballistic models in professional simulation systems.

Multimedia shooting ranges are simulation systems allowing firearms training. Each of the multimedia shooting systems consists of two parts: the real part and the virtual part. The real part consists of the shooter, firearm or replica firearm, the vision system which allows detection of bullet hits, a computer and a screen for projection of shooting targets [11, 12]. Virtual part consists of a generated by computer simulated virtual world with shooting targets. Virtual world generated by the computer is displayed on the screen by the projector. Multimedia shooting ranges can be divided into two categories: using firearms and using replicas of firearms equipped with imu [13]. In the first case, the user of such a shooting range, shots with a firearm into ricochets silencer, which also serves as the screen for display purposes. The algorithm [14] recognizes coordinates where the bullet hits a screen based on the analysis of video stream from the thermal imaging camera. The detected hit point's coordinates in the camera coordinate system are further mapped by a computer algorithm into respective coordinates in a simulation system. In the case of replica firearms, there is a laser operating in the infrared range mounted in the barrel or on picatinny rail. Trigger of the replica firearm is modified in order to wirelessly transmit the information about the press to the computer system. At the moment of a virtual shot a laser pulse is emitted from the replica firearm and special algorithm detects the laser dot on the screen by analyzing the video stream from infrared camera. The detected laser dot's coordinates in the camera coordinate system are further mapped by a computer algorithm into a respective coordinates in a simulation system. Traditional shooting range can be converted into a multimedia shooting range. Transforming traditional shooting range into the multimedia shooting media allows to practice shooting at distances greater than those that provided by traditional, physically limited, indoor shooting ranges. Multimedia shooting ranges should in the most accurate way simulate the flight of the projectile and take into account the rights of external ballistics that affect the flight of the projectile.

Part of the work will be implementation of a mathematical model of projectile motion for multimedia shooting ranges. For this purpose will review the literature related to the external ballistics. Ballistics is a science that studies the process of launching, the behavior of projectile flight and behavior of the impact of projectile into hit targets. A ballistic object is an object which movement is not limited in any way and is subject to forces such as: air resistance, gravity. External ballistics is a section of ballistics researching the movement of propelled ballistic objects and phenomena that have a direct impact on their movement. One of the uses of external ballistics is the study of the movement of rifle's projectile in the Earth's atmosphere. Using mathematical models of projectile motion in the atmosphere it is possible to determine the trajectory of a projectile with knowledge of the initial conditions of the shot. It is also possible to determine the forces and moments acting on the projectile during flight. As part of the work selected mathematical model will be implemented in the MATLAB simulation environment. The results obtained in the simulation will be verified with available ballistic charts. Ballistic charts are supplied by the manufacturers of ammunition and contain information about the flight path of projectile. Along with the trajectory information about the parameters of the atmosphere, setting the iron sights and wind speed, may be provided. The ballistic charts contain information about the bullet weights, muzzle velocity and ballistic coefficients. This information is useful to verify the simulation. Examples of manufacturers of ammunition ballistic tables can be found in [15, 16].

The measurable result of the work will be implementation of ballistics equations in the physical engine. Implementation will reproduce projectile motion in the atmosphere in a manner consistent with the data from the ballistic charts from ammunition manufacturers. Implemented ballistics model could be used for a shooting training e.g. using UAVs [17].

2 External Ballistics

The subject of the external ballistics is the study of the laws of motion of ballistic objects in space around Earth. One of the tasks of external ballistics is modelling of physical and mathematical spatial motion of projectiles. In order to build a physical model of a projectile motion the following phenomena should be taken into account: the basic laws of physics, the structure of the object, external forces acting on an object in flight, the characteristics of the medium in which the object is moving.

External ballistics is used in the study of motion of rifle projectile, artillery shells, rocket ballistic missiles. External ballistics in military applications is described in [18]. The authors of the publication provide general information on mathematical modelling of the motion of ballistic object and phenomena that affect the flight of the projectile: gravity, air resistance, the Coriolis force [19], wind and their mathematical modelling. Mathematical model of projectile motion as a material point and a mathematical model of the atmosphere described in the [18] were selected for implementation. The article [20] contains information on modelling of projectiles motion. It contains mathematical description of the bullet, but the description of phenomena affecting the projectile is not as detailed as in [18]. From the article [20] the equation describing the force of air resistance was used. Mathematical models of projectile motion are limited to the consideration of gravity and air resistance. Publication [21] contains practical information on external ballistics. This position is a textbook for sharpshooters [22, 23]. It contains descriptions of phenomena that affect the flight of the projectile, but does not provide any mathematical models describing these phenomena. Position [21] contains an extensive collection of ballistic tables that provide a lot of information about the projectiles, which can be used to verify the models included in the article [18]. The article [21] did not describe any information to derive a mathematical model of projectile motion. Reference article [24] provides a manual for snipers. It contains a description of the phenomena affecting the bullet: gravity, wind, drag, gyro drift. It contains no mathematical descriptions of phenomena affecting the projectile. It contains a chart of ballistic coefficients of ballistic projectiles that can be used to verify the mathematical models of projectile motion. The article [24] does not contain any information allowing to derive a mathematical model projectile motion.

2.1 External Forces Acting on an Object

Movement of the object takes place under the action of the resultant of external forces. In the case of projectile motion in the atmosphere external forces acting on the projectile resulting from: projectile motion in the gravitational field, the movement of the projectile in the atmosphere, the Earth's rotation, rotational movement of the projectile.

2.1.1 Gravity Force

The process of modelling of the physical projectile motion requires the adoption of the model of gravity field. The choice depends on the firing range. A uniform force field of gravity assumes that in all points of the trajectory the gravity force is constant. Such field is used at ranges up to 20 km, which is sufficient for the needs of multimedia shooting range, because even range of the snipers rarely exceeds 1 km. The central field of gravity assumes that the value of the acceleration varies with the altitude of the projectile. The field is used at ranges from 20 km to 500 km. For ranges longer than 500 km off-center gravity fields are used, which assumes that the Earth is an ellipsoid compressed at the poles.

2.1.2 Aerodynamic Drag

Aerodynamic drag acts on a moving ballistic object in the atmosphere. The value of the force of aerodynamic drag depends on the parameters of the atmosphere, the shape of the projectile, the speed of movement and drag coefficient. Determination of the coefficient of resistance is difficult and complicated, in addition, it changes with the change of velocity of the projectile. For the calculation of the resistance coefficient G1 and G7 resistance models were created. These models define the standard projectile (its mass and shape) for which the value of the drag coefficient is determined depending on the Mach number, the ratio of speed of the object in the medium to the speed of sound in the same medium. For other projectiles with the same shape value of resistance can be calculated with a given model, adjusting it by the value of shape factor of the projectile. The values of the resistance depends on the Mach number for the G7 resistance model (Fig. 1).

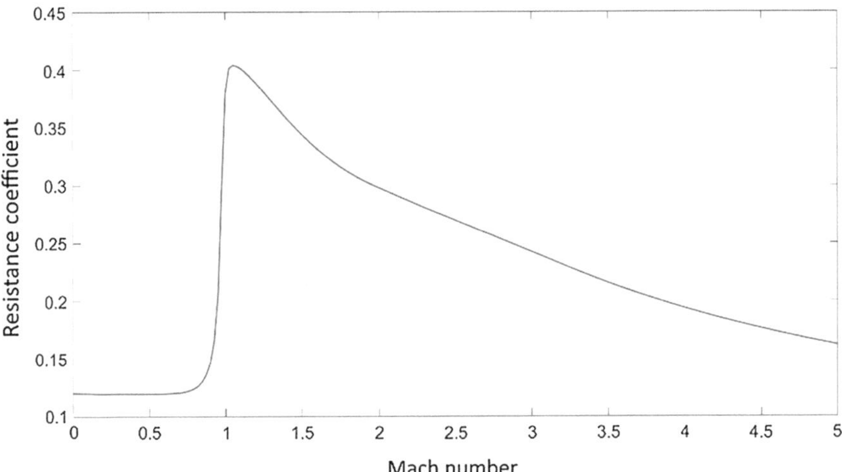

Fig. 1 The influence of Mach number on the value of the drag coefficient for the model G7

It may be noted that for subsonic speeds coefficient is constant at speeds close to the speed of sound value of this ratio is growing rapidly, and at supersonic speed this coefficient decreases.

2.1.3 Wind Force

Wind causes the rotation of the projectile in the direction of the wind, which causes that the drag force is not parallel to the trajectory. Such directed drag force causes the horizontal deflection of the trajectory of the bullet, and is observed by the shooter as a drift of the projectile. Another effect of the wind is change of the speed of the projectile relative to air, which results in change of the aerodynamic force.

2.1.4 Coriolis Effect

Ballistic object moves in a rotating coordinate system of reference, which is the Earth. As a result, on the projectile operates the Coriolis force, which bends the trajectory of the projectile. This force depends on the azimuth and latitude of the shot. Its influence is observed when shots over long distances, for instance with sniper rifles. The Coriolis force has a vertical component and its effect on the object is called Eötvös effect [25]. An object moving in east direction is influenced by vertical force directed upwards. Vertical force directed downwards occurs when object is moving in west direction. This has an impact on the range of shot.

2.1.5 Magnus Effect

Projectiles stabilized by spinning motion are affected by the Magnus effect. When the projectile moves transversely to the air, e.g. under the influence of wind, pressure difference is created on two opposite sides of the projectile which causes the Magnus force. The force acts either up or down, depending on the wind direction and the direction of rotation (Fig. 2). The vertical deviation of the projectile trajectory is smaller than the deviation caused by the wind.

2.1.6 Gyroscope Drift

Projectiles stabilized by spinning motion are affected by the gyro drift. The axis of rotation of the projectile does not coincide with the velocity vector of the projectile, it causes creation of additional force deflecting the trajectory of the bullet in the horizontal and vertical planes. The value of this force depends on the length of the projectile, speed of rotation, range, time of flight, altitude and atmospheric conditions.

2.1.7 Summary

Implementation of mathematical model of projectile motion should meet the following assumptions:

- mathematical model takes into account the effects of projectile motion: gravity, wind, drag, and the Coriolis force,
- it is possible to simulate any type of ammunition for small arms, whose parameters: weight, caliber and ballistic coefficient, the initial conditions of the shot: the initial velocity and the angle of the shot, and ambient conditions: air pressure, air density, wind speed and temperature are known,
- calculations are simple and fast to ensure computations in real time,
- system provides functionality to detect shoots in a single shot mode.

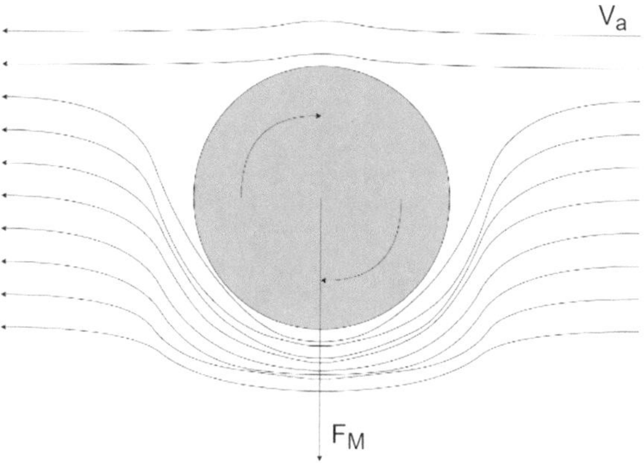

Fig. 2 Example of Magnus force rotating a body in the atmosphere. FM—Magnus force, Va—air velocity

3 Mathematical Model of Projectile Motion

In order to create a mathematical model of projectile motion, it is possible to choose a coordinate system in which a projectile motion will be considered, determine the equation of atmospheric parameters and determine the mathematical model of the object ballistic including selected effects acting on the projectile.

3.1 Coordinate Systems

Using the coordinate system it is possible to unambiguously describe the motion of a ballistic object. Choosing the appropriate coordinate system can facilitate solving equations of external ballistics.

3.2 Fixed Coordinate System Related to Earth

The beginning of the fixed coordinate system associated with the Earth (Fig. 3) shall be the discharge point. Plane $x_g y_g$ is called shooting plane. At discharge point, when the projectile leaves the barrel, the projectile velocity vector lies on a shooting plane. Plane $x_g z_g$ is called the principal plane and is parallel to the plane of the Earth. Point 0_o is both at the origin and the point of departure of the projectile from the barrel (discharge point). Location of object O at any point in time it is given by the coordinates x_g, y_g, z_g.

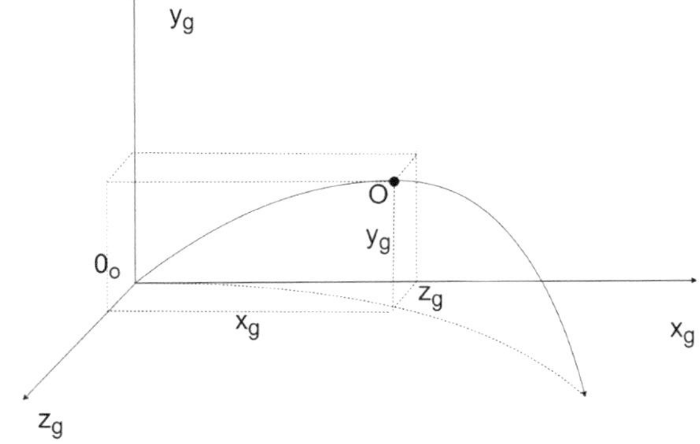

Fig. 3 The coordinate system related to the Earth. O—ballistic object, 0_o—origin of the coordinate system, x_g, y_g, z_g—object's coordinates

3.3 Coordinate System Related to Trajectory

Origin of coordinate system 0_o related to trajectory (Fig. 4) is center of mass of the projectile. The velocity vector of the projectile V lies along the axis x_k. The position of the coordinate system associated with the trajectory of the coordinate system associated with the Earth is defined by:

- angle of the trajectory Ψ,
- angle of inclination of the trajectory Θ.

In order to make the transition from the system associated with the Earth to the coordinate system associated with trajectory rotation of the coordinate system by the angle Ψ along the axis y_g, should be done. Next the coordinate system should be rotated by the angle Θ along the axis $z_{k.}$

3.4 Mathematical Model of Atmosphere

Projectile motion in the atmosphere is influenced by the following parameters of the atmosphere: air pressure, air density, air temperature, gas constant of air, the speed of sound. For calculation of these parameters for any point of projectile's trajectory it is required to know the temperature, air pressure and air density at discharge point. To calculate the parameters there is also required knowledge of the temperature distribution and the acceleration of gravity, depending on the altitude. Equation (1) describes the relationship of air pressure to altitude. To calculate the pressure at a given level the knowledge of the temperature distribution and the acceleration of gravity in the section between the surface and a given height is required.

$$p(h) = p_0 * e^{\left(-\frac{1}{R_p} * \int_0^h \frac{g(h)dh}{T(h)}\right)}, \tag{1}$$

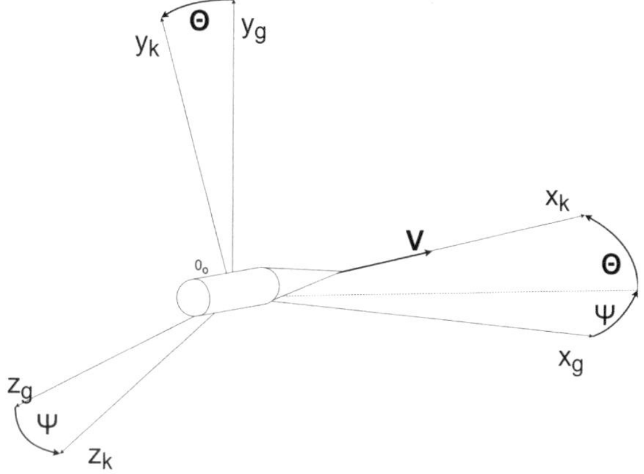

Fig. 4 The position of the coordinate system associated with the trajectory $0x_ky_kz_k$ relative to the coordinate system related to Earth $0x_ky_kz_k$

where:

$p[\mathrm{Pa}]$ air pressure,
$p_0[\mathrm{Pa}]$ air pressure at discharge point,
$h[\mathrm{m}]$ height of the projectile's location,
$T(h)[\mathrm{K}]$ a function of air temperature,
$g(h)\left[\frac{\mathrm{m}}{\mathrm{s}^2}\right]$ function of the gravity acceleration,
$R_p\left[\frac{\mathrm{J}}{\mathrm{kgK}}\right]$ air gas constant.

The resulting air pressure is used to calculate the air density at a given height in Eq. (2).

$$\rho(h) = \frac{\rho_0 * T_0 * p(h)}{p_0 * T(h)}, \tag{2}$$

where:

$\rho\left[\frac{\mathrm{kg}}{\mathrm{m}^3}\right]$ air density,
$\rho_0\left[\frac{\mathrm{kg}}{\mathrm{m}^3}\right]$ air density at discharge point,
$T_0[\mathrm{K}]$ air temperature at discharge point.

To calculate the speed of sound in Eq. (3) uses the calculated value of air pressure and density in the Eqs. (1) and (2).

$$a(h) = \sqrt{k\frac{p(h)}{\rho(h)}}, \tag{3}$$

where:

$a\left[\frac{\mathrm{m}}{\mathrm{s}}\right]$ speed of sound in air,
$k[-]$ heat capacity ratio for air.

The above equations can be used to create a mathematical model of the atmosphere:

$$\begin{cases} T(h) = T_0 + \Delta T(h) \\ g(h) = g_0 + \Delta g(h) \\ p(h) = p_0 * e^{\left(-\frac{1}{R_p} * \int_0^h \frac{g(h)dh}{T(h)}\right)} \\ \rho(h) = \frac{\rho_0 * T_0 * p(h)}{p_0 * T(h)} \\ a(h) = \sqrt{k * \frac{p(h)}{\rho(h)}} \end{cases}, \tag{4}$$

where:

$g_0\left[\frac{\mathrm{m}}{\mathrm{s}^2}\right]$ gravity acceleration at discharge point,
$\Delta g(h)\left[\frac{\mathrm{m}}{\mathrm{s}^2}\right]$ function of gravity acceleration depending on the height,
$\Delta T(h)[\mathrm{K}]$ function of air temperature depending on the height.

3.5 Mathematical Model of Projectile Motion

The aim of projectile motion will be considered as a movement of a material point. The derivation of the mathematical model of projectile motion starts with the vector differential equation of projectile motion:

$$m\frac{d\boldsymbol{V}}{dt} = \boldsymbol{R_A} + \boldsymbol{G} + \boldsymbol{F_C} + \boldsymbol{F_B}, \tag{5}$$

where:

$m[\mathrm{kg}]$ projectile weight,
$\boldsymbol{V}\left[\frac{\mathrm{m}}{\mathrm{s}}\right]$ velocity of the projectile,
$\boldsymbol{R_A}[\mathrm{N}]$ air drag force,
$\boldsymbol{G}[\mathrm{N}]$ gravity force,
$\boldsymbol{F_C}[\mathrm{N}]$ Coriolis force,
$\boldsymbol{F_B}[\mathrm{N}]$ inertia force.

During the formulation of the mathematical model of projectile motion a reference system should be selected for the Eq. (5). For reasons of simplicity, clarity and ease of physical interpretation of equations appropriate coordinate system is a coordinate system associated with the trajectory of the bullet. In the coordinate system, the movement of the projectile is described by three coordinates: velocity V, the trajectory angle Θ and inclination angle Ψ. These coordinates do not give direct information about the location of the projectile with respect to discharge point. You can transform these coordinates to the coordinate system associated with the land where the projectile position is described by coordinates: distance x_g, height y_g, drift z_g. For the transformation the equation system (5) is used.

$$\begin{cases} \frac{dx_g}{dt} = V \cos \Theta \cos \Psi \\ \frac{dy_g}{dt} = V \sin \Theta \\ \frac{dz_g}{dt} = -V \cos \Theta \sin \Psi \end{cases} . \tag{6}$$

Taking into account the force of the wind, it is assumed that the projectile moves at velocity V_{wz} in a coordinate system associated with the air molecules moving with the wind speed V_w in a coordinate system associated with the ground. It is assumed that the origins of the coordinate system associated with the system ground and connected with the air molecule overlap in time of departure (discharge point), and a coordinate system associated with the air molecule will move at the velocity of V_w relative to the coordinate system associated with the earth. This allows the use of Eq. (5) to determine the trajectory of the bullet only by adopting the velocity of the projectile V and the projectile velocity relative to the air V_{wz}. In this case, the transformation from a coordinate system associated with the trajectory of the coordinate system associated with the ground should be transformed to the coordinate system associated with the trajectory of the coordinate system associated with a molecule of air using equation system (6). Then, taking into account wind speed V_w transform the coordinate system associated with the ground using a system of Eq. (7).

$$\begin{cases} \frac{dx_{wz}}{dt} = V_{wz} \cos \Theta \cos \Psi \\ \frac{dy_{wz}}{dt} = V_{wz} \sin \Theta \\ \frac{dz_{wz}}{dt} = -V_{wz} \cos \Theta \sin \Psi \end{cases}, \tag{7}$$

where:

x_{wz}, y_{wz}, x_{wz} the coordinates of the projectile in a coordinate system associated with the air molecule.

$$\begin{cases} \frac{dx_g}{dt} = V_{w_x} + \frac{dx_{wz}}{dt} \\ \frac{dy_g}{dt} = V_{w_y} + \frac{dy_{wz}}{dt} \\ \frac{dz_g}{dt} = V_{w_z} + \frac{dz_{wz}}{dt} \end{cases}, \tag{8}$$

where:

V_{w_x}, V_{w_y}, V_{wz} components of wind speed.

Vector Eq. (5) considered in the coordinate system associated with the trajectory can be described using three scalar equations:

$$\begin{cases} m\frac{dV_{wz}}{dt} = -X_A - G \sin \Theta_{wz} + F_{B_x} \\ mV_{wz}\frac{d\Theta_{wz}}{dt} = -G \cos \Theta_{wz} + F_{C_y} + F_{B_y} \\ -mV_{wz} \cos \Theta_{wz} \frac{d\Psi_{wz}}{dt} = F_{C_z} + F_{B_z} \end{cases}, \tag{9}$$

where:

X_A frontal drag force,

F_{C_y}, F_{C_z} Coriolis force components,
$F_{B_x}, F_{B_y}, F_{B_z}$ inertial forces components.

Components of each of the forces affecting the projectile motion can be substituted into the system of Eq. (9).

The force of gravity G depends on the adopted model of gravity field. In the case of a small distance uniform gravity field model can be assumed, according to which gravitational acceleration is constant. Gravitational acceleration is defined in [25] and is:

$$g_0 = 9.80665\ \frac{\mathrm{m}}{\mathrm{s}^2} \tag{10}$$

The value of the force of gravity G is defined as follows:

$$G = g_0 * \mathrm{m} \tag{11}$$

Leading resistance of the projectile is given by the relationship:

$$X_A = \frac{1}{2} c_x\left(\frac{V_{wz}}{a}\right) V_{wz}^2 S, \tag{12}$$

where:

- S cross-sectional area of the projectile,
- $c_x\left(\frac{V_{wz}}{a}\right)$ function of the ratio of resitance.

The function $c_x\left(\frac{V_{wz}}{a}\right)$ can be written as the product of a bullet shape coefficient i and of function $c_{x_{wz}}\left(\frac{V_{wz}}{a}\right)$. These functions are given in models of air resistance, e.g. Model G7 (Fig. 3). Bullet shape coefficient is defined by the formula:

$$i = \frac{m}{d^2 * BC} 10^{-3}, \tag{13}$$

where:

- d the diameter of the bullet,
- BC ballistic coefficient.

Ballistic coefficients and models of resistance are given by the bullet manufacturers in form of ballistic tables [15, 16]. The sectional area of the projectile can be decomposed as an area of a circle and using Eq. (13) to describe a drag force Eq. (12) can be converted into:

$$X_A = \frac{1}{BC} c_{x_{wz}}\left(\frac{V_{wz}}{a}\right) \frac{V_{wz}^2 \pi m}{8} 10^{-3}. \tag{14}$$

The Coriolis force is defined by the following equation:

$$F_C = -2m * \Omega_z \times V = -2m \begin{bmatrix} i_k & j_k & k_k \\ \Omega_{Z_{xk}} & \Omega_{Z_{yk}} & \Omega_{Z_{zk}} \\ V & 0 & 0 \end{bmatrix} \tag{15}$$
$$= -2mV\Omega_{Z_{zk}} j_k + 2mV\Omega_{Z_{yk}} k_k,$$

where:

Ω_z the angular velocity of rotation of the Earth,
$\Omega_{Z_{xk}}, \Omega_{Z_{yk}}, \Omega_{Z_{zk}}$ components of the angular velocity of the Earth's rotation in the coordinate system associated with the trajectory,
i_k, j_k, k_k coordinate system of versors associated with the trajectory.

After the projection of Eq. (14) on the axes of the coordinate system associated with the trajectory we have:

$$\begin{cases} F_{C_x} = 0 \\ F_{C_y} = 2mV\,\Omega_z \cos B_0 \sin(A_0 - \Psi) \\ F_{C_z} = 2mV\,\Omega_z[\sin B_0 \cos\,\Theta - \,\cos B_0 \sin\,\Theta\,\cos(A_0 - \Psi)] \end{cases}, \tag{16}$$

where:

A_0 aziuth direction of shooting,
B_0 latitude of discharge point.

Inertial forces appearing in Eq. (8) are due to the fact that motion of the projectile in a coordinate system related to molecule of the wind is noninertial. The source of these forces is the variable wind and Coriolis effect. In the system related to the ground forces of inertia are described by system of equations:

$$\begin{cases} F_{B_{xg}} = -m\left[2\left(\Omega_{Z_{yg}} V_{W_z} - \Omega_{Z_{zg}} V_{W_y}\right) + \frac{dV_{W_x}}{dt}\right] \\ F_{B_{yg}} = -m\left[2\left(\Omega_{Z_{zg}} V_{W_x} - \Omega_{Z_{xg}} V_{W_z}\right) + \frac{dV_{W_y}}{dt}\right] \\ F_{B_{zg}} = -m\left[2\left(\Omega_{Z_{xg}} V_{W_y} - \Omega_{Z_{yg}} V_{W_z}\right) + \frac{dV_{W_x}}{dt}\right] \end{cases}, \tag{17}$$

where:

$\Omega_{Z_{xk}}, \Omega_{Z_{yk}}, \Omega_{Z_{zk}}$ components of the angular velocity of the Earth's rotation in the coordinate system related to the earth,
$F_{B_{xg}}, F_{B_{yg}}, F_{B_{zg}}$ component of inertia force in the coordinate system associated with the ground.

To determine the value of the inertial forces in a coordinate system associated with the transformation of the trajectory is used equation:

$$\begin{cases} F_{B_x} = F_{B_{xg}} \cos\Theta_{wz} \cos\Psi_{wz} + F_{B_{yg}} \sin\Theta_{wz} - F_{B_{zg}} \cos\Theta_{wz} \sin\Psi_{wz} \\ F_{B_y} = F_{B_{xg}} \sin\Theta_{wz} \cos\Psi_{wz} + F_{B_{yg}} \cos\Theta_{wz} + F_{B_{zg}} \sin\Theta_{wz} \sin\Psi_{wz} \\ F_{B_z} = F_{B_{xg}} \sin\Psi_{wz} + F_{B_{zg}} \cos\Psi_{wz} \end{cases}. \tag{18}$$

In order to get the full model of projectile motion it is possible to substitute the Eqs. (10), (13), (15) and (17) into system of Eq. (8). The resulting system of equations:

$$\begin{cases} m\frac{dV_{wz}}{dt} = -\frac{1}{BC} c_{x_{wz}} \left(\frac{V_{wz}}{a}\right) \frac{V_{wz}^2 \pi m}{8} 10^{-3} - g_0 m \sin\Theta_{wz} + F_{B_x} \\ mV_{wz}\frac{d\Theta_{wz}}{dt} = -g_0 mc\,os\,\Theta_{wz} + 2mV\,\Omega_z \cos B_0 \sin(A_0 - \Psi) + F_{B_y} \\ -mV_{wz}\cos\Theta_{wz}\frac{d\Psi_{wz}}{dt} = 2m\,V\,\Omega_z[\sin B_0 \cos\Theta - \cos B_0 \sin\Theta \cos(A_0 - \Psi)] + F_{B_z} \end{cases}. \tag{19}$$

Taking into account the transformation between systems (7) and (8), and a mathematical model of the atmosphere (4) the full mathematical model of the projectile motion in the atmosphere is obtained. This model takes into account the gravity, the force of resistance, the Coriolis force, the impact of variable wind and changing parameter values of the atmosphere. It was decided to process a projectile motion as a movement of a material point due to the requirement of simplicity and speed of calculations in the multimedia shooting range. Magnus effect was omitted because it is considered under the assumption that the object is a solid body. Gyroscopic effect was omitted due to lack of a mathematical model of the effect. The effect of Poisson was not included because they are not known mathematical models describing the effect.

4 Implementation of Mathematical Model of Projectile Motion in the Matlab Simulation Environment

Matlab environment is a computer program designed for performing scientific calculations and creating simulations. Mathematical model of projectile motion has been implemented in the form of a script in Matlab environment. In order to implement the model equations they have been discretized using Euler's method [26]. Using the method of Euler's equation defined as $y'(x) = f(x, y)$ it is digitized to form $y_{i+1} = y_i + hf(x_i, y_i)$, where: h—constant calculation step. The selection of the computation step h affects the accuracy of the calculation and computation time. Discretization of the system of Eq. (19), along with a mathematical model of the atmosphere (the system of Eq. (4)) was implemented in the scripting language Matlab.

4.1 Selection of Calculation Step

In order to choose the step value h calculations were performed to compare the time of the calculations of implemented mathematical model of projectile motion taking into account only the force of gravity for different values of h. For comparisons a mathematical model which takes into account only the gravity was used, because such a model can be determined by analytical solution, which allows the calculation of model error. Fixed simulation parameters were:

- bullet weight $m = 15\,\mathrm{g}$,
- bullet caliber d $= 8.6\,\mathrm{mm}$,
- initial velocity of the projectile $v_0 = 920\ \frac{\mathrm{m}}{\mathrm{s}}$,
- shot angle $\Theta_0 = 0°$.

The flight time of the projectile was 1 s. The difference e is calculated:

$$e = \frac{1}{n}\sum_{i=0}^{n}\left((x_s(i) - x_a(ih))^2 + (y_s(i) - y_a(ih))^2\right), \tag{20}$$

where:

n	number of points of simulated trajectory,
$x_s(i), y_s(i)$	simulation result in time i,
$x_a(ih), y_a(ih)$	the analytical solution at the moment ih.

The data from Table 1 indicate that the smaller the calculation step h, the greater the execution time and the smaller the difference. The flight time of the projectile was 1 s, so execution times longer than 1 s should be rejected, because the simulation time takes longer than the actual flight time of the projectile. Calculation step h equal to 0.001 s satisfies the shorter execution time than the flight time of projectile. Trajectory for calculation step h = 0.001 s coincides with the trajectory from the analytical solution (Figs. 5 and 6). For this reason, it was decided to choose the computation step h equal to 0.001 s. The scripting language Matlab is an interpreted language, which results in a long time of execution of the script. After the implementation in the target environment the execution time will be shorter.

Table 1 The differences and distances of simulated shots

	h_1	h_2	h_3	h_4	h_5	h_6	h_7
h (s)	0.1	0.05	0.01	0.005	0.001	0.0005	0.0001
Execution time (s)	0.0068	0.014	0.0673	0.1256	0.627	1.2741	6.3650
Difference e	92.01	46.00	9.201	4.6	0.92	0.46	0.09
Distance (m)	1012	966	929.2	924.6	920.9	920.5	920.1

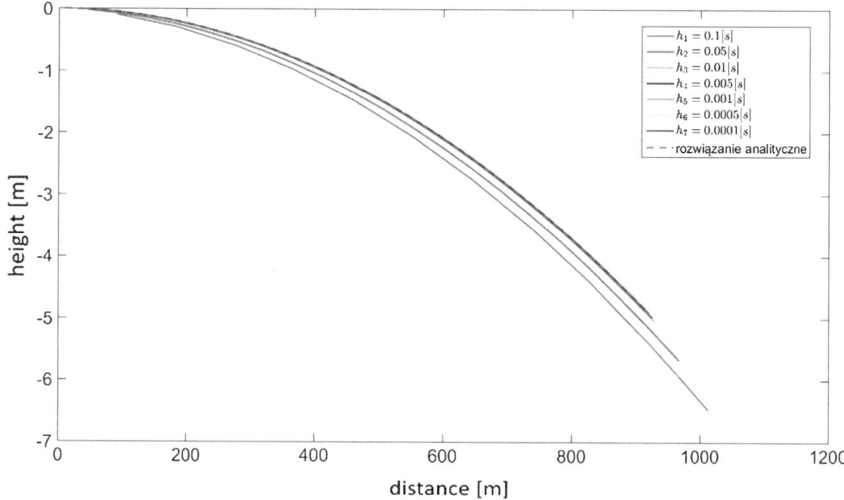

Fig. 5 Obtained projectile trajectories for different values of the computation step h

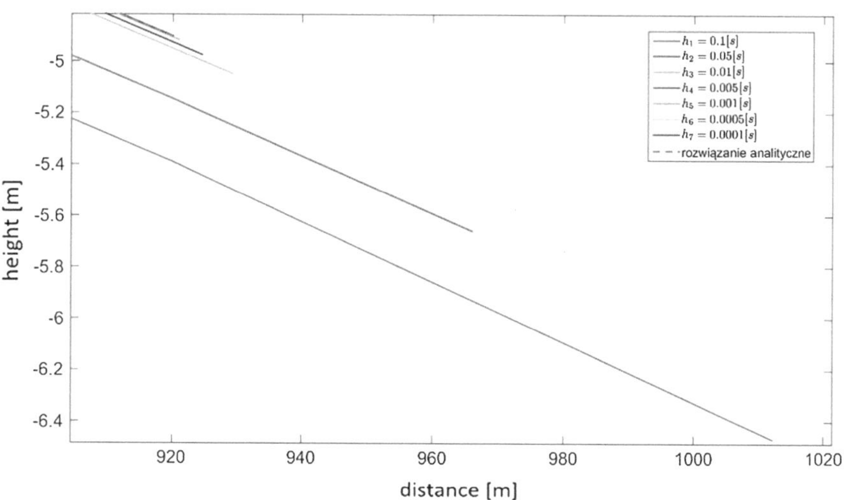

Fig. 6 Approximation of the obtained projectile trajectory for different values of the computation step h

4.2 *Implementation of Aerodynamic Drag Model*

To determine the resistance forces functions of a standard drag coefficient are required to know. A function of the resistance are shown in resistance models such as G1 and G7. These functions are listed in tabular form ([27] for G1, [28] for G7). In order to validate the operation of the implementation a test was performed,

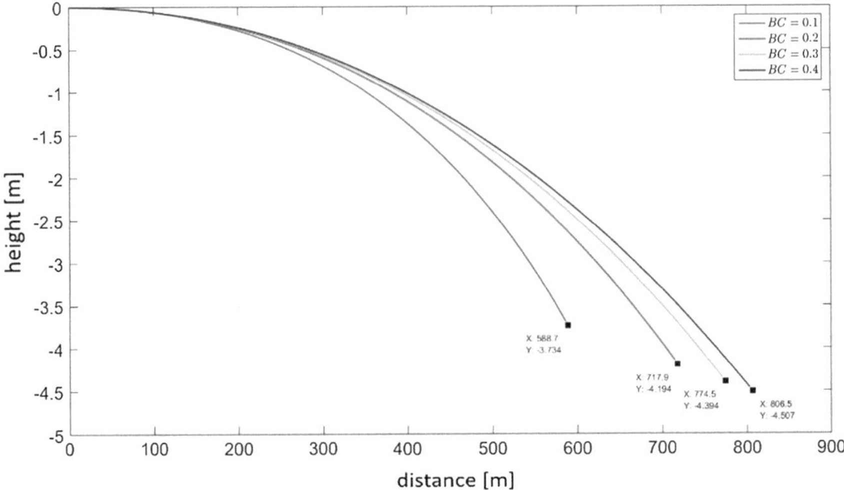

Fig. 7 Comparison of trajectories for different ballistic coefficient

in which the trajectories for different ballistic coefficients of projectile were computed. Fixed simulation parameters were:

- bullet weight m = 15 g,
- bullet caliber $d = 8.6$ mm,
- initial velocity of the projectile $v_0 = 920\ \frac{\text{m}}{\text{s}}$,
- shot angle $\Theta_0 = 0°$.

According to the theory ballistic coefficient determines the ability of a projectile to resist its movement speed reduction caused by air resistance forces. The greatest distance overcame projectile with ballistic coefficient equal to 0.4. The distance was 806.5 m (Fig. 7). Along with the decrease of the ballistic coefficient decreases the distance travelled by projectiles.

4.3 Implementation of Wind Force Model

The aim of checking the implementation of wind force model was based on a comparison of the drift and distance for various directions and wind speeds. The trajectory of the projectile, which was influenced by a variable wind was compared with the trajectory of the projectile, which was affected by a constant wind. Fixed simulation parameters were:

- bullet weight $m = 15$ g,
- bullet caliber $d = 8.6$ mm,
- initial velocity of the projectile $v_0 = 920\ \frac{\text{m}}{\text{s}}$,

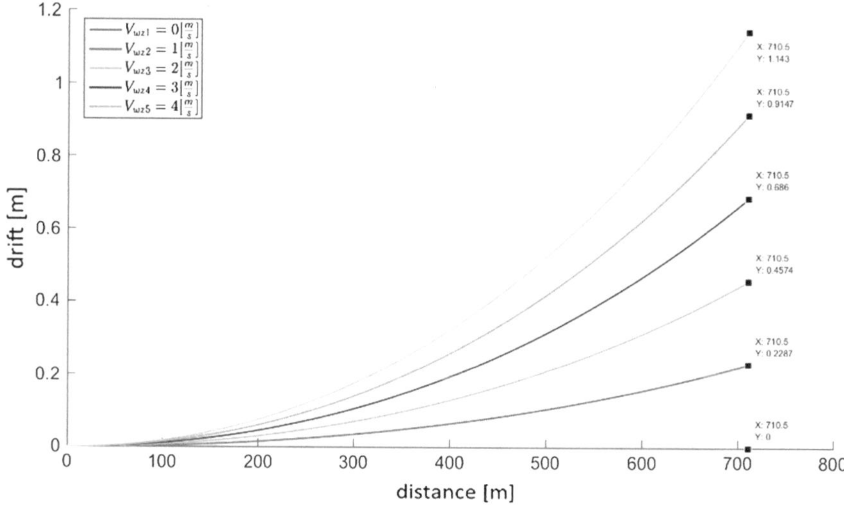

Fig. 8 Comparison of projectile trajectories for different side wind speeds

- shot angle $\Theta_0 = 0°$,
- ballistic coefficient $BC = 0.191$.

The increase in side wind speed causes greater drift of bullet. The range of the shot is not affected by a side wind. The biggest drift has been measured for the wind speed 4 m/s and was 1.143 m (Fig. 8). Along with decreasing wind speed decreases bullet drift.

Projectile flying into the wind is moving at a higher speed relative to the air, which increases air drag forces. This results in a decrease in the projectile range, leading to distance 693.2 m, the wind speed of 10 m/s (Fig. 9). The projectile moving with the wind moves relative to the air at a slower speed, so the force of air resistance is lower and range of the projectile is longer.

Descending variable wind causes smaller drift than the wind with a constant speed value (Fig. 10). In the first case (blue line) wind speed decreases, resulting in less bullet drift in comparison to the second case (red line).

4.4 Implementation of Coriolis Force

In order to check the implementation of the model of the Coriolis force tests were performed which compared the projectile shot trajectory at different time points on the latitude and for different azimuths of the shot. Simulating a shot at the equator in the east and west direction allowed to check the work Eötvös effect. Fixed simulation parameters were:

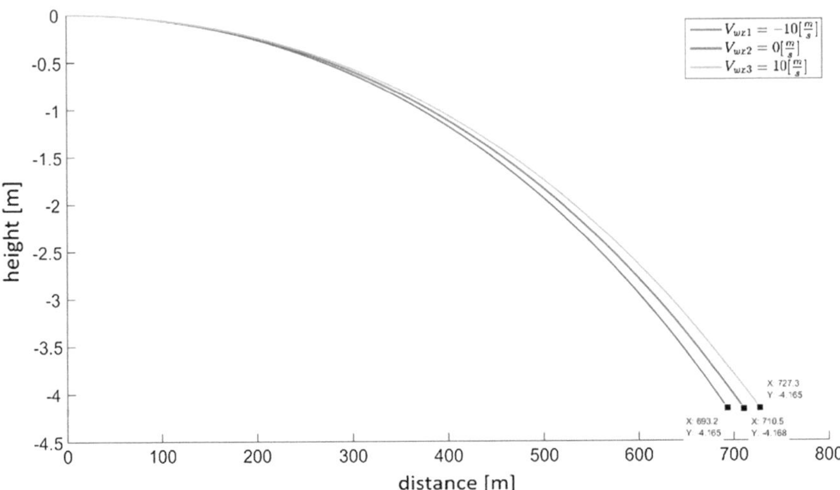

Fig. 9 Comparison of projectile trajectories for different leading wind speeds

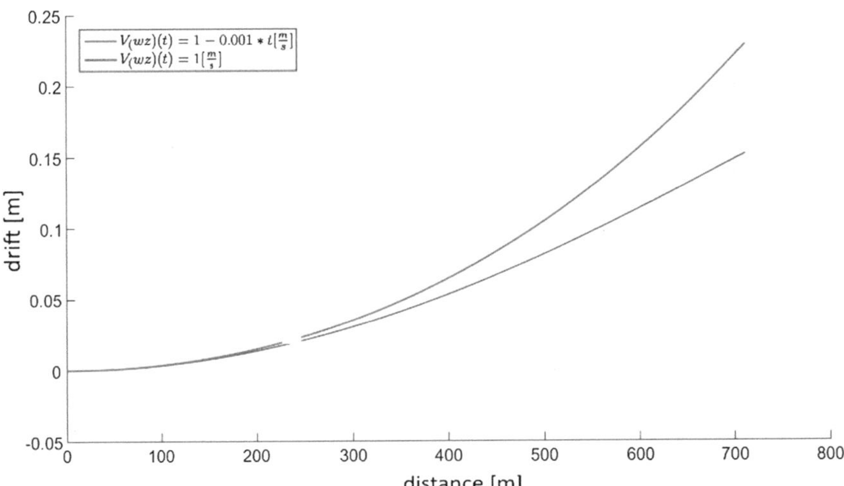

Fig. 10 Comparison of variable and constant wind

- bullet weight $m = 15$ g,
- bullet caliber $d = 8.6$ mm,
- initial velocity of the projectile $v_0 = 920\ \frac{\text{m}}{\text{s}}$,
- shot angle $\Theta_0 = 0°$,
- ballistic coefficient $BC = 0.191$.

At the equator the Coriolis effect causes drift of the bullet equal to 0.0001859 m (Fig. 11). The observed effect of the Coriolis force on the drift of the projectile

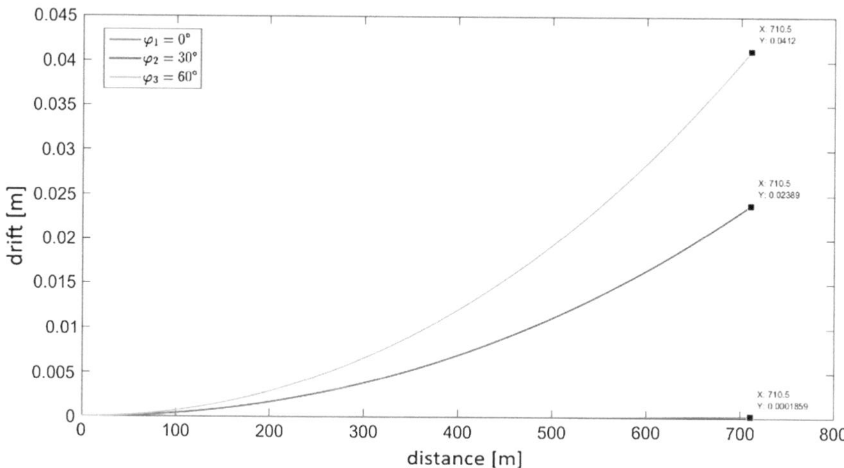

Fig. 11 Comparison of projectile trajectories for shots from different latitudes

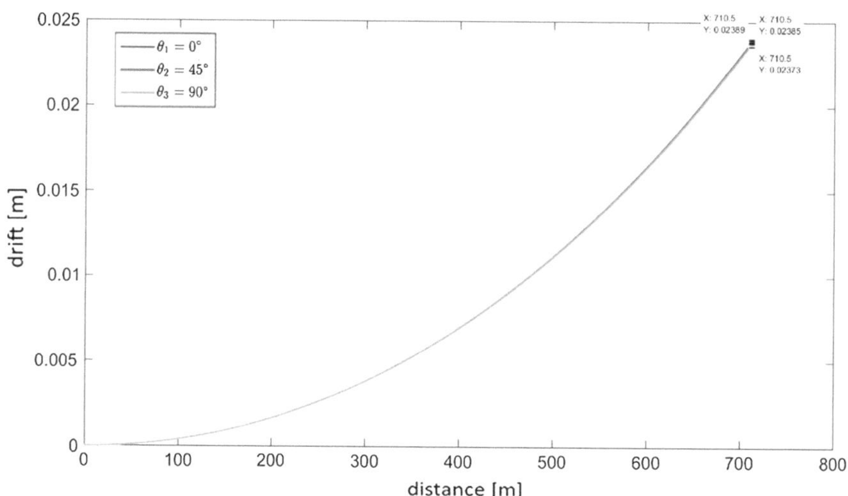

Fig. 12 The impact of changes in azimuth of the shot onto the trajectory of the projectile

increases with latitude of the shot point. Shot at latitude equal to 60° causes bullet drift of 0.0412 m.

For azimuth 0° drift of the bullet is the largest and has a value of 0.02389 m. For azimuth 90° drift is the smallest and equals 0.02373 m (Fig. 12).

Eötvös effect causes that when the projectile is moving towards the east, rotation force around the Earth is acting on it vertically upwards. When the projectile is moving in an easterly direction, force is acting vertically downwards. The effect of

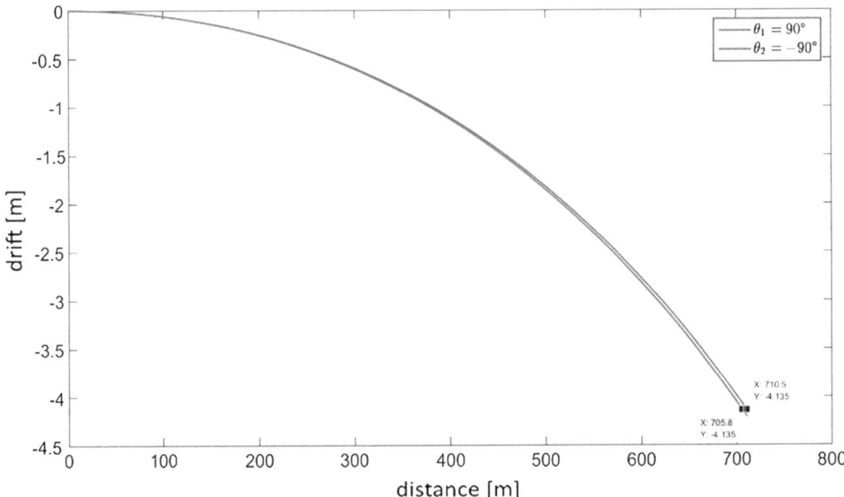

Fig. 13 Eötvös effect

these forces is that the bullet flying in an easterly direction falls more slowly than the bullet flying in a westerly direction (Fig. 13). The bullet moving in an easterly direction (blue line) travelled distance 705.8 m, projectile moving in a westerly direction (red line) travelled distance 710.5 m.

4.5 Comparison of Effects Affecting the Projectile Motion

After implementing the whole mathematical model of projectile motion trajectories were simulated. During simulation multiple effects were applied to verify the impact of these effects on the movement of the projectile. Shot distance is influenced mainly by air resistance (Fig. 14). Other effects: the Coriolis force and the cross wind have negligible effect on the distance.

The projectile trajectory does not deviate in the lateral direction in the case if only the gravity and air resistance forces affect the projectile (Fig. 15). Projectile trajectory deviates in the lateral direction if cross wind or Coriolis force affects the projectile. Wind has a much greater impact on the amount of deviation from the plane of the shot. Coriolis effect deflects the trajectory of 4 cm at the shot distance of 700 m, a cross wind blowing with speed 4 m/s deflects the trajectory of 95 cm.

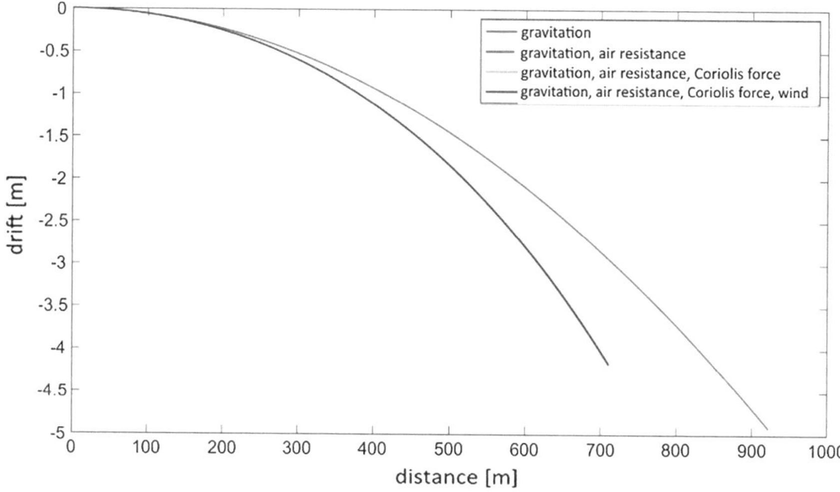

Fig. 14 Comparison of the effects of forces affecting the movement of the projectile

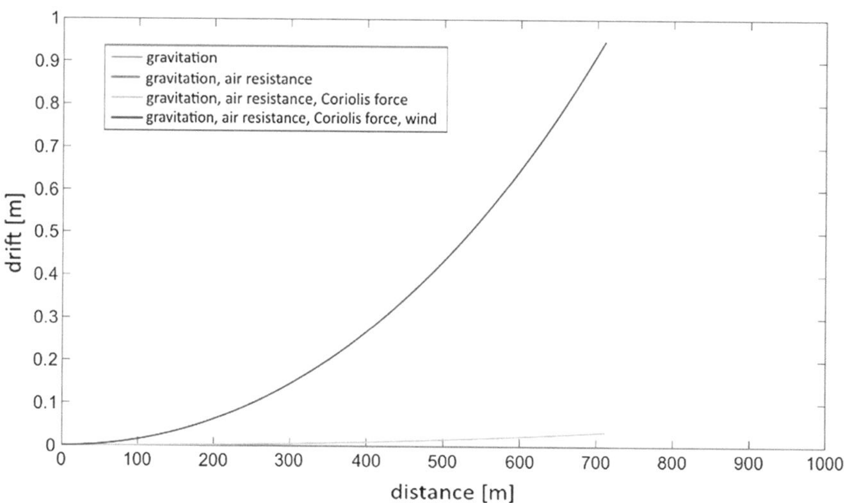

Fig. 15 Comparison of the effects of forces affecting the movement of the projectile

5 Simulation Verification

Verification of the simulation was carried out by comparing the generated trajectory with the results provided by the manufacturers of ammunition and other simulation systems. For verification the following bullets were selected:

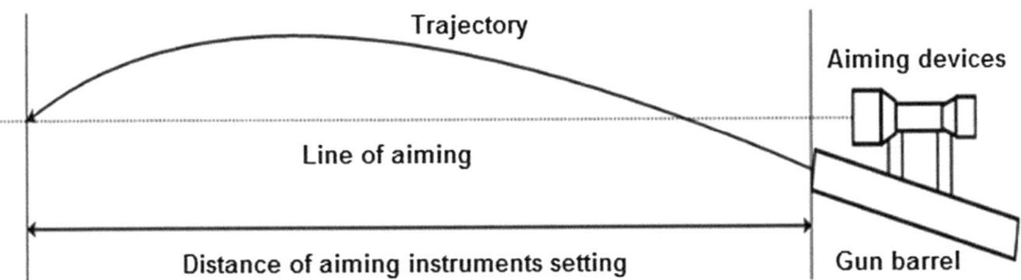

Fig. 16 Diagram showing the distance settings of the iron sights

- Lapua Magnum 0.338 Naturalis,
- Federal American Eagle Ammunition 0.308 Winchester,
- 0.50 BMG M33,
- American Eagle 9 × 19 mm Parabellum.

For all these types of ammunition there are available ballistic tables. In the tables there are placed examples of ballistic trajectories, bullet specification: bullet weight, ballistic coefficient, bullet caliber and firing parameters: atmospheric parameters (temperature, air pressure), and the distance settings of the iron sights. Distance setting of iron sights is the distance at which the projectile will cross the line of sight, established by the sights (Fig. 16).

In case when manufacturer has not specified the parameters of the atmosphere, for the calculation of the parameters Normal Artillery Atmosphere was used:

- $T_0 = 288.15\,\mathrm{K}$,
- $p_0 = 99992\,\mathrm{Pa}$,
- $\rho_0 = 1.2054\,\frac{\mathrm{kg}}{\mathrm{m}^3}$.

The difference between simulation and data from the ballistic sheets is given by the formula:

$$e = \frac{1}{n}\sum_{i=0}^{n}\left((y_s(x_i) - y_t(x_i))^2 + (z_s(x_i) - z_t(x_i))^2\right), \tag{21}$$

where:

n	number of trajectory points,
$y_s(x_i), z_s(x_i)$	simulation results in points x_i,
$y_t(x_i), z_t(x_i)$	Trajectory measurements from ballistic charts in points x_i.

5.1 *Trajectory Verification for Lapua Magnum 0.338 Naturalis*

Producer of Lapua Magnum 0.338 Naturalis specifies the following bullet characteristics [16]:

- bullet weight $m = 15$ g,
- bullet caliber $d = 8.6$ mm,
- ballistic coefficient $BC = 0.191$,
- G7 resistance model.

and the following parameters of the shot:

- sights set on the distance 100 m,
- initial velocity $v_0 = 920\ \frac{\mathrm{m}}{\mathrm{s}}$,
- side wind of speed 4 $\frac{\mathrm{m}}{\mathrm{s}}$.

After travelling 900 ms simulated projectile dropped 2.223 m less than the reference trajectory in the ballistic charts (Fig. 17). After travelling 900 m drift of the simulated projectile was 1.058 m smaller than in reference trajectory in the ballistic charts (Fig. 18 and Table 2).

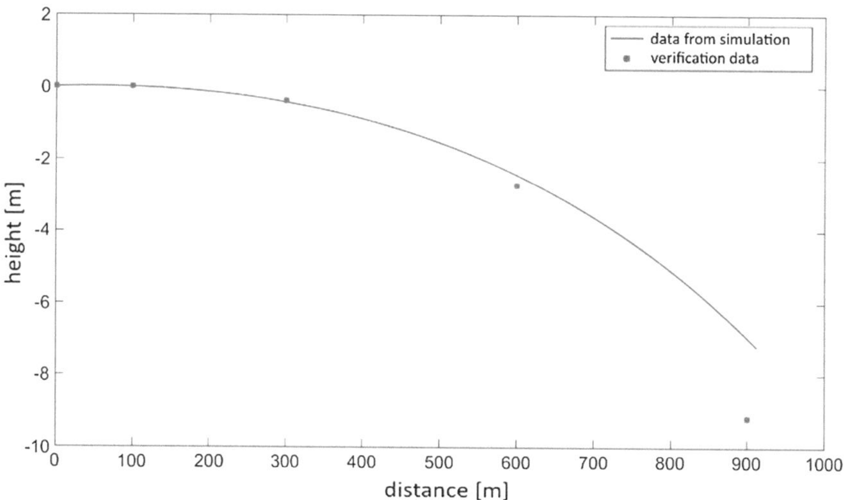

Fig. 17 Comparison of the standard trajectory with the simulated trajectory for the bullet Lapua Magnum 0.338 Naturalis (*side view*)

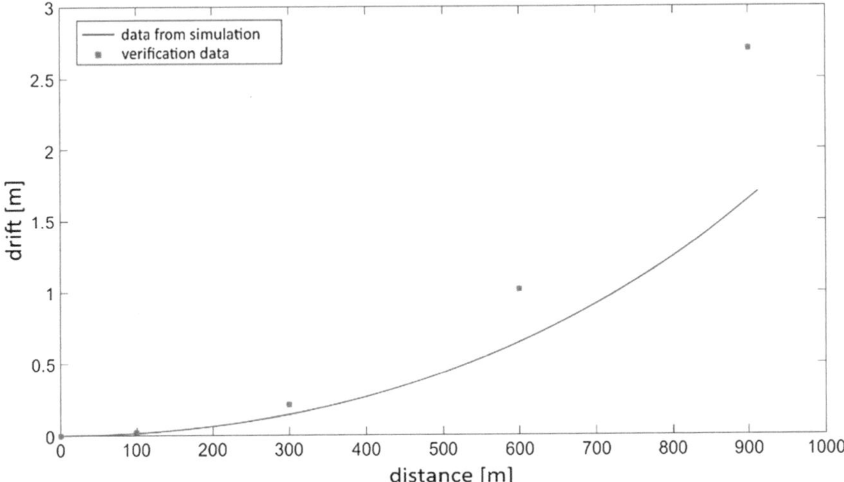

Fig. 18 Rys Comparison of the standard trajectory with the simulated trajectory for the bullet Lapua Magnum 0.338 Naturalis (*top view*)

Table 2 Comparison of the data from the simulation with data from ballistic charts for bullet Lapua Magnum 0.338 Naturalis

Distance (m)		0	100	300	600	900
Height (m)	Verification data	0	0	−0.374	−2.73	−9.176
	Simulation	0	0	0.423	−2.438	−6.953
Drift (m)	Verification data	0	0.021	0.212	1.018	2.703
	Simulation	0	0.015	0.145	0.649	1.645
Difference e_i		0	0.00130	0.76266	2.86415	23.84683

5.2 *Trajectory Verification for Federal American Eagle Ammunition 0.308 Winchester*

Producer of Federal American Eagle Ammunition 0.308 Winchester specifies the following bullet characteristics [29]:

- bullet weight $m = 9.7$ g,
- bullet caliber $d = 7.8$ mm,
- ballistic coefficient $BC = 0.381$,
- G7 resistance model,

and the following parameters of the shot:

- sights set on the distance 150 m,
- initial velocity $v_0 = 860\ \frac{\text{m}}{\text{s}}$.

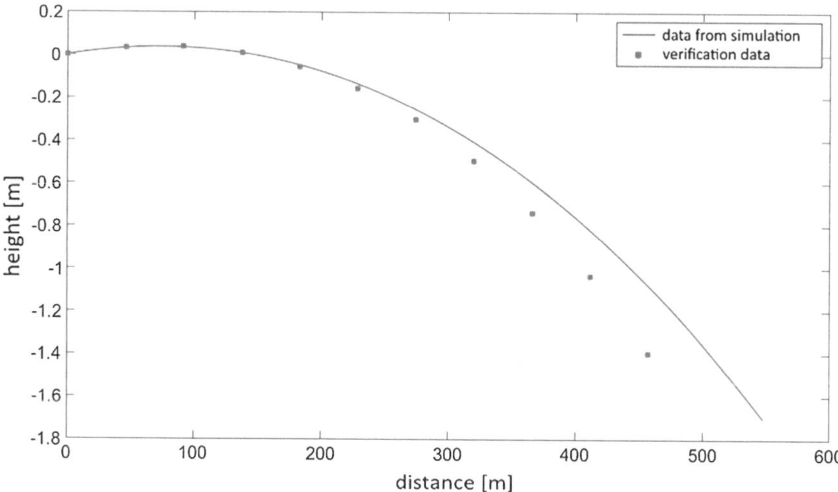

Fig. 19 Comparison of the standard trajectory with the simulated trajectory for the bullet Federal American Eagle Ammunition 0.308 Winchester (*side view*)

Table 3 Comparison of results of simulation with the reference trajectories from ballistic charts for bullet Federal American Eagle Ammunition 0.308 Winchester

Distance (m)		0	45.72	91.44	137.16	182.88
Height (m)	Verification data	0	0.0341	0.0381	0.0094	−0.055
	Simulation	0	0.03181	0.03526	0.0088	−0.0467
Difference e_i		0	0.00001	0.00001	0.00000	0.00007

Distance (m)		228.6	274.3	320	365.7	411.4	457.2
Height (m)	Verification data	−0.1581	−0.3036	−0.4953	−0.7378	−1.0358	−1.395
	Simulation	−0.1344	−0.2544	−0.4077	−0.5945	−0.8175	−1.078
Difference e_i		0.00056	0.0024	0.0076	0.0205	0.1004	0.04765

After travelling 457.2 m drift of the simulated projectile was 0.317 m smaller than in reference trajectory in the ballistic charts (Fig. 19 and Table 3).

5.3 Trajectory Verification for 0.50 BMG M33

Producer of bullet 0.50 BMG M33 specifies the following bullet characteristics [15]:

- bullet weight $m = 42.83$ g,
- bullet caliber $d = 12.7$ mm,
- ballistic coefficient $BC = 0.62$,
- G1 resistance model,

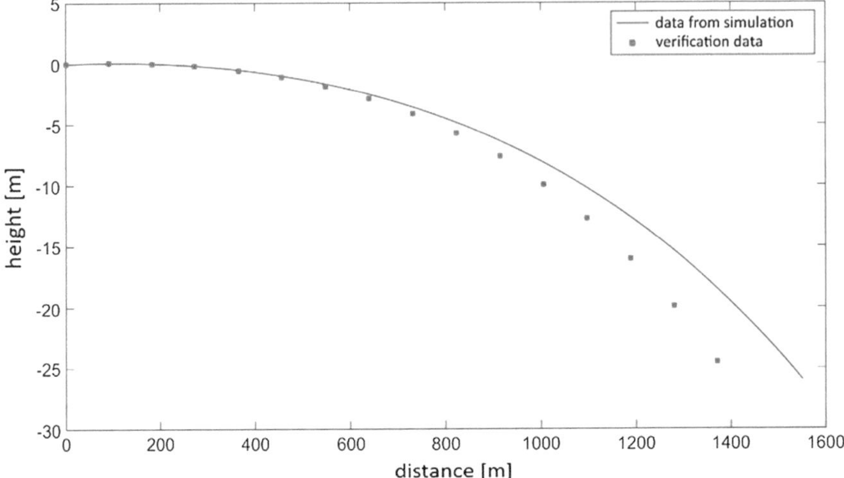

Fig. 20 Comparison of the standard trajectory with the simulated trajectory for the bulle t0.50 BMG M33 (*side view*)

and the following parameters of the shot:

- side wind of speed 0.44 $\frac{\mathrm{m}}{\mathrm{s}}$,
- sights set on the distance 200 m,
- initial velocity $v_0 = 838, 2\ \frac{\mathrm{m}}{\mathrm{s}}$.

After travelling 1371.6 m simulated projectile dropped 5.965 m less than the reference trajectory in the ballistic charts (Fig. 20). After travelling 1371.6 m drift of the simulated projectile was 0.0437 m smaller than in reference trajectory in the ballistic charts (Fig. 21 and Table 4).

5.4 *Trajectory Verification for American Eagle 9 × 19 mm Parabellum*

Producer of bullet American Eagle 9 × 19 mm Parabellum specifies the following bullet characteristics [30]:

- bullet weight $m = 8.03$ g,
- bullet caliber $d = 9$ mm,
- ballistic coefficient $BC = 0.149$,
- G1 resistance model,

and the following parameters of the shot:

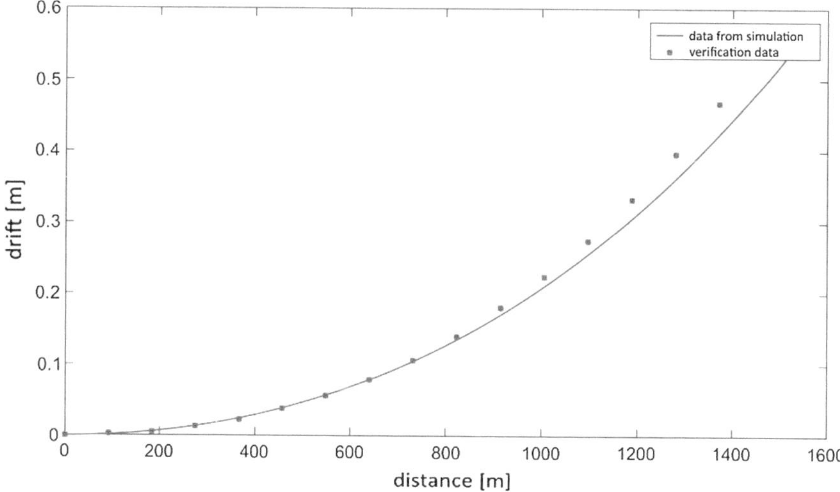

Fig. 21 Comparison of the standard trajectory with the simulated trajectory for the bullet 0.50 BMG M33 (*top view*)

Table 4 Comparison of results of simulation with the reference trajectories from ballistic charts for bullet 0.50 BMG M33

Distance(m)		0	91.4	182.9	274.3	365.8
Height (m)	Verification data	0	0.0762	0.0229	−0.1778	−0.5436
	Simulation	0	0.077	0.0285	−0.1562	−0.4875
Drift (m)	Verification data	0	0.0025	0.0051	0.0127	0.0229
	Simulation	0	0.0014	0.0059	0.0135	0.02447
Difference e_i		0	0.00002	0.00015	0.00115	0.00539

Distance (m)		457.2	548.6	640.1	731.5	823.0
Height (m)	Verification data	−1.0922	−1.8517	−2.8473	−4.1148	−5.6896
	Simulation	−0.9769	−1.64	−2.494	−3.548	−4.829
Drift (m)	Verification data	0.0381	0.0559	0.0787	0.1067	0.1397
	Simulation	0.0388	0.0569	0.0789	0.1051	0.1356
Difference e_i		0.01921	0.05754	0.14966	0.36612	0.81642

Distance (m)		914.4	1005.8	1097.3	1188.7	1280.2	1371.6
Height (m)	Verification data	−7.6149	−9.9416	−12.725	−16.032	−19.928	−24.485
	Simulation	−6.353	−8.129	−10.22	−12.6	−15.38	−18.52
Drift (m)	Verification data	0.1803	0.2235	0.2743	0.3327	0.3962	0.4674
	Simulation	0.1705	0.2102	0.2552	0.3054	0.3623	0.4237
Difference e_i		1.7154	3.4736	6.5554	12.1858	21.2596	36.3752

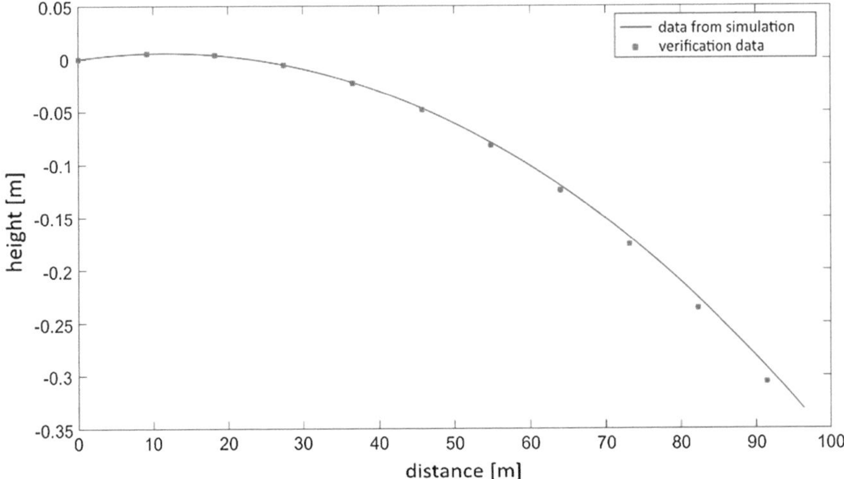

Fig. 22 Comparison of the standard trajectory with the simulated trajectory for the bullet American Eagle 9 × 19 mm Parabellum (*side view*)

Table 5 Comparison of results of simulation with the reference trajectories from ballistic charts for bullet American Eagle 9 × 19 mm Parabellum

Distance (m)		0	9.144	18.288	27.432	36.576
Height (m)	Verification data	0	0.0055	0.0037	−0.0056	−0.0229
	Simulation	0	0.0054	0.0037	−0.0054	−0.0222
Difference e_i		0	1E10^{-8}	0	4E10^{-8}	4.9E10^{-7}

Distance (m)		45.72	54.864	64.008	73.152	82.296	91.44
Height (m)	Verification data	−0.0482	−0.0820	−0.1244	−0.1756	−0.2360	−0.3057
	Simulation	−0.0467	−0.079	−0.12	−0.1688	−0.2262	−0.2923
Difference e_i		2.3E10^{-6}	9E10^{-5}	1.9E10^{-5}	4.6 × 10^{-5}	9.6E10^{-5}	1.8E10^{-4}

- sights set on the distance 25 m,
- initial velocity $v_0 = 341.3\ \frac{\mathrm{m}}{\mathrm{s}}$.

After travelling 91.44 m drift of the simulated projectile was 0.0134 m smaller than in reference trajectory in the ballistic charts (Fig. 22 and Table 5).

5.5 Verification Summary

The presented in the article data were collected as a result of the experiment which involved the comparison of the results of implemented simulation with ballistic charts. The quality of the mathematical model can be assessed mainly by visually

Table 6 Summary of differences between simulation and ballistic data charts for each comparison of ballistic trajectory

Ammunition type	Difference *e*
Lapua Magnum 0.338 Naturalis	5.495
Federal American Eagle Ammunition 0.308 Winchester	0.01631
0.50 BMG M33	5.186
American Eagle 9 × 19 mm Parabellum	0.00032

Fig. 23 Example of a tracer bullet firing in order to analyze the bullet trajectory [31]

comparing the trajectory obtained by the simulation according to the implemented model with the data provided by the manufacturers of ammunition (Fig. 23). For each comparison, the accumulated difference was computed Eq. (21 and Table 6).

For the bullet American Eagle 9 × 19 mm Parabellum we obtained the smallest difference equal to 0.00032. Obtained result of the simulation almost perfectly coincides with the data supplied by the ammo manufacturer. For the bullet Federal American Eagle 0.308 Winchester Ammunition difference is 0.01599 greater than for the bullet American Eagle 9 × 19 mm. The resulting trajectory coincides with the verification data up to 200 m. In the case of simulation of projectile 0.50 BMG M33 the impact of wind on the projectile's motion was included. The resulting trajectory coincides with the reference data used for verification up to 600 m. In the case of simulation of the projectile Lapua Magnum 0.338 Naturalis it can be observed that the difference was equal to 5.495 m and the drift of the simulated projectile was 1.058 m larger than in reference trajectory in the ballistic charts.

6 Conclusions

The aim of the work was the implementation of ballistics equations in the physical engine, which could be used in multimedia shooting ranges. A review of specialized literature on the external ballistics, in order to gather information on the mathematical modelling of projectile motion, was made. The information collected was used for the synthesis of a mathematical model which describes the motion of the projectile in the atmosphere and meet the requirements specified in Sect. 2.1. The selected model was implemented in MATLAB. Using this implementation model was tested in order to verify the correctness of the mathematical model of projectile's motion. The results of the simulation of the projectile's trajectory were compared with the trajectories available in ballistic tables in order to verify the applied mathematical model of projectile motion. The observed problem was that the trajectories in the ballistic tables often are trajectories also obtained by computer simulation.

The manufacturer of ammunition does not provide information which a mathematical model of projectile's motion was used by the manufacturer to develop ballistic charts. Manufacturers do not always also provide information about the parameters of the atmosphere, in which the trajectory of the missile was simulated. Despite the problems, verification of the implemented mathematical model of projectile's motion proved to be accurate enough to be used in the multimedia shooting ranges. The implementation of the presented in the article mathematical model in the physics engine simulates the trajectory of the projectile taking into account the effects acting on the bullet: gravity, air resistance, the Coriolis force and wind in opposition to standard physics engines which takes into account only gravitation and sometimes wind.

Implementation fulfils the following assumptions: takes into account the selected effects affecting the movement of the projectile and takes into account the parameters of the projectile and the parameters of the atmosphere, the calculations are fast enough that they can be carried out in real time. In the article the mathematical model of projectile motion has been extended to a mathematical model of the atmosphere, in order to simulate the atmosphere parameters like temperature, density and pressure at the given time of flight of the projectile.

The use of ballistics equations implemented in the physics engine of the multimedia shooting range will increase the realism of the shooting training. The implemented ballistic model will allow to conduct training on longer distances than allowed by a traditional indoor shooting range, due to a faithful reproduction of projectile's behaviour in the atmosphere. Traditional indoor shooting ranges can be converted into a multimedia shooting ranges, which increase training opportunities through the use of modern computer graphics.

References

1. Retrieved June 15, 2016, from https://upload.wikimedia.org/wikipedia/commons/b/be/Defense.gov_News_Photo_090425-A-2315M-407.jpg.
2. Cymerski J. (2014). Doskonalenie zawodowe funkcjonariuszy Biura Ochrony Rządu, [w:] P. Bogdalski, J. Cymerski, K. Jałoszyński, Bezpieczeństwo osób podlegających ustawowo ochronie wobec zagrożeń XXI wieku, Szczytno s. 382.
3. Cymerski, J. (2015). Współpraca Biura Operacji Antyterrorystycznych KGP z Biurem Ochrony Rządu w ramach struktur przeciwdziałania i zwalczania zagrożeń terrorystycznych [w:] K. Jałoszyński, W Zubrzycki, A. Babiński, Policyjne Siły Specjalne w Polsce, Szczytno, s. 482.
4. Retrieved November 10, 2016, from http://www.specops.pl/vortal/taktyka_czarna/czarna_main.htm.
5. Retrieved November 10, 2016, from http://www.specops.pl/vortal/taktyka_czarna/MOUT/MOUT.htm.
6. Retrieved November 10, 2016, from http://www.specops.pl/vortal/taktyka_zielona/zielona_main.htm.
7. Niezgoda, T., & Barnat, W. (2008). Analysis of protective structures made of various composite materials subjected to impact. *Materials Science and Engineering A, 483,* 705–707.
8. Barnat, W., Dziewulski, P., Niezgoda, T., & Panowicz, R. (2011). Application of composites to impact energy absorption. *Computational Materials Science, 50*(4), 1233–1237.
9. Barnat, W., Panowicz, R., & Niezgoda, T. (2012). Numerical and experimental comparison of combined multilayer protective panels. *Acta Mechanica Et Automatica, 6*(1), 148–153.
10. Sokołowski, D., & Barnat, W. (2016). Numerical and experimental research on the impact of the Twaron T750 fabric layer number on the stab resistance of a body armour package. *Fibres and Textiles in Eastern Europe, 24*(1), 115.
11. Babiarz, A., Bieda, R., Jedrasiak, K., & Nawrat, A. (2013). Machine vision in autonomous systems of detection and location of objects in digital images. In *Vision based systems for UAV applications* (pp. 3–25). Springer.
12. Bereska, D., Daniec, K., Fras, S., Jedrasiak, K., Malinowski, M., & Nawrat, A. (2013). System for multi-axial mechanical stabilization of digital camera. In *Vision based systems for UAV applications* (pp. 177–189). Springer.
13. Daniec, K., Jedrasiak, K., Koteras, R., & Nawrat, A. (2013). Embedded micro inertial navigation system. In *Applied mechanics and materials* (Vol. 249, pp. 1234–1246). Trans Tech Publications.
14. Jedrasiak, K., Andrzejczak, M., & Nawrat, A. (2014). SETh: the method for long-term object tracking. Comput. Vis. Graph. 8671, 302–315. Lecture Notes in Computer Science, 316.
15. Niezgoda, T., Barnat, W., & Ochelski, S. Energy absorption investigation of filling tubes. September 23rd–26th, 2009 Montanuniversität Leoben/Austria, 161.
16. 50 BMG ammunition ballistic chart. Retrieved May 07, 2016, from https://barrett.net/accessories/ammunition/50bmgm33ball/.
17. Jedrasiak, K., Nawrat, A., & Wydmanska, K. (2013) SETh-link the distributed management system for unmanned mobile vehicles. In *Advanced Technologies for Intelligent Systems of National Border Security* (pp. 247–256). Heidelberg: Springer.
18. Prace minerskie i niszczenia, Sztab Generalny, szefostwo Wojsk inżynieryjnych Warszawa 1995r.
19. Persson, A. O. The coriolis effect: Four centuries of conflict between common sense and mathematics, Part I: A history to 1885. Retrieved May 9, 2016, from http://empslocal.ex.ac.uk/people/staff/gv219/classic.d/person_on_coriolis05.pdf (access 9.05.2016).
20. Gacek, J. (1997). Balistyka Zewnętrzna, Część 1, Modelowanie zjawisk balistyki zewnętrznej i dynamiki lotu", Warszawa.

21. Beckenbach E.F. Modern mathematics for the engineer. Retrieved May 7, 2016, from https://books.google.pl/books?hl=pl&lr=&id=E8vDAgAAQBAJ&oi=fnd&pg=PA36&ots=p_TLH2HQYa&sig=mUTFT24yCrT2s-k74xAZGxsHYmQ&redir_esc=y#v=onepage&q&f=false.
22. Ejsmont J., Balistyka dla snajperów. Praktyczny poradnik.
23. 308 Ballistic Chart & Coefficient,Retrieved June 13, 2016. from http://gundata.org/blog/post/308-ballistics-chart/.
24. Litz B., *Applied ballistics for long range shooting: Understanding the elements and application of external ballistics for successful long range target shooting and Hunting.*
25. 338 Lapua Magnum ammunition ballistic chart, Retrieved May 07, 2016, fromhttp://www.lapua.com/en/tactical-ammunition/centerfire-rifle-/-338-lapua-magnum-tactical-ammunition-cartridge.html.
26. The International system of units SI. Retrieved May 4, 2016, from http://physics.nist.gov/Pubs/SP330/sp330.pdf (access 4.05.2016).
27. Klamka, J. (1998). Metody Numeryczne. Wydawnictwo Politechniki Śląskiej, Gliwice.
28. G1 Drag Coefficient vs. Mach Number function. Retrieved June 9, 2016, from http://www.jbmballistics.com/ballistics/downloads/text/mcg1.txt.
29. G7 Drag Coefficient vs. Mach Number function, http://www.jbmballistics.com/ballistics/downloads/text/mcg7.txt Retrieved 9 June 2016 from.
30. x45 mm NATO Ballistic Chart and Coefficient. Retrieved June 13, 2016, from http://www.snipercentral.com/223-remington/.
31. mm NATO Ballistic Chart & Coefficient. Retrieved June 13, 2016, from http://gundata.org/blog/post/9mm-ballistics-chart/.

Manual Calibration of System of the Image Projection Based on DLP Projectors

Dawid Sobel, Karol Jędrasiak, Jarosław Cymerski, Konrad Osiński, Damian Bereska and Aleksander Nawrat

1 Introduction

Systems of the image projection are finding wide applications in the industrial engineering, medicine, army, culture, art, entertainment, learning, education and the business. Such systems are giving the possibility to the presentation and the visualisation of contents in the wide circle of observers [1]. It is possible to single out the number of kinds of systems of the image projection depending on applications.

The shape of the screen of the projection can be any what is giving him the advantage over systems of the visualisation consisting of LCD panels, of which the shape is being dictated by technological limitations. The simplest systems are so-called walls of the video, where the image compound of the projection of many projectors is appearing on the flat surface of the screen [2]. Universally screens cylindrical, surrounding the observer are applicable. Using such a screen allows for surrounding the observer with visional information, what visual feelings are growing thanks to [3]. Nowadays it is possible to observe a growth of the technology of the virtual reality. The user is being immersed in the virtual environment and border between real and with virtual world are fading away. For that purpose such solutions are being created as caves 3D [4], in which user putting on specialist glasses, tracked by the system motion capture can move in the 3D virtual world [5–7]. The knowledge about the position of the observer on the stage [8–11] lets for creating the image projection, correct with regard to a point of view of the observer. Surrounding

D. Sobel (✉) · K. Jędrasiak · K. Osiński · D. Bereska · A. Nawrat
Institute of Automatic Control, Silesian University of Technology,
Akademicka 16, 44-100, Gliwice, Poland
e-mail: dawid.sobel@polsl.pl

J. Cymerski
Government Protection Bureau, Warsaw, Poland
e-mail: j.cymerski@bor.gov.pl

A. Nawrat et al. (eds.), *Advanced Technologies in Practical Applications for National Security*, Studies in Systems, Decision and Control 106,
https://doi.org/10.1007/978-3-319-64674-9_13

the user is another solution hemisphere, how it is taking place e.g. in simulators, air training stimulators or virtual simulators of the warfare [12, 13]. Reconstructing real situations and operations the user can with low cost and without risking life practise piloting the plane [14–17]. In this case, we are also dealing with the system of the image projection, which have to be shown correctly before the user [18, 19].

Creating the system of the image projection they are coming across the row of challenges associated with correct geometrical leveling the image and the unification of the colour and the brightness on the surface of the entire screen of the projection. Applying cheap DLP projects, is giving better visual feeling than applying LCD projectors in the similar price. Unfortunately it is involving appearing of the screen banding, invisible for the observer in the character of the man but can be detected on the stream from camera. That display device is based on optical micro-electro-mechanical technology [20, 21] and color wheel divided into multiple sectors: red, green, blue and in many cases white. The colors are displayed sequentially at a sufficiently high rate that the observer sees a composite "full color" image but camera can see single colors. Performing the automatic calibration, which is a final objective of conducted works, complicates it. In order to get acquainted with challenges of the calibration, a tool of the manual calibration was created and based on previous experience, we plan to conduct scientific research dealing with automatic calibration [22–24].

2 Manual Geometric and Photometric Calibration

In the calibration, of system of the image projection, we distinguish two stages: the geometrical calibration and the photometric calibration. The geometrical registration provides straight lines along the entire visualisation created from many display devices. The photometric calibration provides the continuity for the value on the surface of the chrominance and the brightness of the entire screen of the projection (Fig. 1).

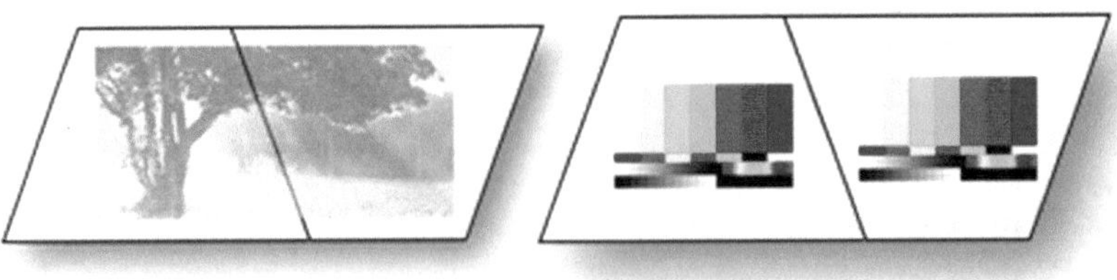

Fig. 1 The geometrical calibration provides the cohesion for the projection under the geometrical account and photometric calibration in terms of levels of the chrominance and the brightness

Transforming the entrance image in this way for the projection of this image falling on the screen of the projection to lie in the earlier designed place is a result of the geometrical calibration. For that purpose, finding the function of such a transformation is needed. The system operator of the image projection can determine such a transformation by hand, changing the geometry of the entrance image and observing changes happening on the screen or perhaps it to be carried out in the automatic way with the feedback in the form of visional information from the camera [25].

Depending on the type of the screen of the projection we can distinguish two with kind of shapes of the screen. The first kind is a screen makes from one or more planar surfaces, a second is a screen of an irregular, unknown shape (Fig. 2). In case of the first type of screens to the calibration it is possible to use linear transformations e.g. homography. In the second case the mapping pixel to pixel is needed (Fig. 3).

An irregular shape of the screen results in it, that change position of the observer, requires retransforming the image. Therefore, when we are dealing with screens of irregular shape, we can distinguish two approaches to the geometric calibration.

The first one assumes that visualization is rigidly "glued" to the projection screen. This approach can be compared to the situation at which the wallpaper is located on the screen. The second approach is to calibrate projection system to the

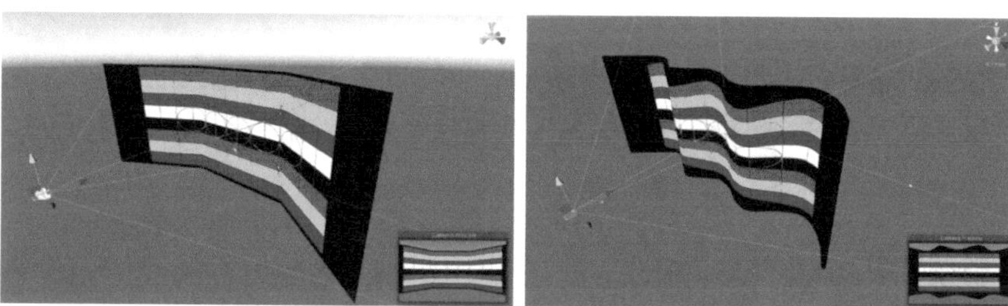

Fig. 2 Screen of the projection consisting of a few plains (*left*) and screen of an irregular shape (*right*)

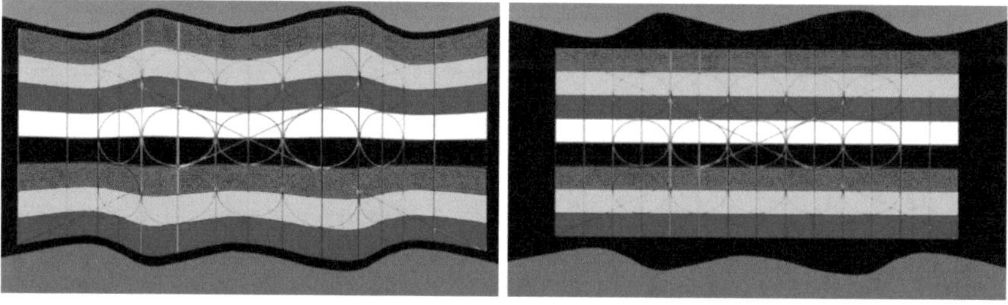

Fig. 3 Two manners of the geometrical calibration at using screens about an irregular shape. First it is "sticking" of projection to the shape of the screen (*left*), second it is a calibration of the projection relative to the chosen position of the observer in the space so that rectangular shape of the projection is preserved (*right*)

desired position of the observer in the space, in such a way that from that location the image is seen correctly. Jeśli obserwator jest ruchomy to zmiana jego położenia wymaga zaktualizowania danych o jego aktualnej pozycji i ponownego przekształcenia rzucanych obrazów.

The next chapter discusses the steps for manual calibration If the projection screen consisting of three interconnected planes.

2.1 Texture Mapping

Having information about how tops of the output image are being transformed in the entrance image of the projector it is possible to make the mapping of the quad texture from one to another. This can be done in several ways, depending on what result you want to achieve [26].

The first of them 1.5 Order of Polynomial Transformation implies a linear transformation between the texture coordinates and the coordinates of the output quadrilateral.

$$U = a_0 + a_1 X + a_2 Y + a_3 XY \tag{1}$$

$$V = b_0 + b_1 X + b_2 Y + b_3 XY \tag{2}$$

Based on knowledge of four points corresponding to each other on both quadrangles coefficients can be determined by Eqs. (1) and (2) on the basis designate new coordinates for each pixel. The use of this method does not preserving these straight lines

The second way is perspective transformation, ensure the maintenance of straight lines in the image but the texture is not distributed evenly inside the quadrangle and its distribution depends on the angle between the opposite edges of the quadrangle (Fig. 5). Transforming can be determined from the four points corresponding to each other and are described by the equations:

$$U = \frac{a_0 + a_1 X + a_2 Y}{1 + c_1 X + c_2 Y} \tag{3}$$

$$V = \frac{b_0 + b_1 X + b_2 Y}{1 + c_1 X + c_2 Y} \tag{4}$$

The last of the presented transformation is bilinear transformation preserving straight lines in the image and the even distribution of texture at the edges (Fig. 5).

The transformation we are setting by found opposite to the equation below:

$$X = a_0 + a_1 U + a_2 V + a_3 UV \tag{5}$$

$$Y = b_0 + b_1 U + b_2 V + b_3 UV \tag{6}$$

After transformations we obtain formulas for the coordinates U and V:

$$U = (X - a_0 - a_2 V)(a_1 + a_3 V) \tag{7}$$

$$V = \frac{-B \pm \sqrt{B^2 - 4AC}}{2A} \tag{8}$$

Each of the above-mentioned mapping methods has been tested in both the virtual and real environment. Description and test results are presented in Chap. 3.

2.2 Image Blending

The geometric calibration challenge is seamless merging of adjacent image projection. Regions projection screen, which overlap projections of adjacent projectors need to be darkened so that the resulting chrominance and luminance color does not differ from the regions where there is no assembly of the image [26]. For this purpose, a method of blending at the edges of each projection is used. Blending is described by the function, which argument is a distance of pixel from the edge of the image. The output value of the pixel to its actual value multiplied by the value of this function and is described by the following formulas:

$$f(x) = \left\{ \begin{array}{ll} 0.5(2x)^p & for\, 0 \le x \le 0.5 \\ 1 - 0.5(2(1-x))^p & for\, 0.5 \le x \le 1 \end{array} \right\} \tag{9}$$

The curvature of the function is controlled by the parameter p Where blending is linear with $p = 1$, the growth of p will increase the degree of curvature of the blending. In the case of image projection systems applies only the function f is no suitable. It is necessary to take account of the gamma function that informs about how the pixel values are mapped to the brightness of the device. The value of gamma G is in the range 1.8–2.2 [27]. The blending after taking into account the gamma is as follows:

$$f(x) = \left\{ \begin{array}{ll} f(x)^{\frac{1}{G}} & for\, 0 \le x \le 0.5 \\ f(1-x)^{\frac{1}{G}} & for\, 0.5 \le x \le 1 \end{array} \right\}. \tag{10}$$

The results of these functions are presented and described in Chap. 3.

2.3 Scaling Colour Manually

After geometric registration and prior to blending regions common to the projectors are required to perform photometric calibration. The method of calibration chrominance and luminance of the screen is to observe the projection screen and manually adjusting the scaling factors of each component color for the projector. Transformations performed for each pixel of the projector can be described by the formulas:

$$I' = I\begin{bmatrix} r_n & 0 & 0 \\ 0 & g_n & 0 \\ 0 & 0 & b_n \end{bmatrix} \tag{11}$$

where I is 1×3 vector in RGB color space of the selected pixel in the input image projector, I' is vector of new pixel values in the image cast by the projector and the r, g, b color scale factors for the projector n.

3 Tests

Tests were performed both in simulation and in reality conditions. They were present at all stages of calibration to verify the correctness of the implementation of the methods and the evaluation results.

Simulation tests were performed in an environment Unity 3D, which enables the deployment of virtual projectors and virtual projection screen, built by the user and in Matlab. They allowed for the preliminary examination of the accuracy of methods used.

The tests were performed on the prepared actual image projection system consisting of three DLP projectors and the screen made up of three perpendicular walls (Fig. 4).

3.1 Texture Mapping in the Virtual and the Real Environment

Mapping method described in Sect. 2.1 have been implemented in Matlab and then tested. Observing the test results, we can conclude that only bilinear and perspective transform keep straight lines in the image. Looking at while submitting the adjacent images can be observed failure to submit images for perspective transformation (Fig. 5). This means that if the images of neighbouring projection better results will apply bilinear transformation.

Fig. 4 The tested system of image projection. Screen and one with 3 DLP projectors

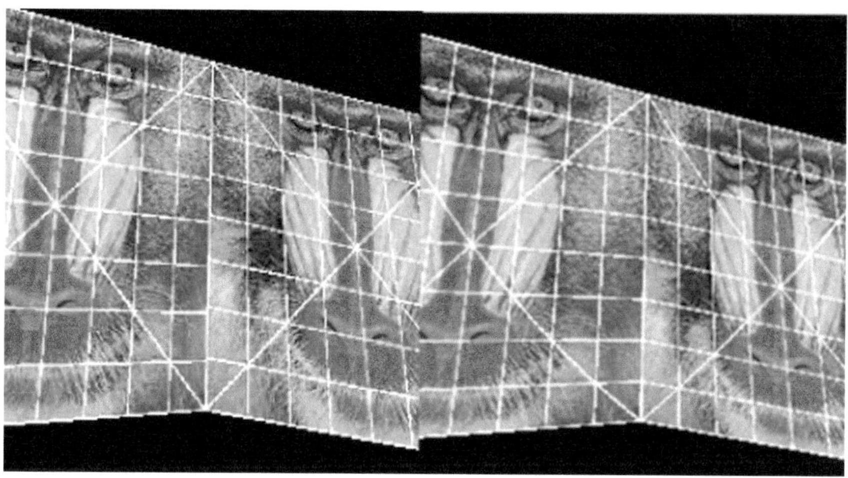

Fig. 5 (*left*) Perspective transformation, (*right*) Bilinear transformation

Tests in real environments have shown that the chosen method works well if you want to combine the two textures in the image from one projector. When combining images of adjacent projectors, there is an additional distortion depends on the angle at which the projector shines onto the plane. The correct geometric calibration was achieved using the alignment angle of projection projectors (Fig. 6).

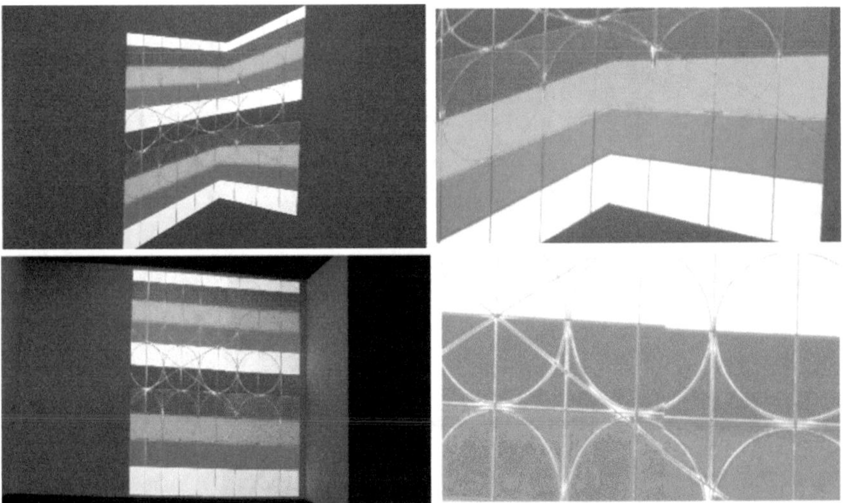

Fig. 6 (*Up*) The correct geometric calibration. (*down*) Incorrect geometric calibration, images of adjacent projectors don't match

Fig. 7 (*Left*) The calibration result for p = 1.9 without correction coefficient G, (*Right*) calibration result for p = 1.9, taking into account the coefficient G

3.2 Blending in Real Application

The table below presents the results of tests involving a change of parameter *p* luminosity function and the pixel values of gamma *G* (Fig. 7). You can be assessed visually, that the inclusion of gamma improves the quality of calibration. Additionally, the proper selection of the curvature ratio *f(x)* also improves the quality of calibration (Fig. 8).

Fig. 8 (*Left*) The calibration result for p = 1 correction factor G (*Right*) calibration result for p = 1.9 correction factor G

4 Summary

The aim of the study was to develop and test methods to create tools for manual calibration image projection system composed of more than one quantity of projectors and the screen in the form of interconnected planes, which can be mapped quadrangular texture. This tool was created using the Unity 3D environment. With its help, a trained operator is able to calibrate the geometry of the system, which results is a visually correct visualization. In the case of calibration appearing challenges of subjective perception of color, brightness and optical illusions what is the sense of sight of the observer which makes it difficult to carry out proper calibration. That is why the subject of further research will be automation of the calibration process and the use of feedback in the form of a camera as an "objective" observer. Another subject of research will also be acquisition the image of the projection system based on DLP projectors due to image distortion occurring during registration (DLP Banding). This task can be solved through the use of better equipment for image acquisition or develop an algorithm filtering data received from the camera.

References

1. Majumder, A., Lai, D. Q., & Tehrani, M. A. (2015). A multi-projector display system of arbitrary shape, size and resolution. In *ACM SIGGRAPH 2015 emerging technologies*. ACM.
2. Sajadi, B., & Majumder, A. (2009). Markerless view-independent registration of multiple distorted projectors on extruded surfaces using an uncalibrated camera. *IEEE Transactions on, Visualization and Computer Graphics, 15*(6), 1307–1316.
3. Raskar, R., et al. (2004). Quadric transfer for immersive curved screen displays. In *Computer graphics forum* (Vol. 23, No. 3). Blackwell Publishing, Inc.
4. Koźlak, M., Kurzeja, A., & Nawrat, A. (2013). Virtual reality technology for military and industry training programs. In *Vision based systems for UAV applications* (pp. 327–334). Springer International Publishing.

5. Daniec, K., Jedrasiak, K., Koteras, R., & Nawrat, A. (2013). Embedded micro inertial navigation system. In *Applied mechanics and materials* (Vol. 249, pp. 1234–1246). Trans Tech Publications.
6. Josinski, H., Switonski, A., Jedrasiak, K., & Kostrzewa, D. (2012). Human identification based on gait motion capture data. In *Proceedings of the 2012 International MultiConference of Engineers and Computer Scientists IMECS* (Vol. 12).
7. Switonski, A., Josinski, H., Jedrasiak, K., Polanski, A., & Wojciechowski, K. (2010). Classification of poses and movement phases, ICCVG 2010. In *Lecture notes in computer science*. Springer.
8. Nawrat, A., & Jedrasiak, K. (2008). Fast colour recognition algorithm for robotics. *Problemy Eksploatacji*, 69–76.
9. Bieda, R., Grygiel, R., & Galuszka, A. (2015). Naive Kalman filtering for estimation of spatial object orientation. In *Methods and models in automation and robotics (MMAR)* (pp. 955–960).
10. Bieda, R., & Grygiel, R. (2014). Wyznaczanie orientacji obiektu w przestrzeni z wykorzystaniem naiwnego filtru Kalmana. *Przeglad Elektrotechniczny, 90,* 34–41.
11. Bieda, R., Jaskot, K., Jędrasiak, K., & Nawrat, A. (2013). Recognition and location of objects in the visual field of a UAV vision system. In *Vision based systems for UAV applications* (pp. 27–45). Springer International Publishing.
12. http://www.fmwconcepts.com/imagemagick/bilinearwarp/index.php.
13. http://www.snieznik.info/pl/gal-pl.
14. Jedrasiak, K., Nawrat, A., & Wydmanska, K. (2013). *SETh-Link the distributed management system for unmanned mobile* (pp. 247–256). Berlin Heidelberg: Springer.
15. Babiarz, A., Bieda, R., Jedrasiak, K., & Nawrat, A. (2013). Machine vision in autonomous systems of detection and location of objects in digital images. In *Vision based systems for UAV applications* (pp. 3–25). Springer International Publishing.
16. Jedrasiak, K., Andrzejczak, M., & Nawrat, A. (2014). SETh: The method for long-term object tracking. In *Computer vision and graphics* (Vol. 8671, pp. 302–315). Lecture Notes in Computer Science, 316.
17. Daniec, K., Iwaneczko, P., Jedrasiak, K., & Nawrat, A. (2013). Prototyping the autonomous flight algorithms using the prepar3D® simulator. In *Vision based systems for UAV applications* (pp. 219–232). Springer International Publishing.
18. Seetzen, H., et al. (2004). High dynamic range display systems. In *ACM transactions on graphics (TOG)* (Vol. 23, No. 3). ACM.
19. Sobel, D., Jedrasiak, K., Daniec, K., Wrona, J., Jurgas, P., & Nawrat, A. (2014). Camera calibration for tracked vehicles augmented reality applications. In *Innovative control systems for tracked vehicle platforms* (pp. 147–162). Springer International Publishing.
20. Bereska, D., Daniec, K., Fras, S., Jedrasiak, K., Malinowski, M., & Nawrat, A. (2013). System for multi-axial mechanical stabilization of digital camera. In *Vision based systems for UAV applications* (pp. 177–189). Springer International Publishing.
21. Jedrasiak, K., Nawrat, A., Daniec, K., Koteras, R., Mikulski, M., & Grzejszczak, T. (2012). A prototype device for concealed weapon detection using IR and CMOS cameras fast image fusion. In *International Conference on Computer Vision and Graphics* (pp. 423–432). Berlin Heidelberg: Springer.
22. Ryt, A., Sobel, D., Kwiatkowski, J., Domzal, M., Jedrasiak, K., & Nawrat, A. (2014). Real-time laser point tracking. In *International Conference on Computer Vision and Graphics* (pp. 542–551). Springer International Publishing.
23. Babiarz, A., Bieda, R., & Jaskot, K. (2013). Vision system for group of mobile robots. In *Vision based systems for UAV applications* (pp. 139–156). Springer International Publishing.
24. Kus, Z., & Nawrat, A. (2013). Object tracking in a picture during rapid camera movements. In *Vision based systems for UAV applications* (pp. 77–91). Springer International Publishing.
25. Bhasker, E., Juang, R., & Majumder, A. (2007). Registration techniques for using imperfect and partially calibrated devices in planar multi-projector displays. *IEEE Transactions on, Visualization and Computer Graphics, 13*(6), 1368–1375.

26. Yang, R., et al. (2001). Pixelflex: A reconfigurable multi-projector display system. In *Proceedings of the Conference on Visualization'01*. IEEE Computer Society.
27. Raskar, R., Welch, G., & Fuchs, H. (1998). Seamless projection overlaps using image warping and intensity blending. In *Fourth International Conference on Virtual Systems and Multimedia*. Gifu, Japan.
28. http://www.lasershot.com.

The Mathematical Model of the Human Arm

Robert Bieda and Krzysztof Jaskot

1 Introduction

The realistic construction of human body is very important in terms of medical applications, as well as entertainment. A very important aspect of the research is to understand the rules of the complex motion of kinematic links occurring during the movement of any part of our body. In the literature we can find some ways that represent human motion [1]. Some authors presents the design process of a gesture control system based on the Microsoft Kinect sensor and model of human arm [2]. In paper [3] authors presents novel approach to gait motion reconstruction based on kinematical loops and functional skeleton features extracted from segmented Magnetic Resonance Imaging (MRI) data.

Kinematics of the human body is concerned with formulating and solving for the translational and rotational position, velocity, and acceleration analysis problems for each human body segment of interest, for various real world motions. Forward kinematics calculates the pose (position and orientation) of each human body segment of interest given the joint angles. The forward kinematics is the problem of finding an end-effector (hand) or tool pose from a set of given joint angles. As a result, statics analysis requires the positions and angles of each segment for static free-body diagrams. In paper [4] author considers a technique for computation of the inverse kinematic model of the human arm for robot based rehabilitation.

Forward dynamics, a much harder mathematical problem, calculates the unknown kinematics terms given the joint torques; this requires the solution of coupled non-linear differential equations. Dynamics requires the translational and rotational

R. Bieda (✉) · K. Jaskot
Institute of Automatic Control, Silesian University of Technology,
Akademicka 16, 44-100 Gliwice, Poland
e-mail: robert.bieda@polsl.pl

K. Jaskot
e-mail: krzysztof.jaskot@polsl.pl

A. Nawrat et al. (eds.), *Advanced Technologies in Practical Applications for National Security*, Studies in Systems, Decision and Control 106,
https://doi.org/10.1007/978-3-319-64674-9_14

position, velocity, and acceleration variables for each human body segment, plus the center-of-mass (CoM) translational accelerations, for dynamic free body diagrams. The paper [5] deals with the dynamic analysis of human arm, determination of shoulder and elbow torques considering isotonic and isokinetic exercise activities.

In this paper full analysis of the kinematics and dynamics of mathematical model of human upper arm is presented. Developing new algorithms [6–9] and computer simulation [10] are classical means to achieve progress of the technology.

2 The Synthesis of a Mathematical Model of the Human Arm

2.1 Kinematics

Kinematics (forward) is a relation of effector position and orientation relative to the base (inertial) frame depending on the position of the joints. In this case, the kinematics of the human arm is easy to position the orientation of the hand (palm) with respect to the base frame associated with shoulder, depending on the position (rotation) in the shoulder and elbow joints. In this paper model/system of m degrees-of-freedom (DoF) consists of n rigid bodies called links (rigid limb) will be considered. One of the arm link is characterized by a few values:

- l_i—the length of the i-th link,
- m_i—mass of the i-th link,
- $l_{c,i}$—the distance of the center-of-mass (CoM) of the i-th link measured from the joint point of reference,
- I_i—moment of inertia (MoI) of the i-th link.

The arm is composed of m rotatable pairs (revolute (rotary) joints). The mobility of the arm joints is characterized by the ability to change the orientation:

- φ_i—rotation angle around the X axis i-th link in the i-th joint,
- θ_i—rotation angle around the Y axis i-th link in the i-th joint,
- ψ_i—rotation angle around the Z axis i-th link in the i-th joint.

One end of the chain forming the model arm is connected to the fixed base (e.g. shoulder) while the other end (hand) remains free.

Pose. In the case of the human arm kinematics position of the i-th link of length l_i with respect to the $(i-1)$-th coordinate frame in which it is fixed, is expressed by the vector $\mathbf{l}_i = \begin{bmatrix} x_i & y_i & z_i \end{bmatrix}^T$. Similarly, the center-of-mass (CoM) of the i-th link relative to the $(i-1)$-th frame is defined as vector $\mathbf{r}_{c,i} = \begin{bmatrix} x_{c,i} & y_{c,i} & z_{c,i} \end{bmatrix}^T$. Accordingly, rotation of the link with respect to i-th joint is expressed by the rotation angles vector $\boldsymbol{\alpha}_i = \begin{bmatrix} \varphi_i & \theta_i & \psi_i \end{bmatrix}^T$ witch defines rotation about the X, Y and Z axis of the $(i-1)$-th frame.

As a result, the position of the k-th link of the arm model relative to the base coordinate frame is determined by relationship:

$$\mathbf{l}_k = \sum_{i=1}^{k} \mathbf{l}_i \tag{1}$$

for $k = 1, 2, \ldots, n$.

Similarly, the pose (CoM position and the orientation) of the k-th link is defined by dependencies:

$$\mathbf{r}_{c,k} = \sum_{i=1}^{k-1} \mathbf{l}_i + \mathbf{l}_{c,k} \tag{2}$$

$$\boldsymbol{\alpha}_k = \sum_{i=1}^{k} \boldsymbol{\alpha}_i \tag{3}$$

In the case of the kinematic chain of the human arm model we consider pairs of rotating (revolute (rotary) joints) only. Because of the fixed length of links sliding pairs do not occur. For the purpose of further analysis of the kinematic model variable values (rotations) in the kinematic chain will be replaced by the generalized variables:

$$\mathbf{q} = \begin{bmatrix} q_1 & q_2 & \cdots & q_m \end{bmatrix}^T \tag{4}$$

Generalized coordinates are the set of independent coordinates explicitly defining position of the system. The number of generalized coordinates is equal to the number of degrees-of-freedom (DoF) of the system. $\mathbf{q}$ is the vector of joints angles.

On human arm model generalized coordinates describing the motion can be selected by specifying the rotation of the first link with respect to the inertial frame and the other link's rotation relative to the previous link.

Velocities. In the case of analysis of direct kinematics of the arm velocity is defined depending on changes in time the position and orientation of the links. Linear velocity k-th link of the arm relative to the base can be expressed as follows [11]:

$$\mathbf{v}_k = \sum_{i=1}^{k} \frac{d\mathbf{l}_i}{dt} = \frac{d\mathbf{l}_k}{dt} = J_k^v(\mathbf{q})\dot{\mathbf{q}} \tag{5}$$

where:

$$J_k^v(\mathbf{q}) = \frac{d\mathbf{l}_k}{dt} \tag{6}$$

is the (position-dependent) Jacobian matrix of the transformation between Cartesian and joint space. $\dot{\mathbf{q}}$ are the joints velocity vector.

Similarly, the linear velocity of the CoM and the angular velocity of k-th link are defined by dependencies:

$$\mathbf{v}_{c,k} = \mathbf{v}_{k-1} + \frac{d\mathbf{l}_{c,k}}{dt} = \frac{d\mathbf{r}_{c,k}}{dt} = J_k^v(\mathbf{q})\dot{\mathbf{q}} \tag{7}$$

$$\boldsymbol{\omega}_k = \sum_{i=1}^{k} \frac{d\boldsymbol{\alpha}_i}{dt} = \frac{d\boldsymbol{\alpha}_k}{dt} = J_k^{\omega}(\mathbf{q})\dot{\mathbf{q}} \tag{8}$$

Velocities of k-th link can be determined using the above definition of generalized variables (4). Jacobian matrix for the k-th link describe change the position and orientation expressed in the system of internal coordinates model $\mathbf{q}$:

$$J_k(\mathbf{q}) = \begin{bmatrix} J_{c,k}^v(\mathbf{q}) \\ J_k^{\omega}(\mathbf{q}) \end{bmatrix} = \begin{bmatrix} \frac{\partial \mathbf{r}_{c,k}}{\partial \mathbf{q}} \\ \frac{\partial \boldsymbol{\alpha}_k}{\partial \mathbf{q}} \end{bmatrix} = \begin{bmatrix} \nabla_{\mathbf{q}} x_{c,k}^T \\ \nabla_{\mathbf{q}} y_{c,k}^T \\ \nabla_{\mathbf{q}} z_{c,k}^T \\ \nabla_{\mathbf{q}} \varphi_k^T \\ \nabla_{\mathbf{q}} \theta_k^T \\ \nabla_{\mathbf{q}} \psi_k^T \end{bmatrix} \tag{9}$$

where:

$$\nabla_{\mathbf{q}} = \begin{bmatrix} \frac{\partial}{\partial q_1} & \frac{\partial}{\partial q_2} & \cdots & \frac{\partial}{\partial q_m} \end{bmatrix}^T \tag{10}$$

is a vector operator gradient calculated on the generalized variables of the arm model. For so defined Jacobian k-th link velocities can be determined from the relationship:

$$\begin{bmatrix} \mathbf{v}_{c,k} \\ \boldsymbol{\omega}_k \end{bmatrix} = J_k(\mathbf{q})\dot{\mathbf{q}} \tag{11}$$

Acceleration. The last group of signals that described the kinematics model of the arm are acceleration. Linear acceleration k-th link of the kinematic chain is defined from the relationship [11]:

$$\mathbf{a}_k = \frac{d\mathbf{v}_k}{dt} = \frac{dJ_k^v(\mathbf{q})\dot{\mathbf{q}}}{dt} = \frac{dJ_k^v(\mathbf{q})}{dt}\dot{\mathbf{q}} + J_k^v(\mathbf{q})\ddot{\mathbf{q}} \tag{12}$$

where:

$$\frac{dJ_k^v(\mathbf{q})}{dt} = \frac{\partial J_k^v(\mathbf{q})}{\partial \mathbf{q}}\dot{\mathbf{q}} = \frac{\partial^2 \mathbf{l}_k}{\partial \mathbf{q}^2}\dot{\mathbf{q}} \tag{13}$$

Similarly, the CoM linear acceleration and angular acceleration k-th link, are defined as:

$$\mathbf{a}_{c,k} = \frac{d\mathbf{v}_{c,k}}{dt} = \frac{dJ_{c,k}^v(\mathbf{q})\dot{\mathbf{q}}}{dt} = \frac{dJ_{c,k}^v(\mathbf{q})}{dt}\dot{\mathbf{q}} + J_{c,k}^v(\mathbf{q})\ddot{\mathbf{q}} \tag{14}$$

$$\varepsilon_k = \frac{d\boldsymbol{\omega}_k}{dt} = \frac{dJ_k^{\omega}(\mathbf{q})\dot{\mathbf{q}}}{dt} = \frac{dJ_k^{\omega}(\mathbf{q})}{dt}\dot{\mathbf{q}} + J_k^{\omega}(\mathbf{q})\ddot{\mathbf{q}} \tag{15}$$

Using the definition of the Jacobian (9) for k-th link and dependence (13) accelerations can be determined using generalized variables:

$$\begin{bmatrix} \mathbf{a}_{c,k} \\ \varepsilon_k \end{bmatrix} = \frac{dJ_k(\mathbf{q})}{d\mathbf{q}}\dot{\mathbf{q}} + J_k(\mathbf{q})\ddot{\mathbf{q}} \tag{16}$$

2.2 Dynamics

In the analysis of the dynamics of the arm equations of motion that describe the motion parameters of the arm and the related forces and torques or external forces applied to the system are considered. The dynamics of the arm refers to the interaction between forces in the system and change of state of the system. A torque acting on the joints causes a change in the joint orientation and velocity.

One way to describe the dynamics of manipulators (arms) is to use the Lagrange equations. They describe the dynamic properties of the system, depending on the kinetic and potential energy expressed as a function of the internal (generalized) coordinate.

Kinetic energy. The kinetic equation is the equation for the energy of motion. The kinetic energy of the k-th link is determined by the relation [12]:

$$K_k = \frac{m_k v_{c,k}^2}{2} + \frac{I_k \omega_k^2}{2} \tag{17}$$

The first component of the above equation is the kinetic energy of translational motion with the velocity of the CoM, the second is the kinetic energy of rotation.

The total kinetic energy for the system can be found by summing the linear and rotational components for each rigid body. A point of interest is that the rotational kinetic energy of the next link depends on the rotational velocity of itself and the previous link. This is because of how the generalized coordinates were selected:

$$K(\mathbf{q}, \dot{\mathbf{q}}) = \sum_{i=1}^{n} K_i = \sum_{i=1}^{n} \left(\frac{m_i \mathbf{v}_{c,i}^T \mathbf{v}_{c,i}}{2} + \frac{I_i \boldsymbol{\omega}_i^T \boldsymbol{\omega}_i}{2} \right) \tag{18}$$

Potential energy. The potential energy equation determines the energy of a rigid body of mass m located in a gravitational field. For the k-th link potential energy is determined by the relation [12]:

$$P_k = -m_k g r_{c,k} \tag{19}$$

where $g \approx 9.81[\mathrm{m/s^2}]$ is the gravitational acceleration of the Earth.

The potential energy for the system is quantified by the change in CoM within a gravitational field. Using the equation (19), the total potential energies for arm model can be expressed as sum of potential energy for each rigid body:

$$P(\mathbf{q}) = \sum_{i=1}^{n} P_i = -\sum_{i=1}^{n} m_i \mathbf{g}^T \mathbf{r}_{c,i} \tag{20}$$

Lagrangian function. The Lagrangian L of arm system is the representation of a system of motion. The Lagrangian can only be used when a system is conservative. Derived from Newton's Laws, the Lagrangian says that if you can find the kinetic equation K and the potential equation P of our complex system in terms of general coordinates and their time derivatives then you can find the equations of the motion of the system in terms of generalized coordinates using the Lagrangian.

If we define total kinetic energy K (18) and total potential energy (20) of the system, we can introduce the concept of the Lagrange function (kinetic potential) in the form [13]:

$$L(\mathbf{q}, \dot{\mathbf{q}}) = K(\mathbf{q}, \dot{\mathbf{q}}) - P(\mathbf{q}) \tag{21}$$

Euler-Lagrange equation of motion. The Lagrangian is defined as the difference between kinetic and potential energy and is used with the Euler-Lagrange equation as follows [13]:

$$\frac{d}{dt}\left(\frac{\partial L(\mathbf{q}, \dot{\mathbf{q}})}{\partial \dot{\mathbf{q}}}\right) - \frac{\partial L(\mathbf{q}, \dot{\mathbf{q}})}{\partial \mathbf{q}} = \boldsymbol{\tau} \tag{22}$$

where $\boldsymbol{\tau}$ represent the m-dimensional vector of generalized torques produced by muscles on the m joints of a limb, and $\mathbf{q}$ is the resulting joint angle trajectories. Torque is applied at each of the joints and as a result we obtain m equations, one for each DoF.

Generalized equation of motion. To a good approximation, human arm dynamics can be modelled as the motion of an open kinematic chain of rigid links, attached together through revolute (rotary) joints, with control torques applied about each joint. Human arm motions can thus be modelled by the same equations used to model revolute (rotary) robot manipulators [11, 14–18]:

$$M(\mathbf{q})\ddot{\mathbf{q}} + C(\mathbf{q}, \dot{\mathbf{q}}) + G(\mathbf{q}) = \boldsymbol{\tau} \tag{23}$$

where $\mathbf{q}$ are the joint angle of the arm. The $M(\mathbf{q})$ is a (position-dependent) symmetric and positive definite inertia matrix. This ensures that the mass is always positive and real valued. The vector $C(\mathbf{q}, \dot{\mathbf{q}})$ contains the Coriolis and centripetal torques and the vector $G(\mathbf{q})$ contains the gravitational torques. The generalized forcing input $\boldsymbol{\tau}$ represents the control torques applied at each arm joint.

2.3 Non Linear Model of Arm

The design of the arm dynamics model is based on the system of m differential equations describing the motion of the system in the space of generalized coordinate:

$$\frac{d}{dt}\left(\frac{\partial L}{\partial \dot{\mathbf{q}}}\right) - \frac{\partial L}{\partial \mathbf{q}} - \boldsymbol{\tau} = \mathbf{0} \tag{24}$$

or in the form of generalized systems manipulating robots [19]:

$$M(\mathbf{q})\ddot{\mathbf{q}} + C(\mathbf{q}, \dot{\mathbf{q}}) + G(\mathbf{q}) - \boldsymbol{\tau} = \mathbf{0} \tag{25}$$

In the general case determination of changes in the value of generalized coordinates reduced to the solution of the equation:

$$F(\mathbf{q}, \dot{\mathbf{q}}, \ddot{\mathbf{q}}, \boldsymbol{\tau}) = \mathbf{0} \tag{26}$$

In the present case, the model of the human arm, having regard to the form of the equations of motion (25), angular acceleration of joints determines the relationship [19]:

$$\ddot{\mathbf{q}} = f(\mathbf{q}, \dot{\mathbf{q}}, \boldsymbol{\tau}) = M(\mathbf{q})^{-1}(\boldsymbol{\tau} - C(\mathbf{q}, \dot{\mathbf{q}}) - G(\mathbf{q})) \tag{27}$$

Finally, the form of non-linear model describing the dynamics of changes in the system of generalized variables is defined as follows:

$$\begin{cases} \ddot{\mathbf{q}} = f(\mathbf{q}, \dot{\mathbf{q}}, \boldsymbol{\tau}) \\ \dot{\mathbf{q}} = \int \ddot{\mathbf{q}} dt \\ \mathbf{q} = \int \dot{\mathbf{q}} dt \end{cases} \tag{28}$$

2.4 Linear Model of the Arm

A common problem in determining the dynamics of the mathematical model is to determine the function $f(\mathbf{q}, \dot{\mathbf{q}}, \boldsymbol{\tau})$ which is the solution of the Eq. (26). An alternative approach is to define a model of the linear nature. This model provides a significant simplification of the calculation of the cost of approximation properties of the non-linear model in a range of actions.

Due to the nature of this problem it was decided that the state-space representation was a good way to approach the problem. Thus, the proposed linear model is described by two equations: the equation of state changes and output equation [19]:

$$\begin{cases} \dot{\mathbf{x}} = \mathbf{A}\mathbf{x} + \mathbf{B}\mathbf{u} \\ \mathbf{y} = \mathbf{C}\mathbf{x} + \mathbf{D}\mathbf{u} \end{cases} \tag{29}$$

State, output and control variables. Having selected state-space formulation for representing the model, it is now important to choose the state vector that will better describe the system. For the proposed model of human arm state variables $\mathbf{x}$, signal output $\mathbf{y}$ and control $\mathbf{u}$ are defined as follows:

$$\begin{aligned} \mathbf{x} &= [\mathbf{q} \quad \dot{\mathbf{q}}]^T \\ \mathbf{y} &= \ddot{\mathbf{q}} \\ \mathbf{u} &= \boldsymbol{\tau} \end{aligned} \tag{30}$$

Linearisation process takes around an equilibrium point. In the case of a model composed of n rotatable joints there are two equilibrium points: stable and unstable. In the presented system stable equilibrium point corresponds to a situation of free over-hang of the human arm.

In the general case, however, we consider the operating point:

$$\rho_0 = [\mathbf{q}_0 \quad \dot{\mathbf{q}}_0 \quad \ddot{\mathbf{q}}_0 \quad \boldsymbol{\tau}_0]^T \tag{31}$$

For such defined operating point linearisation process equations of motion (26) leads to a linear relationship:

$$\left.\frac{\partial F}{\partial \mathbf{q}}\right|_{\rho_0} \Delta\mathbf{q} + \left.\frac{\partial F}{\partial \dot{\mathbf{q}}}\right|_{\rho_0} \Delta\dot{\mathbf{q}} + \left.\frac{\partial F}{\partial \ddot{\mathbf{q}}}\right|_{\rho_0} \Delta\ddot{\mathbf{q}} - \mathbf{I}\Delta\boldsymbol{\tau} = \mathbf{0} \tag{32}$$

Given the initial conditions of the arm system, and bearing in mind that $\mathbf{z} = \Delta\mathbf{z} + \mathbf{z}_0$, for $\mathbf{z} = \{\mathbf{q}, \dot{\mathbf{q}}, \ddot{\mathbf{q}}, \boldsymbol{\tau}\}$ the above linear equation of motion can be written in the general form:

$$\mathbf{A}_0\mathbf{q} + \mathbf{B}_0\dot{\mathbf{q}} + \mathbf{C}_0\ddot{\mathbf{q}} - \boldsymbol{\tau} = \mathbf{0} \tag{33}$$

where $\mathbf{A}_0 = \frac{\partial F}{\partial \mathbf{q}}|_{\rho_0}$ corresponds to joint stiffness, $\mathbf{B}_0 = \frac{\partial F}{\partial \dot{\mathbf{q}}}|_{\rho_0}$ represents joints damping (viscosity), and $\mathbf{C}_0 = \frac{\partial F}{\partial \ddot{\mathbf{q}}}|_{\rho_0}$ is the inertia matrix.

Taking into account the definition (30) of the state variables, output and control signal the above equation becomes:

$$[\mathbf{A}_0 \quad \mathbf{B}_0]\mathbf{x} + \mathbf{C}_0\mathbf{y} - \mathbf{u} = \mathbf{0} \tag{34}$$

Finally, from the Eq. (34) can be derived equations of state changes and output of the linear arm model:

$$\begin{cases} \dot{\mathbf{x}} = -\mathbf{C}_0^{-1}\begin{bmatrix} \mathbf{0} & -\mathbf{C}_0 \\ \mathbf{A}_0 & \mathbf{B}_0 \end{bmatrix}\mathbf{x} - \mathbf{C}_0^{-1}\begin{bmatrix} \mathbf{0} \\ -\mathbf{I} \end{bmatrix}\mathbf{u} \\ \mathbf{y} = -\mathbf{C}_0^{-1}\begin{bmatrix} \mathbf{A}_0\ \mathbf{B}_0 \end{bmatrix}\mathbf{x} + \mathbf{C}_0^{-1}\mathbf{u} \end{cases} \tag{35}$$

Substituting for the variables of the model in state space signals directly related to the change in the joints of the arm angles and torques, we get a linear model (29):

$$\begin{cases} \begin{bmatrix} \dot{\mathbf{q}} \\ \ddot{\mathbf{q}} \end{bmatrix} = \mathbf{A} \begin{bmatrix} \mathbf{q} \\ \dot{\mathbf{q}} \end{bmatrix} + \mathbf{B}\tau \\ \ddot{\mathbf{q}} = \mathbf{C} \begin{bmatrix} \mathbf{q} \\ \dot{\mathbf{q}} \end{bmatrix} + \mathbf{D}\tau \end{cases} \tag{36}$$

where:

$$\begin{aligned} \mathbf{A} &= \begin{bmatrix} \mathbf{0}\ \mathbf{I} \\ \mathbf{C} \end{bmatrix} \qquad \mathbf{B} = \begin{bmatrix} \mathbf{0} \\ \mathbf{D} \end{bmatrix} \\ \mathbf{C} &= \begin{bmatrix} -\mathbf{C}_0^{-1}\mathbf{A}_0 & -\mathbf{C}_0^{-1}\mathbf{B}_0 \end{bmatrix} \ \mathbf{D} = \mathbf{C}_0^{-1} \end{aligned} \tag{37}$$

3 2link-3DoF Arm Model

The proposed mathematical model of the human arm assumes the possibility of bending the arm at the elbow [20]. As a result, this model is defined by the kinematic structure consisting of two rigid connections and two joints [21]. This model takes into account the fact that the upper arm consists of one bone and is connected to the shoulder in the gleno-humeral joint (GH-joint) [22]. The GH-joint is well modelled as a ball-and-socked joint with three DoF. In proposed model of human arm this GH-joint is modelled as two degrees-of-freedom joint. The second elbow joint is modelled as a hinge joint. Elbow joint has one DoF. Consequently, the presented variant of human arm model has three DoF. The idea of the construction of the proposed model with two links and three DoF (2link-3DoF) is shown in Fig. 1. Taking into account the definition of the generalized coordinates in the proposed model the rotation angles of the shoulder and elbow joints $(\psi_1, \varphi_1, \psi_2)$ are labelled as variables (q_1, q_2, q_3). The range of motion of the arm is shown in Table 1.

4 Kinematics

Pose. Using the definition of the positions (1) and (2) and orientation (3) the transformation of joint angles to the position of links, their CoM and the orientation are given by:

$$[\mathbf{l}_1 \quad \mathbf{l}_2] = \begin{bmatrix} l_1 s_1 c_2 & l_1 s_1 c_2 + l_2 s_{13} c_2 \\ -l_1 c_1 c_2 & -l_1 c_1 c_2 - l_2 c_{13} c_2 \\ -l_1 s_2 & -l_1 s_2 - l_2 s_2 \end{bmatrix} \tag{38}$$

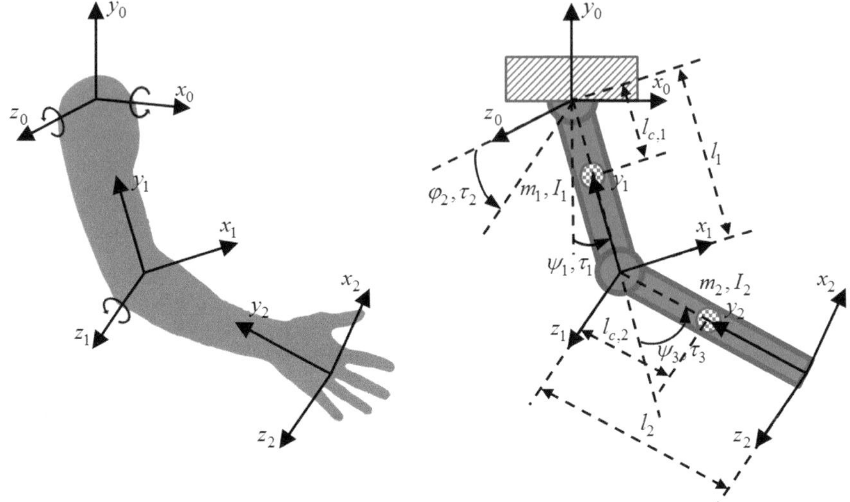

Fig. 1 Diagram of a model of human arm by two link and three degree-of-freedom

Table 1 Parameters of healthy right arm

Joint	Direction	Range (°)	
Shoulder	Flexion	180	ψ_1
	Hyperextension/extension	150	
	Abduction	180	φ_1
	Adduction	10	
Elbow	Flexion	145	ψ_2
	Hyperextension/extension	0	

$$[\mathbf{r}_{c,1} \quad \mathbf{r}_{c,2}] = \begin{bmatrix} l_{c,1}s_1c_2 & l_1s_1c_2 + l_{c,2}s_{13}c_2 \\ -l_{c,1}s_1c_2 & -l_1s_1c_2 - l_{c,2}s_{13}c_2 \\ -l_{c,1}s_2 & -l_1s_2 - l_{c,2}s_2 \end{bmatrix} \tag{39}$$

$$[\boldsymbol{\alpha}_1 \quad \boldsymbol{\alpha}_2] = \begin{bmatrix} 0 & 0 \\ q_2 & q_2 \\ q_1 & q_1 + q_3 \end{bmatrix} \tag{40}$$

where: $s_i = sin(q_i)$; $c_i = cos(q_i)$ and $s_{ij} = sin(q_i + q_j)$; $c_{ij} = cos(q_i + q_j)$.

Velocity. Defining the Jacobian matrices (9) for two rigid links of the arm model:

$$J_1 = \begin{bmatrix} l_{c,1}c_1c_2 & -l_{c,1}s_1s_2 & 0 \\ l_{c,1}s_1c_2 & l_{c,1}c_1s_2 & 0 \\ 0 & -l_{c,1}c_2 & 0 \\ 0 & 0 & 0 \\ 0 & 1 & 0 \\ 1 & 0 & 0 \end{bmatrix} \tag{41}$$

and for second link:

$$J_2 = \begin{bmatrix} c_2(l_1c_1 + l_{c,2}c_{13}) & -s_2(l_1s_1 + l_{c,2}s_{13}) & l_{c,2}c_{13}c_2 \\ c_2(l_1s_1 + l_{c,2}s_{1,3}) & s_2(l_1c_1 + l_{c,2}c_{13}) & l_{c,2}s_{13}c_2 \\ 0 & -c_2(l_1 + l_{c,2}) & 0 \\ 0 & 0 & 0 \\ 0 & 1 & 0 \\ 1 & 0 & 1 \end{bmatrix} \tag{42}$$

we can determine the velocity vectors of CoM's of links and their angular velocities:

$$\begin{bmatrix} \mathbf{v}_{c,1} & \mathbf{v}_{c,2} \\ \boldsymbol{\omega}_1 & \boldsymbol{\omega}_2 \end{bmatrix} = \begin{bmatrix} l_{c,1}(c_1c_2\dot{q}_1 - s_1s_2\dot{q}_2 & y_{xc,2} \\ l_{c,1}(s_1c_2\dot{q}_1 - c_1s_2\dot{q}_2 & y_{yc,2} \\ -l_{c,1}c2\dot{q}_2 & -c_2(l_1 + l_{c,2})\dot{q}_2 \\ 0 & 0 \\ \dot{q}_2 & \dot{q}_2 \\ \dot{q}_1 & \dot{q}_1 + \dot{q}_3 \end{bmatrix} \tag{43}$$

with coefficients:

$$y_{xc,2} = c_2\left(l_{c,2}c_{13} + l_1c_1\right)\dot{q}_1 + l_{c,2}c_{13}c_2\dot{q}_3 - s_2\left(l_1s_1 + l_{c,2}s_{13}\right)\dot{q}_2$$

$$y_{yc,2} = c_2\left(l_{c,2}s_{13} + l_1s_1\right)\dot{q}_1 + l_{c,2}s_{13}c_2\dot{q}_3 + s_2\left(l_1c_1 + l_{c,2}c_{13}\right)\dot{q}_2.$$

Acceleration. Using the relationship (16) and Jacobian matrices of each links we determine the acceleration in the system:

$$\begin{bmatrix} \mathbf{a}_{c,1} & \mathbf{a}_{c,2} \\ \varepsilon_1 & \varepsilon_2 \end{bmatrix} = \begin{bmatrix} a_{xc,1} & a_{xc,2} \\ a_{yc,1} & a_{yc,2} \\ l_{c,1}(s_2\dot{q}_2^2 - c_2\ddot{q}_2) & (l_1 + l_{c,2})(\ddot{q}_2^2s_2 - \ddot{q}_2c_2) \\ 0 & 0 \\ \ddot{q}_2 & \ddot{q}_2 \\ \ddot{q}_1 & \ddot{q}_1 + \ddot{q}_3 \end{bmatrix} \tag{44}$$

with coefficients:

$$a_{xc,1} = l_{c,1}(-s_1c_2\dot{q}_1^2 - 2\dot{q}_1c_1s_2\dot{q}_2 - s_1c_2\dot{q}_2^2 + c_1c_2\ddot{q}_1 - s_1s_2\ddot{q}_2)$$

$$a_{yc,1} = l_{c,1}(c_1c_2\dot{q}_1^2 - 2\dot{q}_1s_1s_2\dot{q}_2 + c_1c_2\dot{q}_2^2 + s_1c_2\ddot{q}_1 + c_1s_2\ddot{q}_2)$$

$$\begin{aligned}a_{xc,2} = &-c_2\dot{q}_1^2l_1s_1 - c_2\dot{q}_1^2l_{c,2}s_{13} - 2\dot{q}_1s_2\dot{q}_2l_1c_1\\ &- 2\dot{q}_1s_2\dot{q}_2l_{c,2}c_{13} - 2\dot{q}_1l_{c,2}s_{13}c_2\dot{q}_3 - c_2\dot{q}_2^2l_1s_1\\ &- c_2\dot{q}_2^2l_{c,2}s_{13} - 2\dot{q}_2l_{c,2}c_{13}s_2\dot{q}_3 - l_{c,2}s_{13}c_2\dot{q}_3^2\\ &- s_2\ddot{q}_2l_{c,2}s_{13} + c_2\ddot{q}_1l_1c_1 + c_2\ddot{q}_1l_{c,2}c_{13} - s_2\ddot{q}_2l_1s_1\\ &+ l_{c,2}c_{13}c_2\ddot{q}_3\end{aligned}$$

$$\begin{aligned}a_{yc,2} = &\,c_2\dot{q}_1^2l_1c_1 + c_2\dot{q}_1^2l_{c,2}c_{13} - 2\dot{q}_1s_2\dot{q}_2l_1s_1\\ &- 2\dot{q}_1s_2\dot{q}_2l_{c,2}s_{13} + 2\dot{q}_1l_{c,2}c_{13}c_2\dot{q}_3 + c_2\dot{q}_2^2l_1c_1\\ &+ c_2\dot{q}_2^2l_{c,2}c_{13} - 2\dot{q}_2l_{c,2}s_{13}s_2\dot{q}_3 + l_{c,2}c_{13}c_2\dot{q}_3^2\\ &+ c_2\ddot{q}_1l_1s_1 + c_2\ddot{q}_1l_{c,2}s_{13} + s_2\ddot{q}_2l_1c_1 + s_2\ddot{q}_2l_{c,2}c_{13}\\ &+ l_{c,2}s_{13}c_2\ddot{q}_3.\end{aligned}$$

5 Dynamics

For the presented model of the human arm Lagrange function takes the form:

$$\begin{aligned}L(\mathbf{q},\dot{\mathbf{q}}) = &\,0.5\left(m_1l_{c,1}^2c_2^2 + m_2c_2^2l_{c,2}^2 + I_2 + I_1\right.\\ &\left.+ m_2c_2^2l_1^2\right)\dot{q}_1^2 + m_2\dot{q}_2^2l_1s_1l_{c,2}s_{13} + gc_2m_2l_1c_1\\ &+ 0.5\left(I_2\dot{q}_2^2 + m_1\dot{q}_2^2l_{c,1}^2 + m_2\dot{q}_2^2l_1^2 + I_1\dot{q}_2^2 + I_2\dot{q}_3^2\right.\\ &\left.+ m_2\dot{q}_2^2l_{c,2}^2 + m_2\dot{q}_3^2l_{c,2}^2c_2^2\right) + m_2\dot{q}_2^2l_1c_2^2l_{c,2}\\ &+ m_2\dot{q}_2^2l_1c_1l_{c,2}c_{13} + \left(m_2c_2^2l_1s_1\dot{q}_3l_{c,2}s_{13} + I_2\dot{q}_3\right.\\ &\left.+ m_2c_2^2l_{c,2}^2\dot{q}_3 + m_2c_2^2l_1c_1l_{c,2}c_{13}\dot{q}_3\right)\dot{q}_1\\ &+ gc_2m_1l_{c,1}c_1 - m_2\dot{q}_2^2l_1c_1l_{c,2}c_{13}c_2^2\\ &- m_2\dot{q}_2^2l_1s_1l_{c,2}s_{13}c_2^2 + m_2c_2^2l_{c,2}s_{13}l_1s_1\dot{q}_1^2\\ &+ gc_2m_2l_{c,2}c_{13} - m_2\dot{q}_2l_1s_1s_2l_{c,2}c_{13}c_2\dot{q}_3\\ &+ m_2\dot{q}_2l_1c_1s_2\dot{q}_3l_{c,2}s_{13}c_2 + m_2c_2^2l_{c,2}c_{13}l_1c_1\dot{q}_1^2\end{aligned} \tag{45}$$

Euler-Lagrange equation of motion. Then using the Lagrangian formula (24) we compute three equations of motion:

$$
\begin{aligned}
&- \dot{q}_2\dot{q}_1 m_2 l_1^2 s_{22} - \dot{q}_2\dot{q}_1 m_1 l_{c,1}^2 s_{22} + \ddot{q}_1 I_2 + \ddot{q}_1 I_1 + \ddot{q}_3 I_2 \\
&- \dot{q}_3\dot{q}_1 m_2 l_{c,2} l_1 s_3 - \dot{q}_2 m_2 l_{c,2}^2 \dot{q}_3 s_{22} - \tau_1 + \ddot{q}_1 m_2 l_{c,2} l_1 c_3 \\
&+ \dot{q}_2\dot{q}_1 m_2 l_{c,2} l_1 s_{3.22} - \dot{q}_2\dot{q}_1 m_2 l_{c,2} l_1 s_{223} - \dot{q}_2\dot{q}_1 m_2 l_{c,2}^2 s_{22} \\
&+ 0.5\left(m_2\dot{q}_2 l_1 \dot{q}_3 l_{c,2} s_{3.22} - m_2\dot{q}_2 l_1\dot{q}_3 l_{c,2} s_{223} + g m_2 l_1 s_{12}\right. \\
&- m_2 l_1 \dot{q}_3^2 l_{c,2} s_3 + g m_2 l_{c,2} s_{123} + g m_2 l_{c,2} s_{13.2} \\
&+ \ddot{q}_1 m_2 l_{c,2} l_1 c_{3.22} + \ddot{q}_1 m_2 l_{c,2} l_1 c_{223} + g m_2 l_1 s_{1.2} \\
&+ \ddot{q}_1 m_2 l_{c,2}^2 + \ddot{q}_1 m_2 l_{c,2}^2 c_{22} + \ddot{q}_3 m_2 l_{c,2} l_1 c_3 + \ddot{q}_3 m_2 l_{c,2}^2 \\
&+ \ddot{q}_3 m_2 l_{c,2}^2 c_{22} + g m_1 l_{c,1} s_{1.2} + g m_1 l_{c,1} s_{12} + \ddot{q}_1 m_1 l_{c,1}^2 \\
&+ \ddot{q}_1 m_1 l_{c,1}^2 c_{22} + \ddot{q}_1 m_2 l_1^2 + \ddot{q}_1 m_2 l_1^2 c_{22} - \dot{q}_3\dot{q}_1 m_2 l_{c,2} l_1 s_{223} \\
&\left.- \dot{q}_3\dot{q}_1 m_2 l_{c,2} l_1 s_{3.22}\right) + 0.25\left(\ddot{q}_3 m_2 l_{c,2} l_1 c_{3.22}\right. \\
&\left.- m_2 l_1 \dot{q}_3^2 l_{c,2} s_{223} - m_2 l_1 \dot{q}_3^2 l_{c,2} s_{3.22} + \ddot{q}_3 m_2 l_{c,2} l_1 c_{223}\right) = 0
\end{aligned}
$$

$$
\begin{aligned}
&- \tau_2 + \ddot{q}_2 m_1 l_{c,1}^2 + \ddot{q}_2 m_2 l_1^2 + \ddot{q}_2 m_2 l_1 l_{c,2} c_{22} \\
&- m_2 \dot{q}_2^2 l_1 l_{c,2} s_{22} + \ddot{q}_2 m_2 l_1 l_{c,2} + \dot{q}_1 m_2 l_{c,2}^2 \dot{q}_3 s_{22} \\
&+ \ddot{q}_2 I_2 + \ddot{q}_2 I_1 + \ddot{q}_2 m_2 l_1 l_{c,2} c_3 - m_2 \dot{q}_2 l_1 \dot{q}_3 l_{c,2} s_3 \\
&+ \ddot{q}_2 m_2 l_{c,2}^2 + 0.5\left(m_2 \dot{q}_3^2 l_{c,2}^2 s_{22} + m_2 \dot{q}_2^2 l_1 l_{c,2} s_{223}\right. \\
&- \dot{q}_1^2 m_2 l_{c,2} l_1 s_{3.22} + \dot{q}_1^2 m_2 l_{c,2} l_1 s_{223} - \ddot{q}_2 m_2 l_1 l_{c,2} c_{3.22} \\
&+ \dot{q}_1^2 m_2 l_1^2 s_{22} - \ddot{q}_2 m_2 l_1 l_{c,2} c_{223} + m_2 \dot{q}_2 l_1 \dot{q}_3 l_{c,2} s_{3.22} \\
&+ m_2 \dot{q}_2 l_1 \dot{q}_3 l_{c,2} s_{223} + \dot{q}_1^2 m_2 l_{c,2}^2 s_{22} + g m_2 l_{c,2} s_{123} \\
&- g m_2 l_{c,2} s_{13.2} + g m_2 l_1 s_{12} - g m_2 l_1 s_{1.2} + \dot{q}_1^2 m_1 l_{c,1}^2 s_{22} \\
&- g m_1 l_{c,1} s_{1.2} + g m_1 l_{c,1} s_{12} + \dot{q}_3 \dot{q}_1 m_2 l_{c,2} l_1 s_{223} \\
&\left.- \dot{q}_3 \dot{q}_1 m_2 l_{c,2} l_1 s_{3.22} - m_2 \dot{q}_2^2 l_1 l_{c,2} s_{3.22}\right) \\
&+ 0.25\left(m_2 l_1 \dot{q}_3^2 l_{c,2} s_{223} - m_2 l_1 \dot{q}_3^2 l_{c,2} s_{3.22}\right. \\
&\left.+ \ddot{q}_3 m_2 l_{c,2} l_1 c_{3.22} - \ddot{q}_3 m_2 l_{c,2} l_1 c_{223}\right) = 0
\end{aligned}
$$

$$
\begin{aligned}
&\ddot{q}_1 I_2 + \ddot{q}_3 I_2 - \dot{q}_2 m_2 l_{c,2}^2 \dot{q}_3 s_{22} - \tau_3 - \dot{q}_2 \dot{q}_1 m_2 l_{c,2}^2 s_{22} \\
&+ 0.5\left(m_2 \dot{q}_2^2 l_1 l_{c,2} s_3 + \dot{q}_1^2 m_2 l_{c,2} l_1 s_3 + g m_2 l_{c,2} s_{123}\right) \\
&+ 0.5\left(g m_2 l_{c,2} s_{13.2} + \ddot{q}_1 m_2 l_{c,2}^2 + \ddot{q}_1 m_2 l_{c,2}^2 c_{22}\right. \\
&+ \ddot{q}_1 m_2 l_{c,2} l_1 c_3 + \dot{q}_2 \dot{q}_1 m_2 l_{c,2} l_1 s_{3.22} - \dot{q}_2 \dot{q}_1 m_2 l_{c,2} l_1 s_{223} \\
&\left.+ \ddot{q}_3 m_2 l_{c,2}^2 + \ddot{q}_3 m_2 l_{c,2}^2 c_{22}\right) + 0.25\left(m_2 \dot{q}_2^2 l_1 l_{c,2} s_{3.22}\right. \\
&+ m_2 \dot{q}_2^2 l_1 l_{c,2} s_{223} + \dot{q}_1^2 m_2 l_{c,2} l_1 s_{3.22} + \dot{q}_1^2 m_2 l_{c,2} l_1 s_{223} \\
&+ \ddot{q}_2 m_2 l_1 l_{c,2} c_{3.22} - \ddot{q}_2 m_2 l_1 l_{c,2} c_{223} + \ddot{q}_1 m_2 l_{c,2} l_1 c_{3.22} \\
&\left.+ \ddot{q}_1 m_2 l_{c,2} l_1 c_{223}\right) = 0
\end{aligned}
\tag{46}
$$

where: $s_{ij} = sin(q_i - q_j)$; $c_{ij} = cos(q_i - q_j)$ and $s_{ijk} = sin(q_i + q_j + q_k)$; $c_{ijk} = cos(q_i + q_j + q_k)$ and $s_{i.jk} = sin(q_i - q_j - q_k)$; $c_{i.jk} = cos(q_i - q_j - q_k)$ and $s_{ij.k} = sin(q_i + q_j - q_k)$; $c_{ij.k} = cos(q_i + q_j - q_k)$.

Generalized equation of motion. We obtained the generalized equation of motion (25) by conversion of equations (46) with corresponding coefficients:

$$M(\mathbf{q}) = \begin{bmatrix} M_{11} & 0 & M_{13} \\ 0 & M_{22} & M_{23} \\ M_{31} & M_{32} & M_{33} \end{bmatrix} \tag{47}$$

where:

$$\begin{aligned}
M_{11} &= m_1 l_{c,1}^2 c_2^2 + I_2 + m_2 l_1^2 c_2^2 + m_2 l_{c,2}^2 c_2^2 + I_1 \\
&\quad + 2 l_{c,2} \left(m_2 l_1 s_1 c_2^2 s_{13} + m_2 l_1 c_1 c_2^2 c_{13} \right) \\
M_{13} &= M_{31} = I_2 + m_2 l_1 c_1 c_2^2 l_{c,2} c_{13} + m_2 l_{c,2}^2 c_2^2 \\
&\quad + m_2 l_1 s_1 c_2^2 l_{c,2} s_{13} \\
M_{22} &= m_2 l_{c,2}^2 + m_1 l_{c,1}^2 + I_1 + I_2 + m_2 l_1^2 \\
&\quad + 2 l_{c,2} \left(m_2 l_1 s_1 s_{13} - m_2 l_1 c_1 c_2^2 c_{13} + m_2 l_1 c_2^2 \right. \\
&\quad \left. + m_2 l_1 c_1 c_{13} - m_2 l_1 s_1 c_2^2 s_{13} \right) \\
M_{23} &= M_{32} = -m_2 l_1 s_1 s_2 l_{c,2} c_{13} c_2 + m_2 l_1 c_1 s_2 l_{c,2} s_{13} c_2 \\
M_{33} &= I_2 + m_2 c_2^2 l_{c,2}^2
\end{aligned}$$

$$C(\mathbf{q}, \dot{\mathbf{q}}) = \begin{bmatrix} C_1 \\ C_2 \\ C_3 \end{bmatrix} \tag{48}$$

where:

$$\begin{aligned}
C_1 &= \dot{q}_2 \dot{q}_1 m_2 l_{c,2} l_1 s_{3.22} - \dot{q}_2 \dot{q}_1 m_2 l_1^2 s_{22} - \dot{q}_2 \dot{q}_1 m_1 l_{c,1}^2 s_{22} \\
&\quad - \dot{q}_3 \dot{q}_1 m_2 l_{c,2} l_1 s_3 - \dot{q}_2 m_2 l_{c,2}^2 \dot{q}_3 s_{22} - \dot{q}_2 \dot{q}_1 m_2 l_{c,2} l_1 s_{223} \\
&\quad - \dot{q}_2 \dot{q}_1 m_2 l_{c,2}^2 s_{22} + 0.5 \left(m_2 \dot{q}_2 l_1 \dot{q}_3 l_{c,2} s_{3.22} \right. \\
&\quad - m_2 l_1 \dot{q}_3^2 l_{c,2} s_3 - m_2 \dot{q}_2 l_1 \dot{q}_3 l_{c,2} s_{223} - \dot{q}_3 \dot{q}_1 m_2 l_{c,2} l_1 s_{223} \\
&\quad \left. - \dot{q}_3 \dot{q}_1 m_2 l_{c,2} l_1 s_{3.22} \right) + 0.25 \left(-m_2 l_1 \dot{q}_3^2 l_{c,2} s_{23} \right. \\
&\quad \left. - m_2 l_1 \dot{q}_3^2 l_{c,2} s_{3.22} \right) \\
C_2 &= \dot{q}_1 m_2 l_{c,2}^2 \dot{q}_3 s_{22} - m_2 \dot{q}_2^2 l_1 l_{c,2} s_{22} - m_2 \dot{q}_2 l_1 \dot{q}_3 l_{c,2} s_3 \\
&\quad + 0.5 \left(m_2 \dot{q}_3^2 l_{c,2}^2 s_{22} + m_2 \dot{q}_2^2 l_1 l_{c,2} s_{223} - m_2 \dot{q}_2^2 l_1 l_{c,2} s_{3.22} \right. \\
&\quad - \dot{q}_1^2 m_2 l_{c,2} l_1 s_{3.22} + \dot{q}_1^2 m_2 l_{c,2} l_1 s_{223} + \dot{q}_1^2 m_2 l_1^2 s_{22} \\
&\quad + m_2 \dot{q}_2 l_1 \dot{q}_3 l_{c,2} s_{3.22} + m_2 \dot{q}_2 l_1 \dot{q}_3 l_{c,2} s_{223} + \dot{q}_1^2 m_2 l_{c,2}^2 s_{22} \\
&\quad \left. + \dot{q}_1^2 m_1 l_{c,1}^2 s_{22} + \dot{q}_3 \dot{q}_1 m_2 l_{c,2} l_1 s_{223} - \dot{q}_3 \dot{q}_1 m_2 l_{c,2} l_1 s_{3.22} \right) \\
&\quad + 0.25 \left(m_2 l_1 \dot{q}_3^2 l_{c,2} s_{223} - m_2 l_1 \dot{q}_3^2 l_{c,2} s_{3.22} \right) \\
C_3 &= -\dot{q}_2 \dot{q}_1 m_2 l_{c,2}^2 s_{22} - \dot{q}_2 m_2 l_{c,2}^2 \dot{q}_3 s_{22} \\
&\quad + 0.5 \left(\dot{q}_1^2 m_2 l_{c,2} l_1 s_3 + m_2 \dot{q}_2^2 l_1 l_{c,2} s_3 - \dot{q}_2 \dot{q}_1 m_2 l_{c,2} l_1 s_{223} \right.
\end{aligned}$$

$$+\dot{q}_2\dot{q}_1 m_2 l_{c,2} l_1 s_{3.22}\big) + 0.25\left(\dot{q}_1^2 m_2 l_{c,2} l_1 s_{223} + \dot{q}_1^2 m_2 l_{c,2} l_1 s_{3.22} + m_2\dot{q}_2^2 l_1 l_{c,2} s_{223} + m_2\dot{q}_2^2 l_1 l_{c,2} s_{3.22}\right)$$

$$G(\mathbf{q}) = \begin{bmatrix} gc_2(m_1 l_{c,1} s_1 + m_2 l_1 s_1 + m_2 l_{c,2} s_{13}) \\ gs_2(m_1 l_{c,1} c_1 + m_2 l_1 c_1 + m_2 l_{c,2} c_{13}) \\ gc_2 m_2 l_{c,2} s_{13} \end{bmatrix} \quad (49)$$

6 Non Linear Model of Arm

The solution of the above system of equations of motion allows to define a non-linear model of the arm. However, due to the high complexity of solving of equations (46) only the general form is presented:

$$\begin{bmatrix} \ddot{q}_1 \\ \ddot{q}_2 \\ \ddot{q}_3 \end{bmatrix} = M(\mathbf{q})^{-1}\left(\begin{bmatrix} \tau_1 \\ \tau_2 \\ \tau_3 \end{bmatrix} - C(\mathbf{q}, \dot{\mathbf{q}}) - G(\mathbf{q})\right) \quad (50)$$

in which matrices $M(\mathbf{q})$, $C(\mathbf{q}, \dot{\mathbf{q}})$ and $G(\mathbf{q})$ are described by the formulas (47), (48) and (48).

7 Linear Model of Arm

Linearised equation of motion (45) takes the form of (33) with matrices as below:

$$\mathbf{A}_0 = \begin{bmatrix} A_{11} & A_{12} & A_{13} \\ A_{21} & A_{22} & A_{23} \\ A_{31} & A_{32} & A_{33} \end{bmatrix} \quad (51)$$

where:

$$A_{11} = gc_2\left(m_2 l_{c,2} c_1 c_3 - m_2 l_{c,2} s_1 s_3 + m_2 l_1 c_1 + l_{c,1} m_1 c_1\right)$$

$$\begin{aligned} A_{12} = {} & g(-m_2 l_{c,2} c_1 s_3 s_2 - m_2 l_{c,2} s_1 c_3 s_2 - m_2 l_1 s_1 s_2 \\ & - l_{c,1} m_1 s_1 s_2) + 2\left(m_2 l_1 l_{c,2}\dot{q}_3^2 s_3 s_2 c_2 + \dot{q}_2\dot{q}_1 m_1 l_{c,1}^2\right. \\ & - \ddot{q}_3 m_2 l_{c,2} l_1 c_3 s_2 c_2 + \dot{q}_2\dot{q}_1 m_2 l_{c,2}^2 - \ddot{q}_3 m_2 l_{c,2}^2 s_2 c_2 \\ & - \ddot{q}_1 m_2 l_{c,2}^2 s_2 c_2 - \ddot{q}_1 m_1 l_{c,1}^2 s_2 c_2 - \ddot{q}_1 m_2 l_1^2 s_2 c_2 \\ & \left. + m_2 l_{c,2}\dot{q}_2 l_1\dot{q}_3 c_3 + \dot{q}_2\dot{q}_1 m_2 l_1^2 + \dot{q}_2 m_2 l_{c,2}^2\dot{q}_3\right) \\ & + 4\left(\dot{q}_3\dot{q}_1 m_2 l_{c,2} l_1 s_3 s_2 c_2 - m_2\dot{q}_2 l_1 l_{c,2}\dot{q}_3 c_3 c_2^2\right. \\ & - \dot{q}_2\dot{q}_1 m_1 l_{c,1}^2 c_2^2 - \dot{q}_2\dot{q}_1 m_2 l_{c,2}^2 c_2^2 - \ddot{q}_1 m_2 l_{c,2} l_1 c_3 s_2 c_2 \\ & \left. + \dot{q}_2\dot{q}_1 m_2 l_{c,2} l_1 c_3 - \dot{q}_2\dot{q}_1 m_2 l_1^2 c_2^2 - \dot{q}_2 m_2 l_{c,2}^2\dot{q}_3 c_2^2\right) \\ & - 8\dot{q}_2\dot{q}_1 m_2 l_{c,2} l_1 c_3 c_2^2 \end{aligned}$$

$$
\begin{aligned}
A_{13} &= -m_2 l_{c,2} c_2 (l_1 \dot{q}_3^2 c_3 c_2 + \ddot{q}_3 l_1 s_3 c_2 - 2\dot{q}_2 l_1 \dot{q}_3 s_3 s_2 \\
&\quad + 2\dot{q}_3 \dot{q}_1 l_1 c_3 c_2 - 4\dot{q}_2 \dot{q}_1 l_1 s_3 s_2 - g c_1 c_3 + g s_1 s_3 \\
&\quad + 2\ddot{q}_1 l_1 s_3 c_2) \\
A_{21} &= -g s_2 \left(m_2 l_{c,2} c_1 s_3 + m_2 l_{c,2} s_1 c_3 + m_2 l_1 s_1 + l_{c,1} m_1 s_1 \right) \\
A_{22} &= g m_2 l_1 c_1 c_2 + g l_{c,1} m_1 c_1 c_2 + g m_2 l_{c,2} c_1 c_3 c_2 \\
&\quad - g m_2 l_{c,2} s_1 s_3 c_2 - m_2 l_{c,2} l_1 \dot{q}_3^2 c_3 - m_2 l_{c,2} \ddot{q}_3 l_1 s_3 \\
&\quad - \dot{q}_1^2 m_2 l_{c,2}^2 - m_2 \dot{q}_3^2 l_{c,2}^2 - \dot{q}_1^2 m_2 l_1^2 - \dot{q}_1^2 m_1 l_{c,1}^2 \\
&\quad + 2 \Big(\ddot{q}_3 m_2 l_{c,2} l_1 s_3 c_2^2 - m_2 l_{c,2} \dot{q}_3 \dot{q}_1 l_1 c_3 + \dot{q}_1^2 m_1 l_{c,1}^2 c_2^2 \\
&\quad + m_2 l_1 l_{c,2} \dot{q}_3^2 c_3 c_2^2 + m_2 \dot{q}_2^2 l_1 l_{c,2} + \dot{q}_1^2 m_2 l_1^2 c_2^2 \\
&\quad + \dot{q}_1^2 m_2 l_{c,2}^2 c_2^2 + m_2 \dot{q}_3^2 l_{c,2}^2 c_2^2 - \dot{q}_1 m_2 l_{c,2}^2 \dot{q}_3 \\
&\quad - m_2 l_{c,2} \dot{q}_1^2 l_1 c_3 - m_2 l_{c,2} \dot{q}_2^2 l_1 c_3 \Big) \\
&\quad + 4 \left(\dot{q}_3 \dot{q}_1 m_2 l_{c,2} l_1 c_3 c_2^2 - m_2 \dot{q}_2 l_1 l_{c,2} \dot{q}_3 s_3 s_2 c_2 \right. \\
&\quad + \dot{q}_1 m_2 l_{c,2}^2 \dot{q}_3 c_2^2 + m_2 \dot{q}_2^2 l_1 l_{c,2} c_3 c_2^2 + \dot{q}_1^2 m_2 l_{c,2} l_1 c_3 c_2^2 \\
&\quad \left. + \ddot{q}_2 m_2 l_1 l_{c,2} c_3 s_2 c_2 - \ddot{q}_2 m_2 l_1 l_{c,2} s_2 c_2 - m_2 \dot{q}_2^2 l_1 l_{c,2} c_2^2 \right) \\
A_{23} &= m_2 l_{c,2} \left(-2\ddot{q}_2 l_1 s_3 - 2\dot{q}_2 l_1 \dot{q}_3 c_3 - l_1 \dot{q}_3^2 s_3 s_2 c_2 \right. \\
&\quad + \ddot{q}_3 l_1 c_3 s_2 c_2 + 2\dot{q}_2 l_1 \dot{q}_3 c_3 c_2^2 - 2\dot{q}_3 \dot{q}_1 l_1 s_3 s_2 c_2 \\
&\quad - g s_1 c_3 s_2 - g c_1 s_3 s_2 + 2\ddot{q}_2 l_1 s_3 c_2^2 - 2\dot{q}_1^2 l_1 s_3 s_2 c_2 \\
&\quad \left. - 2\dot{q}_2^2 l_1 s_3 s_2 c_2 \right) \\
A_{31} &= g m_2 l_{c,2} c_2 (c_1 c_3 - s_1 s_3) \\
A_{32} &= -m_2 l_{c,2} (\ddot{q}_2 l_1 s_3 + 4\dot{q}_2 \dot{q}_1 l_1 c_3 c_2^2 - 2\dot{q}_2 \dot{q}_1 l_1 c_3 \\
&\quad + g s_1 c_3 s_2 + g c_1 s_3 s_2 + 2\ddot{q}_3 l_{c,2} s_2 c_2 + 2\ddot{q}_1 l_1 c_3 s_2 c_2 \\
&\quad - 2\ddot{q}_2 l_1 s_3 c_2^2 + 2\dot{q}_1^2 l_1 s_3 s_2 c_2 + 2\ddot{q}_1 l_{c,2} s_2 c_2 - 2\dot{q}_2 l_{c,2} \dot{q}_3 \\
&\quad + 4\dot{q}_2 l_{c,2} \dot{q}_3 c_2^2 - 2\dot{q}_2 \dot{q}_1 l_{c,2} + 2\dot{q}_2^2 l_1 s_3 s_2 c_2 + 4\dot{q}_2 \dot{q}_1 l_{c,2} c_2^2) \\
A_{33} &= m_2 l_{c,2} c_2 \left(2\dot{q}_2 \dot{q}_1 l_1 s_3 s_2 + g c_1 c_3 - g s_1 s_3 + \right. \\
&\quad \left. - \ddot{q}_1 l_1 s_3 c_2 + \ddot{q}_2 l_1 c_3 s_2 + \dot{q}_1^2 l_1 c_3 c_2 + \dot{q}_2^2 l_1 c_3 c_2 \right)
\end{aligned}
$$

$$
\mathbf{B}_0 = \begin{bmatrix} B_{11} & B_{12} & B_{13} \\ B_{21} & B_{22} & B_{23} \\ B_{31} & B_{32} & B_{33} \end{bmatrix} \tag{52}
$$

where:

$$
\begin{aligned}
B_{11} &= -2c_2 \Big(\dot{q}_2 m_1 l_{c,1}^2 s_2 + \dot{q}_3 m_2 l_{c,2} l_1 s_3 c_2 \\
&\quad + 2\dot{q}_2 m_2 l_{c,2} l_1 c_3 s_2 + \dot{q}_2 m_2 l_1^2 s_2 + \dot{q}_2 m_2 l_{c,2}^2 s_2 \Big) \\
B_{12} &= -B_{21} = -2 s_2 c_2 \Big(\dot{q}_1 m_2 l_{c,2}^2 + 2\dot{q}_1 m_2 l_{c,2} l_1 c_3 \\
&\quad + m_2 l_{c,2}^2 \dot{q}_3 + \dot{q}_3 m_2 l_{c,2} l_1 c_3 + \dot{q}_1 m_2 l_1^2 + \dot{q}_1 m_1 l_{c,1}^2 \Big) \\
B_{13} &= -2 m_2 l_{c,2} c_2 \left(\dot{q}_3 l_1 s_3 c_2 + \dot{q}_2 l_1 c_3 s_2 + \dot{q}_1 l_1 s_3 c_2 \right. \\
&\quad \left. + \dot{q}_2 l_{c,2} s_2 \right) \\
B_{22} &= 2 m_2 l_{c,2} l_1 \left(\dot{q}_3 s_3 c_2^2 - \dot{q}_3 s_3 + 2\dot{q}_2 c_3 s_2 c_2 - 2\dot{q}_2 s_2 c_2 \right) \\
B_{23} &= 2 m_2 l_{c,2} \left(\dot{q}_3 l_1 c_3 s_2 c_2 + \dot{q}_2 l_1 s_3 c_2^2 \right.
\end{aligned}
$$

$$- \dot{q}_2 l_1 s_3 + \dot{q}_1 l_{c,2} s_2 c_2 + l_{c,2} \dot{q}_3 s_2 c_2 + \dot{q}_1 l_1 c_3 s_2 c_2)$$
$$B_{31} = -2m_2 l_{c,2} c_2 \left(-\dot{q}_1 l_1 s_3 c_2 + \dot{q}_2 l_1 c_3 s_2 + \dot{q}_2 l_{c,2} s_2\right)$$
$$B_{32} = -2m_2 l_{c,2} c_2 \left(\dot{q}_1 l_1 c_3 s_2 - \dot{q}_2 l_1 s_3 c_2 + l_{c,2} \dot{q}_3 s_2 + \dot{q}_1 l_{c,2} s_2\right)$$
$$B_{33} = -2\dot{q}_2 m_2 l_{c,2}^2 s_2 c_2$$

$$\mathbf{C}_0 = \begin{bmatrix} C_{11} & 0 & C_{13} \\ 0 & C_{22} & C_{23} \\ C_{31} & C_{32} & C_{33} \end{bmatrix} \tag{53}$$

where:

$$C_{11} = 2m_2 l_1 l_{c,2} c_3 c_2^2 + I_1 + I_2 + m_2 l_1^2 c_2^2 + m_1 l_{c,1}^2 c_2^2 + m_2 l_{c,2}^2 c_2^2$$
$$C_{13} = C_{31} = m_2 l_1 l_{c,2} c_3 c_2^2 + I_2 + m_2 l_{c,2}^2 c_2^2$$
$$C_{22} = I_1 + 2m_2 l_1 l_{c,2} c_2^2 + I_2 + m_2 l_{c,2}^2 + m_2 l_1^2 + 2m_2 l_{c,2} l_1 c_3 + m_1 l_{c,1}^2 - 2m_2 l_1 l_{c,2} c_3 c_2^2$$
$$C_{23} = C_{32} = m_2 l_1 l_{c,2} s_3 s_2 c_2$$
$$C_{33} = I_2 + m_2 l_{c,2}^2 c_2^2.$$

However due to the high complexity of the matrix equations of state changes and outputs only their general form are presented:

$$\mathbf{A} = \begin{bmatrix} \mathbf{0} \; \mathbf{I} \\ \mathbf{C} \end{bmatrix} \qquad \mathbf{B} = \begin{bmatrix} \mathbf{0} \\ \mathbf{D} \end{bmatrix}$$
$$\mathbf{C} = \left[-\mathbf{C}_0^{-1}\mathbf{A}_0 \; -\mathbf{C}_0^{-1}\mathbf{B}_0\right] \; \mathbf{D} = \mathbf{C}_0^{-1} \tag{54}$$

8 Results of the Simulation

For the presented model of the human arm simulation analysis of the properties of non-linear model and the corresponding linear model was carried out.

In order to examine the dynamic properties of the models information of the physical features of the individual parts of the human arm was used. According to [20] representative anatomical and physical value of the arm are shown in Table 2. They contain the average values of the arm for an adult male with a height of 1.829 [m] and a weight of 90.7 [kg]. Based on Table 2 parameter values of individual models of human arm were defined. These parameters are used in simulation studies are presented in Table 3.

As an experiment, showing the behavior of linear and non-linear system with different initial conditions. For first experiment presented on Figs. 2 and 3 initial conditions was—free movement $\boldsymbol{\tau} = \mathbf{0}$, operating point $\mathbf{q} = \mathbf{0}, \dot{\mathbf{q}} = \mathbf{0}, \ddot{\mathbf{q}} = \mathbf{0}$, starting point $\mathbf{q} = [30, -45, 15]^T$ [deg] and $\dot{\mathbf{q}} = \mathbf{0}, \ddot{\mathbf{q}} = \mathbf{0}$. Second experiment presented on Figs. 4 and 5 initial conditions was taken from linear model $\boldsymbol{\tau} = [9, -12, 3]^T$, operating

Table 2 Anatomical physical parameters of adult human arm

	Male		
	Upper arm	Forearm	Hand
Length [m]	0.315	0.287	0.105
Weigth [N]	28.91	16.64	5.78
Mass [kg]	2.947	1.696	0.589
CoM [m]	0.137	0.127	0.049
MoI [kgm^2]	0.0303	0.0128	0.0005

Table 3 The numerical values of model arm parameters

	2L-3DoF
l_1 [m]	0.315
$l_{c,1}$ [m]	0.137
m_1 [kg]	2.947
I_1 [kgm^2]	0.0303
l_2 [m]	0.392
$l_{c,2}$ [m]	0.176
m_2 [kg]	2.285
I_2 [kgm^2]	0.0133

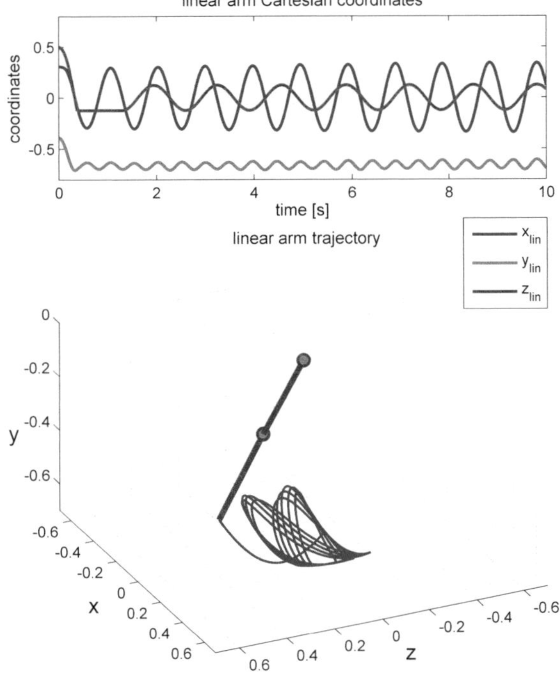

Fig. 2 Linear model—free movement

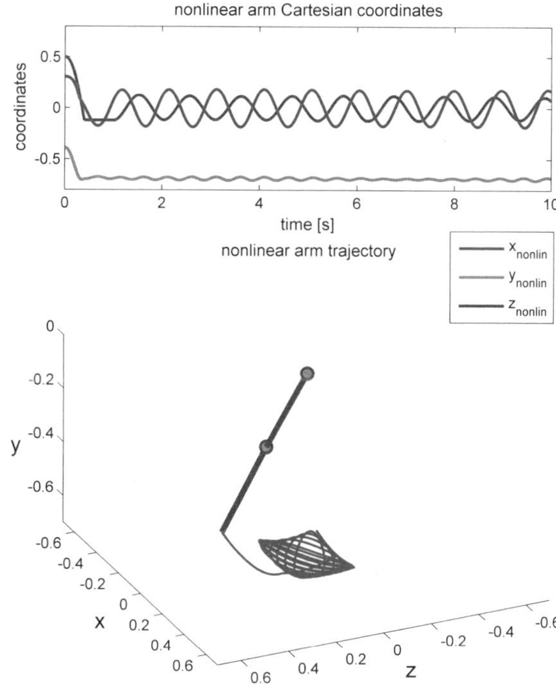

Fig. 3 Non-linear model—free movement

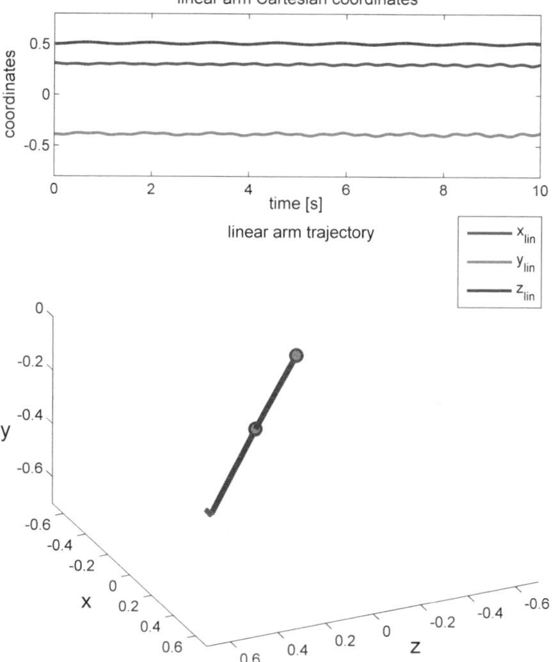

Fig. 4 Linear model—$\boldsymbol{\tau} = [9, -12, 3]^T$

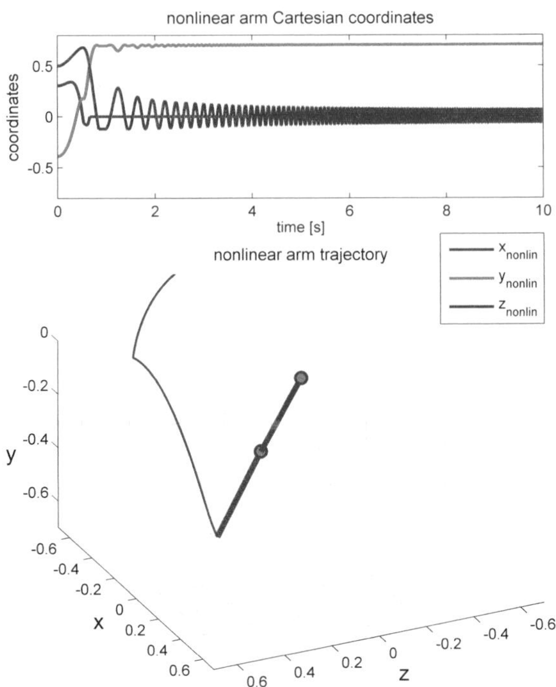

Fig. 5 Non-linear model—$\boldsymbol{\tau} = [9, -12, 3]^T$

point $\mathbf{q} = \mathbf{0}, \dot{\mathbf{q}} = \mathbf{0}, \ddot{\mathbf{q}} = \mathbf{0}$, starting point $\mathbf{q} = [30, -45, 15]^T$ [deg] and $\dot{\mathbf{q}} = \mathbf{0}, \ddot{\mathbf{q}} = \mathbf{0}$. Third experiment presented on Figs. 6 and 7 initial conditions was—free movement $\boldsymbol{\tau} = \mathbf{0}$, operating point $\mathbf{q} = [30, -45, 15]^T$ [deg], $\dot{\mathbf{q}} = \mathbf{0}, \ddot{\mathbf{q}} = \mathbf{0}$, starting point $\mathbf{q} = [30, -45, 15]^T$ [deg] and $\dot{\mathbf{q}} = \mathbf{0}, \ddot{\mathbf{q}} = \mathbf{0}$. Fourth experiment presented on Figs. 8 and 9 initial conditions was taken from linear model $\boldsymbol{\tau} = [0, -3, 0]^T$, operating point $\mathbf{q} = [30, -45, 15]^T$ [deg], $\dot{\mathbf{q}} = \mathbf{0}, \ddot{\mathbf{q}} = \mathbf{0}$, starting point $\mathbf{q} = [30, -45, 15]^T$ [deg] and $\dot{\mathbf{q}} = \mathbf{0}, \ddot{\mathbf{q}} = \mathbf{0}$.

Fifth experiment presented on Figs. 10 and 11 initial conditions was—free movement $\boldsymbol{\tau} = \mathbf{0}$, operating point $\mathbf{q} = \mathbf{0}, \dot{\mathbf{q}} = \mathbf{0}, \ddot{\mathbf{q}} = \mathbf{0}$, starting point $\mathbf{q} = [-45, -1, 90]^T$ [deg] and $\dot{\mathbf{q}} = \mathbf{0}, \ddot{\mathbf{q}} = \mathbf{0}$.

Sixth experiment presented on Figs. 12 and 13 initial conditions was taken from linear model $\boldsymbol{\tau} = [-6, 0, -3]^T$, operating point $\mathbf{q} = \mathbf{0}, \dot{\mathbf{q}} = \mathbf{0}, \ddot{\mathbf{q}} = \mathbf{0}$, starting point $\mathbf{q} = [-45, -1, 90]^T$ [deg] and $\dot{\mathbf{q}} = \mathbf{0}, \ddot{\mathbf{q}} = \mathbf{0}$.

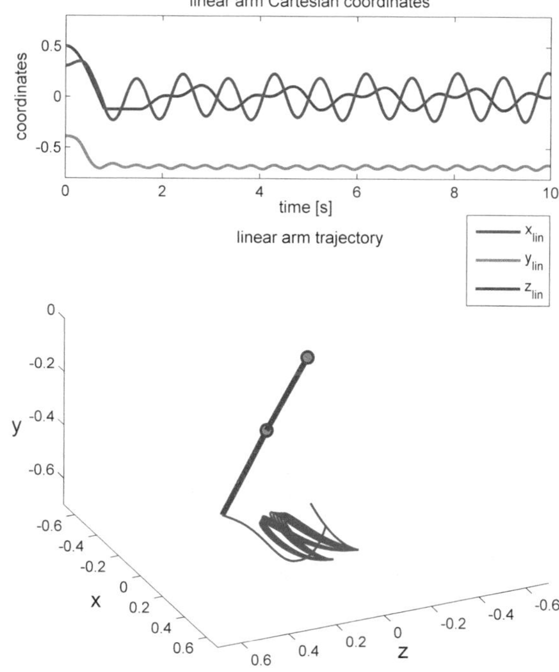

Fig. 6 Linear model—free movement

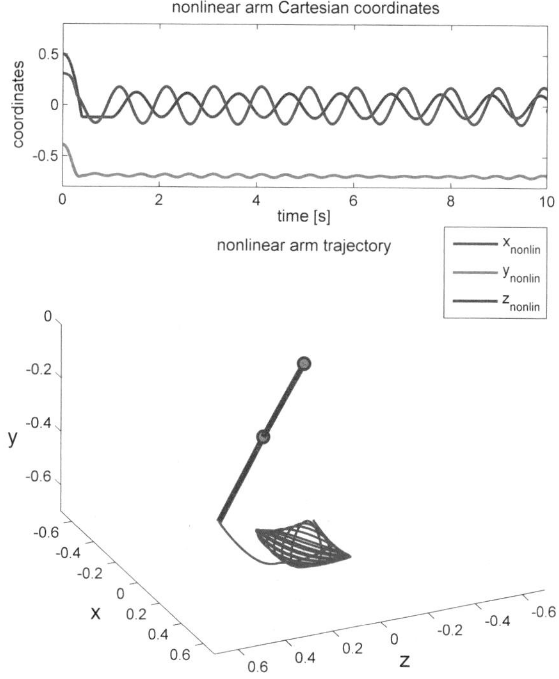

Fig. 7 Non-linear model—free movement

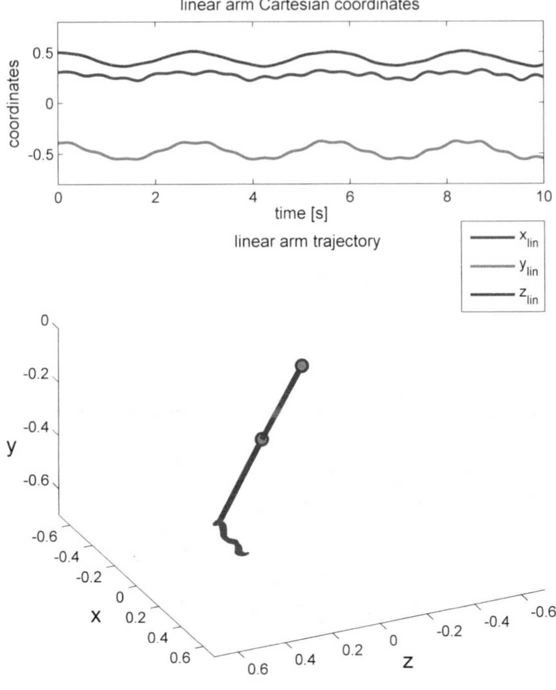

Fig. 8 Linear model—$\boldsymbol{\tau} = [0, -3, 0]^T$

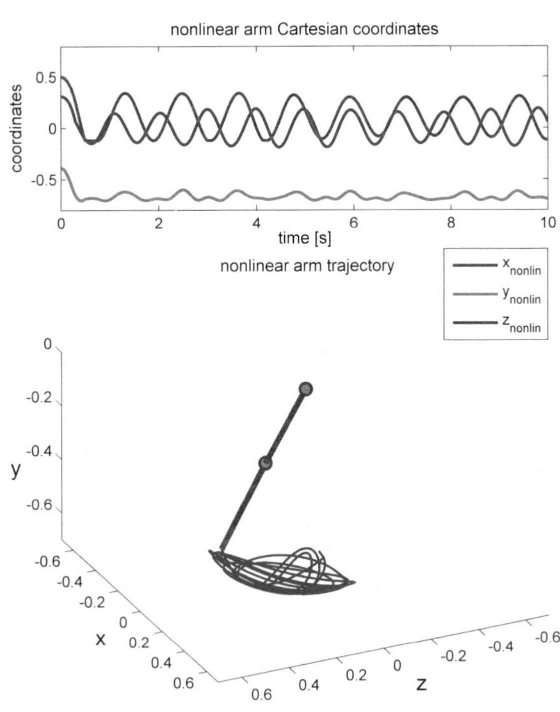

Fig. 9 Non-linear model—$\boldsymbol{\tau} = [0, -3, 0]^T$

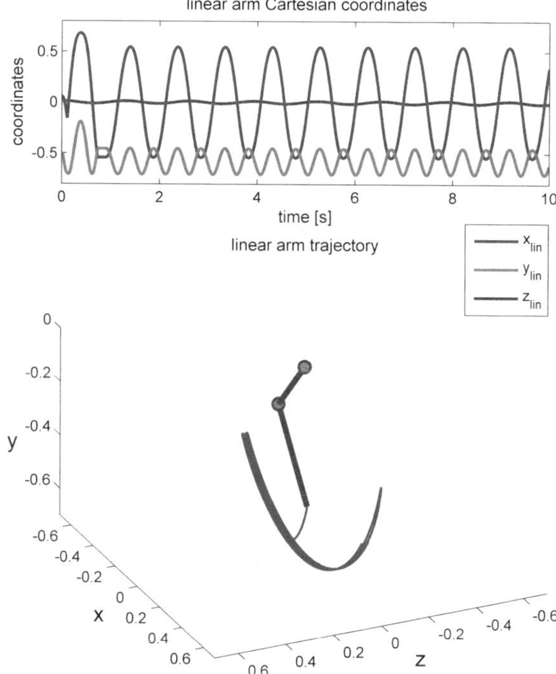

Fig. 10 Llinear model—free movement

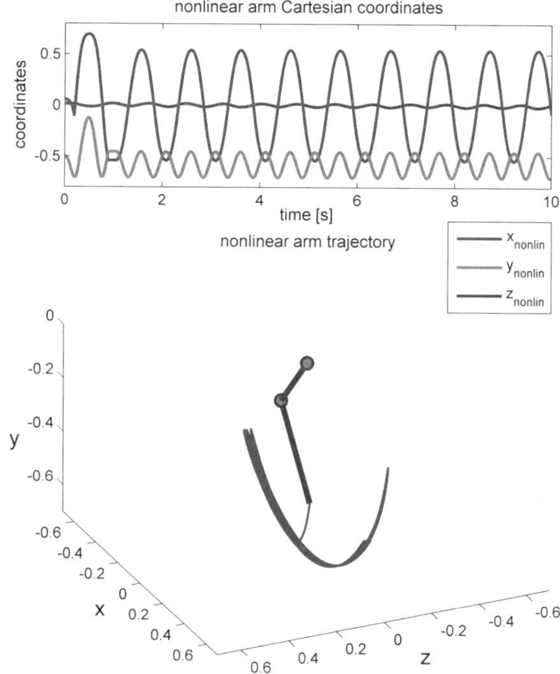

Fig. 11 Non-linear model—free movement

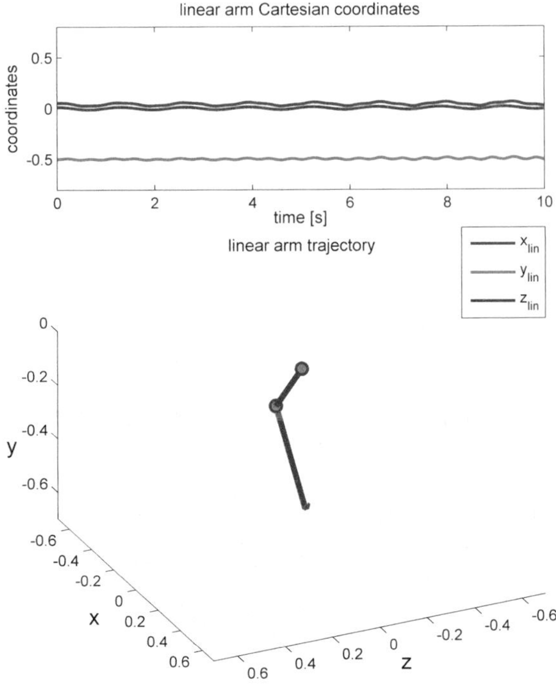

Fig. 12 Linear model—$\boldsymbol{\tau} = [-6, 0, -3]^T$

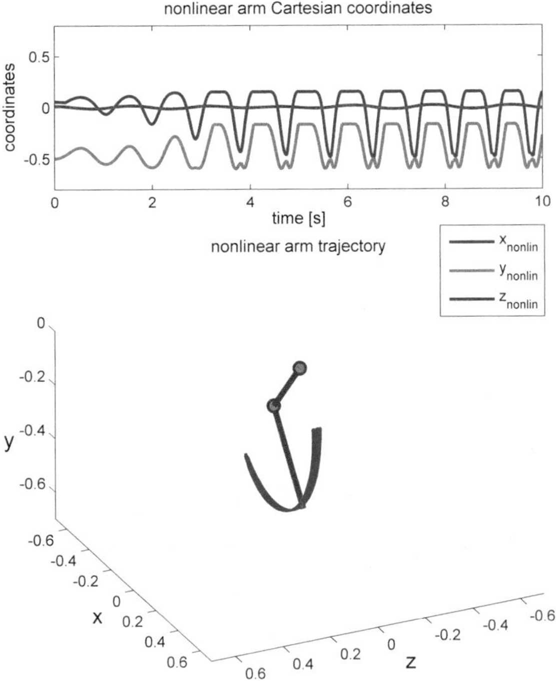

Fig. 13 Non-linear model—$\boldsymbol{\tau} = [-6, 0, -3]^T$

9 Summary

The results confirm the effectiveness of the use of linear models. This approach simplifies the synthesis of model. Properties of the resulting solutions show a high similarity in the behaviour of the non-linear model. They are retained even when operating point is strongly different than point of linearisation.

The simulations show the effectiveness of linear model of the dynamics of the human arm. Through the process of linearisation synthesis of linear dynamic model is reduced significantly compare to the non-linear model. At the same time their dynamic properties allow the analysis and synthesis of close-loop control systems.

In terms of future research authors aim to verify linear and non-linear models with using IMU sensors [23–26] and physical 2link-3DoF arm model.

References

1. Babiarz, A., Bieda, R., Jaskot, K., & Klamka, J. (2013). The dynamics of the human arm with an observer for the capture of body motion parameters. *Bulletin of the Polish Academy of Sciences: Technical Sciences, 61*(4), 955–971.
2. Parzych, M., Dabrowski, A., & Cetnarowicz, D. (2014). Aspects of microsoft kinect sensor application to servomotor control. *Bulletin of the Polish Academy of Sciences: Technical Sciences, 62*(3), 595–601.
3. Tändl, M., Stark, T., Erol, N. E., Löer, F., & Kecskeméthy, A. (2009). An object-oriented approach to simulating human gait motion based on motion tracking. *International Journal of Applied Mathematics and Computer Science, 19*(3), 469–483.
4. Mihelj, M. (2009). Human arm kinematics for robot based rehabilitation. *Robotica, 24*(3), 377–383.
5. Nagarsheth, H., Savsani, P. S., & Patel, M. (2008). Modeling and dynamics of human arm. In *IEEE International Conference on Automation Science and Engineering, 2008 (CASE 2008)*, Arlington, VA.
6. Switonski, A., Josinski, H., Jedrasiak, K., Polanski, A., & Wojciechowski, K. (2010). Classification of poses and movement phases. *Lecture Notes in Computer Science*.
7. Ryt, A., Sobel, D., Kwiatkowski, J., Domzal, M., Jedrasiak, K., & Nawrat, A. (2015). Real-time laser point tracking. *International Conference on Computer Vision and Graphics* (pp. 542–551).
8. Sobel, D., Jedrasiak, K., Daniec, K., Wrona, J., Jurgas, P., & Nawrat, A. (2014). Camera calibration for tracked vehicles augmented reality applications. *Innovative control systems for tracked vehicle platforms* (pp. 147–162).
9. Nawrat, A., & Jedrasiak, K. (2008). Fast colour recognition algorithm for robotics. *Problemy Eksploatacji* 69–76.
10. Daniec, K., Iwaneczko, P., Jedrasiak, K., & Nawrat, A. (2013). Prototyping the autonomous flight algorithms using the prepar3D® simulator. *Vision based systems for UAV applications* (pp. 219–232).
11. Wang, T. (2000). *Control force change due to adaptation of forward model in human motor control*. Master's thesis, The Johns Hopkins University Baltimore, Maryland.
12. Jankowski, K. (2011). *Dynamics of double pendulum with parametric vertical excitation*. Master's thesis, Technical Univesity of Lodz.
13. Driver, J., & Thorpe, D. (2004). *Design, build and control of a single/double rotational inverted pendulum*. Technical report, The University of Adelaide, Australia.

14. Tee, K. P., Burdet, E., Chew, C. M., & Milner, T. E. A model of force and impedance in human arm movements. In *Biological Cybernetics*, vol. 90, pp. 368–375, Springer.
15. Burdet, E., Tee, K. P., Mareels, I., Milner, T. E., Chew, C. M., Franklin, D., et al. (2006). Stability and motor adaptation in human arm movements. *Biological Cybernetics*, *94*(94), 20–32.
16. Grecu, V., Dumitru, N., & Grecu, L. (2009). Analysis of human arm joints and extension of the study to robot manipulator. In *Proceedings of the International MultiConference of Engineers and Computer Scientists*, vol. II, Hong Kong.
17. Omar, N. (2012). Mathematical model of the arms during Kayaking using gordon's method. *International Journal of Modern Physics: Conference Series*, *9*, 174–177.
18. Sanner, R. M., & Kosha, M. (1999). A mathematical model of the adaptive control of human arm motions. *Biological Cybernetics*, *80*, 369–382.
19. Li, W. (2006). *Optimal control for biological movement systems*. Ph.D. thesis, University of California, San Diego.
20. Williams, R. (2014). *Engineering biomechanics of human motion*. Technical report, Ohio University.
21. Klopčar, N., & Lenarčič, J. (2005). Kinematic model for determination of human arm reachable workspace. *Meccanica*, *40*(2), 203–219.
22. Moeslund, T. B. (2002). *Modeling the human arm*. Technical report, Laboratory of Computer Vison and Media Technology, Aalborg University, Denmark.
23. Jaskot, K., & Babiarz, A. (2010). The inertial measurement unit for detection of position. *Przegląd Elektrotechniczny*, *86*(11a), 323–333.
24. Bieda, R. (2013). Determining the IMU orientation in 3D space using tensor matrix rotation and non-stationary Kalman filter. *Przegląd Elektrotechniczny*, *89*(12), 68–78.
25. El-Gohary, M., & McNames, J. (2015). Human joint angle estimation with inertial sensors and validation with a robot arm. *IEEE Transactions on Biomedical Engineering*, *62*(7), 1759–1767.
26. Bieda, R., & Jaskot, K. (2016). Determinig of an object orientation in 3D space using direction cosine matrix and non-stationary Kalman filter. *Archives of Control Sciences*, *26*(2), 147–168.

Interactive Application Using Augmented Reality and Photogrammetric Scanning

Karol Jędrasiak, Natalia Hładczuk, Krzysztof Daniec and Aleksander Nawrat

1 Introduction

The aim of the article is to present augmented reality as an innovative technology used to create interactive applications. As part of the work there was created a computer game based on the three-dimensional image reconstruction technique. Although the application serves entertainment, its purpose is to present the next steps of the transition from virtual reality to the augmented reality. There were discussed issues related to the methodology of the reconstruction of the 3D models, image processing and interaction between the reconstructed objects and real objects [1–3]. The results were analyzed for possible future improvements and development.

Presented as part of the work is augmented reality technology. It is a transition state between virtual reality and the real world. Virtual Reality is created using computer technology and allows the user to immerse himself in the synthetic virtual world [4]. It can imitate reality, but does not contain its actual elements. The laws of physics such as gravity, time and properties of matter may, but may not be maintained [5]. Virtual reality is currently the most popularized the technology used to create the various simulators, computer games and as a tool for digital modeling [6–8].

Between virtual reality and the real world there is a spectrum of technologies called mixed reality. Depending on the number of real elements they can be classified as the world closer to real or virtual, but the distribution is often ambiguous. Technology, which is closest to the virtual reality is called augmented virtuality (AV). Environment is still synthetic, however certain elements of reality are added [9]. Applications classified as virtuality extended can contain static elements of reality as texture, and object models. They may also allow the user to manipulate virtual objects, e.g. using hands [10, 11]. An interesting example is the Swiss Birdly

K. Jędrasiak (✉) · N. Hładczuk · K. Daniec · A. Nawrat
Institute of Automatic Control, Silesian University of Technology,
Akademicka 16, 44-100 Gliwice, Poland
e-mail: karol.jedrasiak@polsl.pl

A. Nawrat et al. (eds.), *Advanced Technologies in Practical Applications for National Security*, Studies in Systems, Decision and Control 106,
https://doi.org/10.1007/978-3-319-64674-9_15

project [12], a flight simulator in which the user move across the virtual space by utilizing hand movements to move virtual wings attached. The system is equipped with a fan and a spring, which mimic the sensation of floating in the air. The whole scenery, however, is computer-generated.

The closest to the real world in terms of number of real elements is augmented reality. This technology is based on the co-existence of a real world and a three-dimensional virtual objects in the same space in real time [13]. The idea of augmented reality is to broaden the perception of the user and can also apply to other senses like touch, smell and hearing. An example would be a project Aura developed by Google in 2016, which with the help of bone conduction allow to send audio signals directly to hearing center of the user. Augmented reality is characterized by the inverse proportion between the number of virtual and real objects, than in the case of virtuality expanded.

There is no clearly visible boundary between these technologies, therefore it is difficult to assign many applications clearly to one of the classes. The concept of augmented reality is based on real-life scene with additional virtual objects, but the process can also be reversed. Technology acting similar to augmented reality in which the number of elements of the space is reduced, reduced is called Diminished Reality. This method allows you to not only to ignore, but also replace unwanted landscape elements with other [14]. The biggest problem when using this solution is appropriate approximation of the background color for the liquidation of the object, which must correspond to the rest of the scene [15]. Many applications utilize hybrid technology, combining the two aforementioned technologies, eliminating some irrelevant elements and expanding others.

Achieving of the effect of augmented reality in application requires the use of techniques for mapping real objects into their virtual, three-dimensional representation. There are several methods to accomplish this goal. One of the most accurate is laser scanning, based on measuring the distance between the pulses of light, and the elements of the object [16]. The advantage is the independence of the quality of the scan performed on the weather and lighting. This method, however, requires the imposition of texture on the restored objects or integrate the scanner with the camera still at the stage of shooting. Due to the equipment used, the cost of making three-dimensional scans is extremely high. In addition, this method does not allow to reconstruct for instance water surface, because the laser beams are not reflected by it.

Another approach to the reconstruction of the 3D scene is one of the techniques of photogrammetry-stereovision. Photogrammetry is an area dedicated to the reconstruction of the position, shape, orientation in space and the size of the objects based on images [17]. Stereovision is a method of mimicking the natural mechanism that occurs in a healthy human eye, or binocular vision (stereoscopic). It involves playing and apply for the position [18–20] of 3D objects while using at least two sources of perception [21]. Stereovision as a technique is more readily available and less expensive than laser scanning, and the models resulting from its use keep their original colors and textures. But the main reason that determines the choice of this method is that it is based on the equipment used in the technology of augmented reality.

The purpose of the stereovision techniques as part of the work was to obtain three-dimensional models for the application. On the basis of a series of images photographed objects were reconstructed and analyzed for detection and matching characteristic point clouds algorithms. The quality of the obtained models was compared and a proposal for the proper methodology of photographing objects according to their properties was made. The reconstructed three-dimensional models were then subjected to graphic processing including correction of the size, shape, texture and lighting. At the same time the real and fully virtual objects were modeled, and then was arcade video game based on the coexistence of both types of models was implemented. The transition mechanism between global coordinate systems and coordinates of existing objects in the virtual world was analyzed. Scripts that support dynamic properties, behavior and interactions between objects were implemented in the application. Both stereovision algorithms and the implemented code have been tested and analyzed for optimality, effectiveness.

Implemented application can be widely used in various types of systems and training simulators [22, 23]. Reconstructed using stereovision scenery allows the user to explore places difficult to reach or dangerous in the real world. An example of created the game, it could be a simulator to carry out anti-terrorist action. The advantages of such an application would have an opportunity to examine the topography of the area before carrying out the raid without arranging real exercise. In addition, it would allow training of different operational variants and evacuation scenarios.

Augmented reality is a technology with educational value. The background of created games, could serve as models of buildings of historical value, such as e.g. Acropolis. Player (student) could be moving in the reconstructed space, where he could see the antique buildings restored with the help of computer graphics. The idea of "learning by playing" would take place through user interactions with the figures of Greek gods, associated with particular temples and so on. The game might contain references to mythology, showing the attributes and stories of individual deities. The idea of application combining combines real scenes with virtual characters, would allow students to more easily associate and remember historical facts.

The above examples show the range of possible applications and potential applications based on 3D reconstruction technology and augmented reality.

2 Existing Solutions

The theme carried out in the framework of the work is to raise the issue of methods of 3D reconstruction in the context of augmented reality technology. Created game is to show stereovision, as a basic tool in the development of applications based on complete immersion of the real world and the virtual. Augmented reality is a technology assuming synchronization of real and virtual elements in real time. A simple example of a two-dimensional field can be a mechanism to detect faces while taking pictures with a digital camera. An essential element of a three-dimensional augmented reality are Head Mounted Displays (HMD). They allow the application of computer-generated elements, onto the actual scenery. Depending on the manner in which images are combined there can be distinguished HMD types: "optical see-through" and "video see-through" [24]. The first type of device allows the user to see the virtual objects placed directly on the actual scenery. Glasses HMD "video see-through" cut off the user from the visual stimuli from the outside world. Generating a scene based on the image recording by the camera located on the glasses [25]. This solution allows precise integrate virtual elements in the real scenery, eliminate sight of existing objects or modify their shapes. Glasses should not, however, be used when driving, or in the open, because the registration error scene could expose to the accident. Therefore, use of the glasses type HMD "video see-through" is still very limited. Augmented reality technology is currently used in many fields [26–28]. Medicine could be an example, where it is possible to apply a scan of the patient's body tissue, which facilitates the identification of lesions during surgical procedures. One of the application carries out the idea of using HDM in medicine is Even: Medical (Fig. 1). The device allows to visualize the vein directly on the skin and precisely determine the position of the needle puncture. At the same time they do not interfere with vision glasses other objects, allowing the wearer to them to make eye contact with the patient [29].

HMD displays can also be used to create virtual sketches. The user with the help of hand gestures draws, keep visualized shape in the air. Created this way virtual object can be 3D printed.

Augmented reality offers a variety of solutions dedicated to architects and designers. Examples are applications which goal is to simulate the reconstruction of damaged buildings, visualization, interior design and projection in real-time virtual models of buildings, to places where they will be built. Application designed for architectural visualization is the product of Urbansee [30]. Using your tablet or phone, you can play on the screen three-dimensional structure based on flat sketch (Fig. 2).

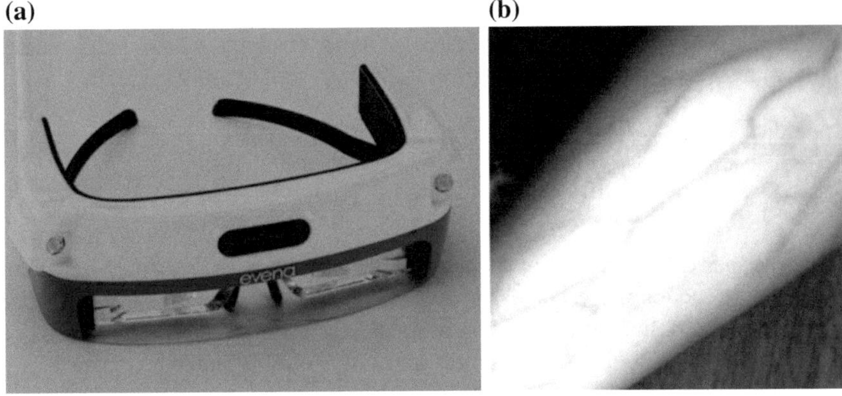

Fig. 1 **a** Evena medica glasses as an example of the "optical see-through" HMD, **b** scan of the veins in real time using the glasses

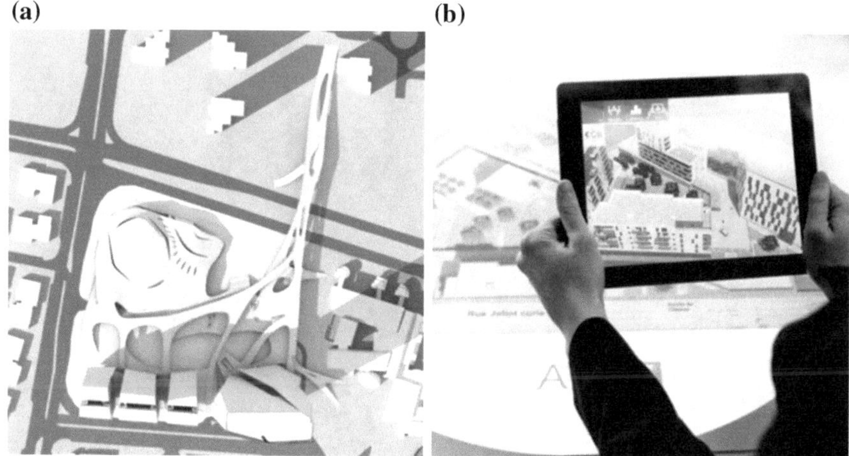

Fig. 2 **a** 2D architectural model used for reconstruction, **b** three-dimensional visualization of the model on the tablet

Relatively new technology is currently being tested by NASA—Fused Reality. It combines the concept of augmented reality with the capabilities of ground-based flight simulators. The pilot controls the actual machine which is in the air. It has, however, wearing special glasses generating virtual objects. The principle of operation is similar as in the case of augmented reality, the additional feature is the fact that the actual objects can affect those computer generated [31]. An example of such technology is a maneuver of air refueling taking place during real flight, however the function of the air tank is performed by a virtual plane.

Apart from the above examples the augmented reality technology is used in simulations of military operations, aviation, automotive, and computer games. An integral part of the application is based on the technology of augmented reality using stereovision. Glasses HMD type "video see-through" use stereoscopic algorithms generated in real-time to restore objects. The aim of the work is to show how important for the development of augmented reality is to find effective methods to reconstruct the three-dimensional scene (Fig. 3).

The most common use of stereovision is to create three-dimensional maps. In addition to the traditional cartography ground reconstruction using this method can also be used in outer space. An example would be the rover Curiosity, which performs stereoscopic images of the Martian surface using infrared camera [32]. One of the interesting applications of 3D reconstruction are flight simulators. The market offers a variety of both professional and amateur applications that allow playback trajectory on your home computer [33]. Stereoscopic pictures of the terrain are most often acquired using drones and then reconstructed into three-dimensional scene. An example of such an application may be Mega-SceneryEarth containing tens of reconstructed locations, in which the user can move virtual plane (Fig. 4). To the player's choice there are several types of air

Fig. 3 The virtual plane on the background of the actual scene, during maneuver

Fig. 4 Screen shots from MegaSceneryEarth: **a** view a virtual plane on the reconstructed scene, **b** the reconstructed city from the window of a virtual plane

vehicles of different parameters and different mode of control. The downside of the application is the ability to interact with the scenery modeled only at the time of takeoff and landing aircraft. During the game the user is reconstructed landscape with a bird's eye view, but there is no possibility of dynamic interaction with real objects.

The project that uses stereovision to reconstruct the premises, written using the Unity game engine is the IC-CRIME. This application is based on the reconstruction of the premises of the actual crime scenes, using laser scanning. Reconstructed in this way the scenery is covered with texture and transferred to the environment Unity. The application is dedicated for people investigating the circumstances of the offense and allows them an insight into the crime scene unaltered, long after its commission. In the space they are placed the items crucial to the case, along with descriptions useful in the investigation. Although the shape of the room was completed with very high precision, color quality and light and shade effect it is

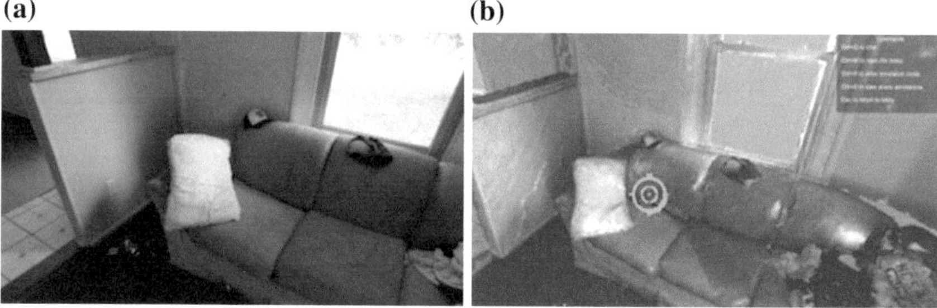

Fig. 5 **a** photo of the crime scene, **b** reconstructed 3D model using the IC-CRIME app

much worse than using stereovision techniques. Texture objects do not look realistic, despite the correctness of the stored geometry (Fig. 5).

The above examples can be stated that the quality of the applications based on the stereovision technology include: precision of reproducing geometric model and putting the right color and lighting. The quality of the model consists of a number of captured images and the variety used for the reconstruction techniques such as laser scanning, stereovision from the ground stations, or photos using drones. In addition, the application should allow interaction with the world recreated by the physical interaction between the player and reconstructed objects. Most of the applications available on the market focuses on the realization of only one criterion for your project. If the action game is set in space, as in the case MegaSceneryEarth, this criterion is precisely recreate the 3D scenery from the air. Despite the realistic models of the landscape. Hitting the building will break down the virtual plane, but does not imply the collapse of the building. A drawback in many applications based on 3D reconstruction is too large discrepancy between the appearance of real and virtual objects. In order to maintain coherence computer graphics modeled objects should be indistinguishable from those reconstructed. Many applications use computer modeling to create imitations of real objects (e.g. The control panel of the aircraft), as compared with actual 3D objects gives the impression of a lack of consistency.

Presented as part of the work application takes into account the above-mentioned objectives allowing interaction of elements of the real world with modeled computer. The user can navigate on the surface of objects and move with them. Falling from an aerial point of setting the character controlled results in losing life of the player. Applications such as IC-CRIME and flight simulators are based solely on exploration reconstructed scenery and do not assume fictional twists for instance: plane crash, natural disasters, or changes in weather conditions. Created as part of the work application is interactive, that is a decision taken by the user impinge on its course and determine victory or defeat. The game is set in one room with clearly defined limits. So constructed scenery make it easy to transfer the application to the field of real-time gameplay and using the HMD, which would not be possible for a large open space with chaotic construction. Table 1 presents the criteria to be met made within the application work, the methods used to meet these criteria, and the result of their use.

Table 1 The criteria, methods and the effect of their use

Criteria	Applied method	Results
Photorealistic	Stereovision	Acquiring reconstructed 3D objects
Visual diversity	Computer modeling	Creating virtual objects and correcting the real models
Graphical consistency	Unity engine	Unification of scale and details of created objects
Interactivity	Mesh process	The possibility of collisions and interactions between real and virtual objects
Dynamics	Script programming	Realistic kinematics of created objects

Table 2 Analysis of functional requirements and non-functional requirements

Functional requirements	Nonfunctional requirements
Obtaining a photo of the object of interest	Bright background
	Pattern background
	Good photo quality
	Possibility to transfer photos to computer
3D Reconstruction	No more than 70 photographs
	The amount of detail matching to the size of the object
	The proportions of colors and models relevant to the real objects
World integrity	Predictability
	Uniform scale
	Interaction between objects
Responding to the player's behavior	Collisions between player and objects
	Consistent moves
Responding to player commands	Response to the commands from the keyboard
	Compliance of the movements with the commands from the player via keyboard
	No response for undefined keys

3 Photogrammetry Scanning

Goal of the created game is to combine the advantages of visual entertainment. It should reconstruct space with an accuracy of flight simulators, while maintaining the possibility of user interaction with the elements of scenery. To achieve this goal, it was necessary to meet by the application the specific functional and nonfunctional requirements summarized in Table 2.

3.1 Stereovision as a Method of 3D Reconstruction

In order to reconstruct real objects the stereovision technique was used. It allows for reliable reconstruction of the scene without the use of specialized equipment and large financial outlays. It allows you to recreate a three-dimensional scene from a series of photos. The object to reconstruct be first photographed under strict rules. Knowledge of the different steps of the algorithm is helpful in the taking correct photographs.

The reconstruction of the world with the help of stereovision can be divided into two stages:

- Designation of common points, which occurred in both images and measurement of the distance (in pixels) between them.
- Carrying out triangulation, which determine the coordinates (x, y, z) of analyzed pixels [34].

The first stage is to determine for each image point cloud of characteristic points. Detection algorithms of these points may be based on both color information (e.g. histogram) and geometric (e.g. detected at edges and corners) [35]. Properly photographed object should be characterized thus:

- clarity contours,
- distinction of the object color of the background,
- a small number of rounded edges,
- non-uniform, preferably patterned background to provide a large number of characteristic points,
- no shaded, reflective or transparent areas.

Estimating the depth using the stereovision is based on disparition, which is the difference in the position of the object of interest caused by the change in perspective of the observer [36] (Fig. 6). Finding the coordinates of a point in space (triangulation) based on disparition is only possible in the case of the coplanar camera system.

A helpful tool in finding common points on two photographs (correspondence problem) made from a variety of perspectives is the epipolar geometry. Each photo-shooting for the purpose of this work was characterized by change in not only translation, but also in the rotation of the camera. If there are known the relative positions of cameras and a point x2 projected onto the shooting plane it is possible to determine the course of the epipolar line x1e1 [37]. Such defined matching of points is reduced to the search for solutions in the field of one-dimensional (point x1 must in fact be on the epipolar line *x1e1*. Figure 7 shows that the described features for the non-canonical system, the one in which the axes of the cameras are not parallel, or they have different focal lengths to [38].

X—point in space,
x1, *x2*—images of a point in the right and left camera,
O1, *O2*—focal of left and right camera,
e1, *e2*—epipolar poles,
x1, *e1*, *x1*, *e2*—epipolar lines.

Fig. 6 Value of disparition d using bottle as an example

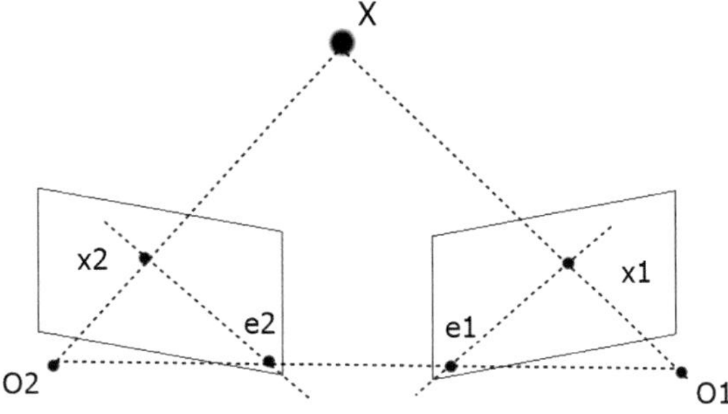

Fig. 7 Epipolar geometry for non-canonical camera system

Concepts inherent geometry epipolarną are matrices: essential and fundamental. Essential matrix is a matrix of dimensions describing the relative position of the cameras in the coordinates of the scene:

$$E = RS, \tag{1}$$

where:

$$R = \begin{pmatrix} r_{11} & r_{12} & r_{13} \\ r_{21} & r_{22} & r_{23} \\ r_{31} & r_{23} & r_{33} \end{pmatrix} \tag{2}$$

is the rotation matrix, and

$$S = \begin{pmatrix} 0 & -T_z & T_y \\ T_z & 0 & -T_x \\ -T_y & T_x & 0 \end{pmatrix} \tag{3}$$

Is a matrix operator of vector product for the translation vector. Such expressed essential matrix allow formulating an equation defining the limit of points x_1 and x_2:

$$x_2^T E x_1 = 0, \tag{4}$$

where x_1, x_2 are the projections of points lying on a line connecting the actual point of the center of projection cameras. Fundamental matrix described in Eq. (5) determines the relative positions of the cameras is not in the coordinates of the scene, but in the image plane, based on knowledge of their internal parameters:

$$F = A^{-T} E A'^{-1}, \tag{5}$$

where A is the matrix of the internal parameters of the camera.

$$A = \begin{pmatrix} f_x & 0 & c_x \\ 0 & f_y & c_y \\ 0 & 0 & 1 \end{pmatrix}. \tag{6}$$

f_x, f_y are horizontal and vertical focal length of the camera and c_x, c_y are horizontal and vertical offset of the optical axis relative to the center of the matrix. Fundamental matrix F is expressed analogously to the essential matrix equation:

$$m_2^T F m_1 = 0, \tag{7}$$

where m_1, m_2 are homogeneous coordinates of points in both images.

On the basis of images identified that the for smooth running of this phase the following rules are important:

- shoot from the constant spatial intervals, to give its shape with equal accuracy, from all perspectives,
- ensuring the existence of common points each image with at least two photographs neighbors. None, or illegibility in common stereo pair resulting in difficulty in reconstructing the model. It is necessary to manually identify corresponding points between the image and the two adjacent (Fig. 8). However, if

Fig. 8 Manual solution to the problem of detection of corresponding points between the image *1*, and *2* and *3*, which are adjacent to it

this problem affects many pictures and is more time effective to redo the entire series.

The above described steps are performed similarly for all the stereoscopic pairs. With the increase in the number of images increases the waiting time to reconstruct the model. The purpose of the application was to obtain a satisfactory mapped 3D objects, while minimizing run-time of algorithms. Compatible with the expectations of the quality of the reconstruction portion of the room, has been obtained for a series of 11 images. In order to correct detection points corresponding stereo pair, while taking more shots should ensure the existence of fragments of images overlap with the previous picture (Fig. 9).

One of the components of the application are modeled geometric primitives. Depending on the quality of the model, these objects were left subject to adjustment graphic, or placed directly in the scenery of the game. For each element local coordinate system was defined to facilitate subsequent location in space and giving kinematic properties. Reconstruction of the room took place by making the rotation camera for subsequent shots, the invariance of its position. This way of shooting allows the user to find the center of the room. Reconstruction of blocks based on a translation of the camera for subsequent shots without changing the angle of inclination with respect to the lens element (Fig. 10).

When taking pictures in this way we get a convex model (Table 3), which then is isolated from the background, cutting elements of scenery that were not the subject of reconstruction Table 4.

One of the most frequent problems were defects in the mesh and texture of the reconstituted model and stereoscopic distortions, or shape differences between the real object and the reconstructed object [39]. These adverse effects can be corrected by using the tools offered by the program Blender (Fig. 11). To carry out the reconstruction of the most commonly used grid-based vertex scaling and carving a

Fig. 9 A sample set of images used for 3D reconstruction of room

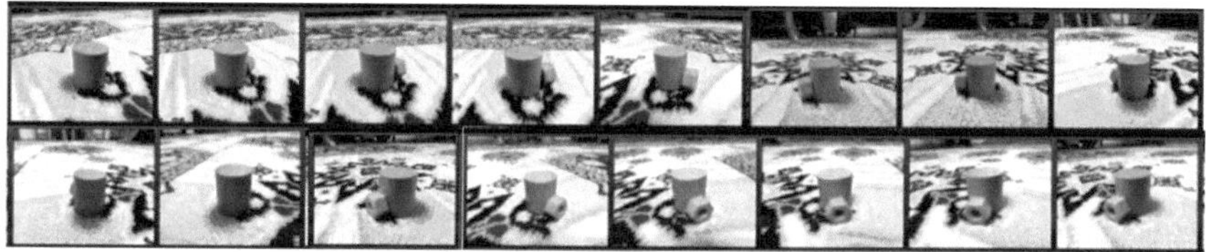

Fig. 10 A series of 16 shots of photographed object used for 3D reconstruction

block. Correction textures required while UV mapping, which plays a grid object on a two dimensional plane. Minor corrections can also be done using a color palette and brush.

3.2 System Architecture

Architecture of the systems includes: the subject, the photographic station and computer, which is used for computation of all the algorithms of reconstructing the scene. The photographic station met the functional requirements necessary to obtain the correct photos. The background to create photo was a patterned carpet, enabling easy detection of characteristic points. The pictures were taken at a resolution of

Table 3 3D objects used in the game and the number of images used for the reconstruction

Rigid body	Number of photos	Rigid body	Number of photos
Cylinder 	17	Cuboid 	17
Cylinder and cap 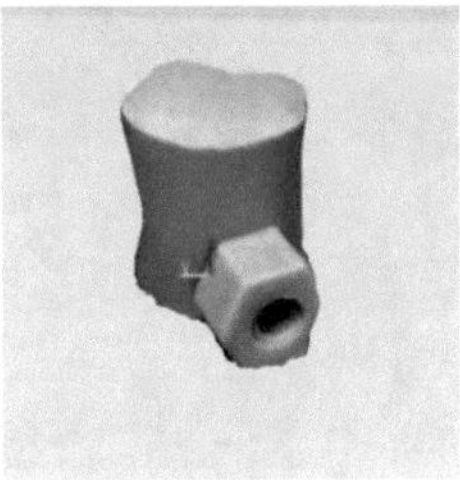	16	Polyhedra 	21
Cube 	16	Polyhedra 	19

1024 × 726 pixels, in the bright day light. The number of photographs depending on the subject were in the interval starting from 10 to 70 shots and the time they in which they were generated was in interval starting from 4 to 20 min.

The functional requirements for the integrity of the worlds and to interact with the player, were implemented using Unity engine. Using the appropriate scripts reconstructed objects were scaled, given the dynamic properties and allowed to collisions. The interaction between objects and the player and control of the game was also implemented in Unity.

After transferring to a computer, the objects were automatically reconstructed, using photogrammetric techniques. Then, depending on the quality, 3D models were transferred to Unity, or were subject to prior correction graphics in Blender. Some of the objects used in the application has been fully modeled by computer, without the reconstruction methods (Fig. 12).

Table 4 Comparison of geometry of reconstructed and reference objects

Researched object	Parameter	Object (cm)	Conclusions
Diameter	3,3	5,5	The scale factor of the diameter of the model to the real object is 0.6
Height	3,25	6	Height scale factor of the model is 0.54. This means that the object has been stretched along. To ratios between the diameter and height were correct the model should have a height of 5.42 [j]
Ratio	1,06	0,92	Reconstructed model does not reflect the actual proportion of real rigid body. When using the application these differences, however, are imperceptible for the human observer

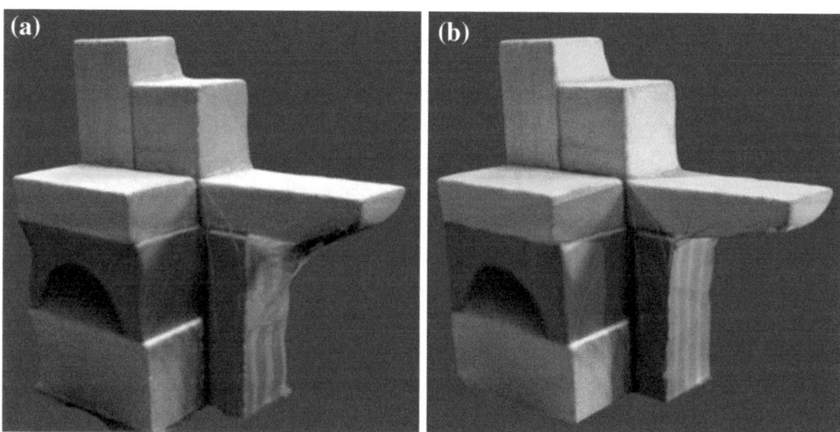

Fig. 11 3D models rendered in Blender software: **a** raw object without correction, **b** model with enhanced edges, mesh and texture

3.3 Application Goal

The developed application is a skill game located in a reconstructed scenery using stereovision techniques. During the game the user's goal is to travel to the end point

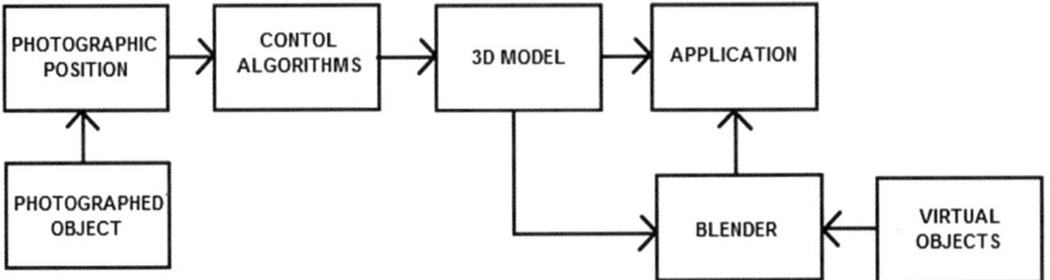

Fig. 12 The individual components of the system

without losing all live points and collecting as many coins. Arcade element is the need to overcome the moving platforms, which requires reflexes and proper estimating distances. On the way to the final point locations are also enemies, which the player must avoid. If the player succeeds points are counted otherwise the whole game starts from the beginning. Platforms, structures with wooden blocks, silhouette player and enemies modeled as chess pieces, the background in the game were all reconstructed using stereovision techniques. Coins and bonuses have been modeled in Blender.

4 Tests

To achieve the goal the Unity environment was used. Unity allows to create three-dimensional applications, moreover, is compatible program Blender, allowing to import files from this environment. All code is implemented in object-oriented programming language C#.

The application starts by displaying the start menu, after which the user navigates a cursor. Depending on the decision, it is possible to load various scenes from the game, or exit the program. If the user choose to exit, he is asked again to confirm his selection. After loading the game board it is possible to pause, which is visible as a screen freeze. There is the option to return to the game, or go to the Start menu. There are two variants of game completion. If you succeed, automatically go to the start menu otherwise there is an additional option to restart the board (Fig. 13).

Graphics created an application are based on a combination of reconstructed 3D objects with virtual elements. On the board there are placed reconstructed 3D models and bonuses: coins and life symbols modeled in Blender (Fig. 14). Virtual objects are blended into real scene, the player must overcome to get them (Fig. 15).

The background of the game is an example room, and therefore on the board objects appear blended in the convention, e.g. books and laptop. On the computer screen there was placed a screenshot of the game in design mode, which allows the player to see the whole board before its exploration (Fig. 16).

During the game the player must avoid contact with bullets, shot by black chess knights (Fig. 17). After each contact with the ball, the player loses one life. The loss

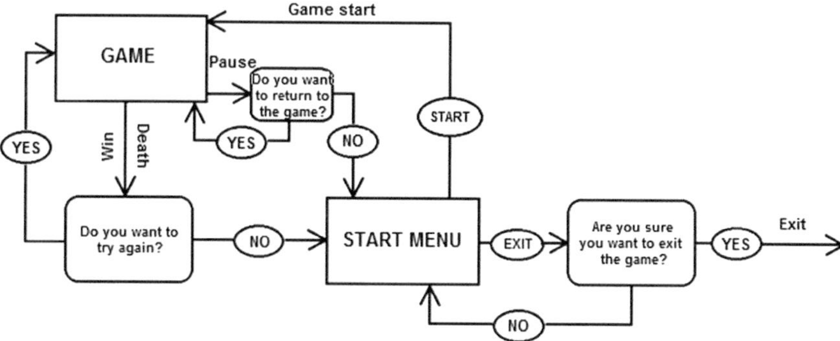

Fig. 13 Application logic schema

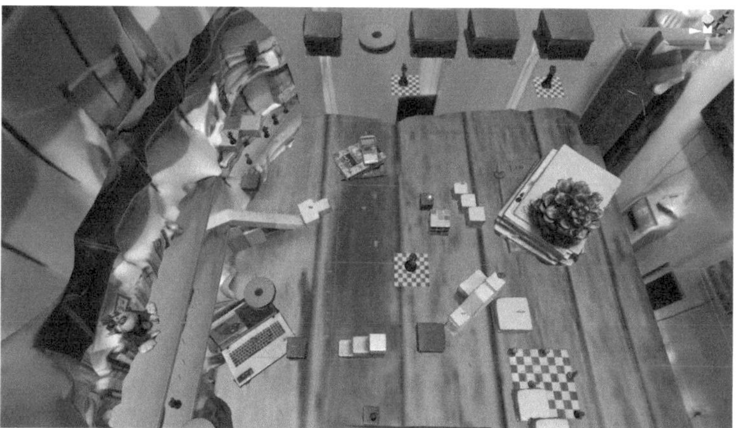

Fig. 14 The view from the top of the scene (level)

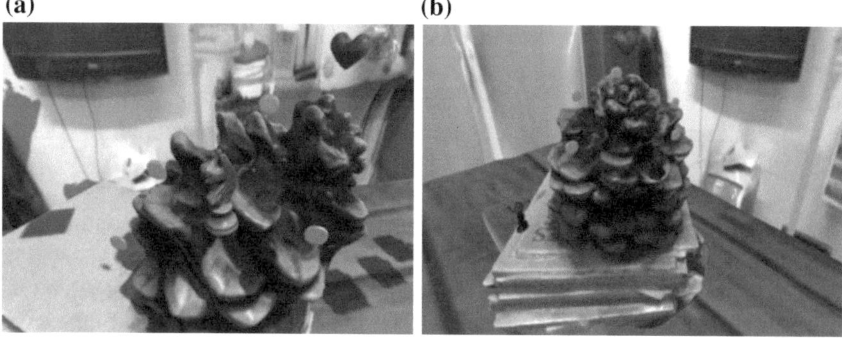

Fig. 15 Model of pine cones reconstructed using a series of 67 images: **a** player climbing on the block to collect bonus points, **b** view of the model in edit mode

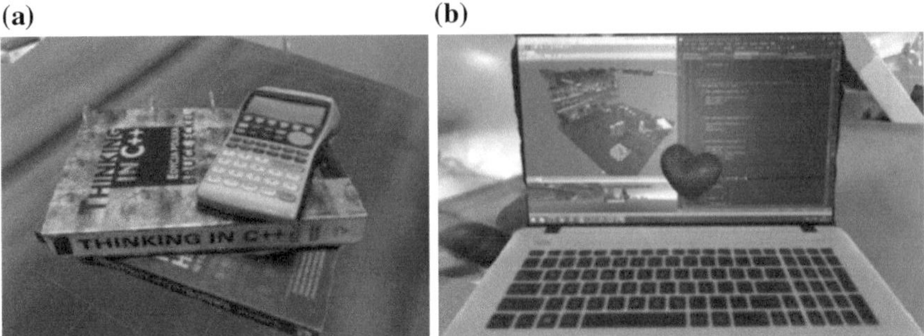

Fig. 16 The reconstructed 3D models, along with virtual elements: **a** books, calculator and coins, **b** laptop and bonus

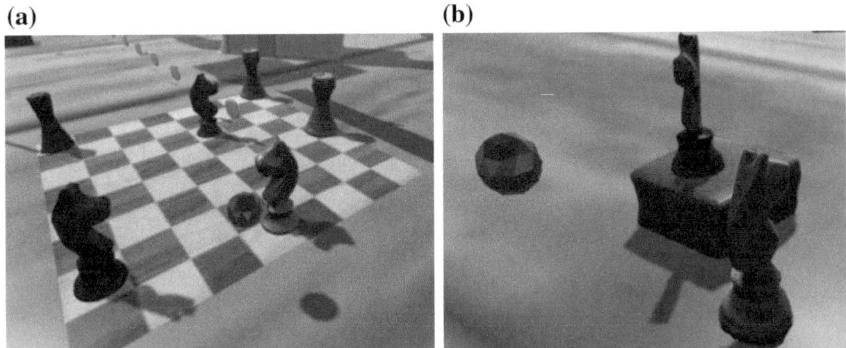

Fig. 17 Enemies appearing on the board: **a** black chess knight aiming ball in the player, **b** chess knight placed on the platform

of all life points is synonymous with the end of the game. Black figures were reconstructed on the basis of light ones and repainted Blender. The bullets, which are shoot enemies are the computer created polyhedras.

The board on which moves the player includes both static and dynamic elements. An example of still images are: shelves, base, walls, books and part of the platform. These objects do not affect the player's moving along them. Part of the solids had losses in the grid, or a texture and required prior correction in Blender (Fig. 18).

In addition to static objects in space there are arranged also dynamically moving parts e.g. platforms. A player on the surface of such bodies is moving on them. To ensure proper player interaction with moving objects on the platform there were imposed additional grids generating collisions. These elements are arranged on the platform, and have a symmetrical bodies. This allows the correct player interaction with the platform, even if the grid were irregularities. Additional elements generating collisions are not visible in game mode (Fig. 19).

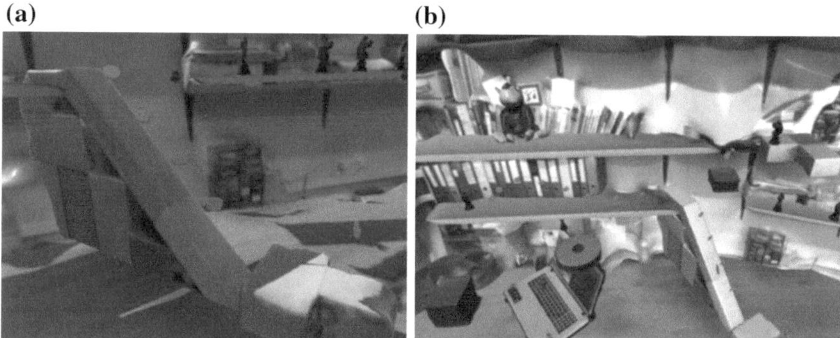

Fig. 18 The corrected static platform using Blender: **a** platform on which the player can climb, **b** incorporation of the platform in the scenery

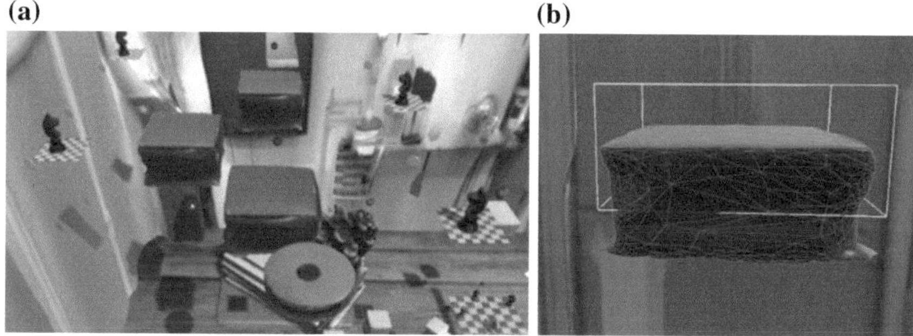

Fig. 19 **a** System of movable platforms, **b** platform with invisible for the player bounding box used for collision detection

During work a few three-dimensional objects of different shape, color, texture, and reflected light level of complexity were reconstructed. The experience allowed the selection of the best parameters of the position of the photographic and establish a reasonable methodology for capturing images.

The application included a large number of items shown on a small scale and with a lot of detail. The optimal solution therefore must have sought minimizing time and maximizing the quality of the 3D model. Into account must have also been taken versatility and availability of environment. One of the tests was to examine the influence of the number of photos on-time and accurate reconstruction of the object. The test results are shown in Table 5.

Reconstruction of the solid body based on a series of 67 images was completed during 17 min. It can be concluded that the looks of the objects is very realistic, there are no defects in the mesh and texture and proper proportions are maintained. However on the edges of books there is still minimal distortion, which can be caused by poor lighting or mistakes in framing (Fig. 20).

Table 5 Reconstruction quality comparison

No. photos	Time (min)	3D quality	No. photos	Time (min)	3D quality
7	2	There are significant gaps in the grid model, the object is flattened and merges with the ground. The proportions are not retained	15	4	Preserved are essential proportions of the model. There are minor defects, details are blurred together, and the edges blurred
20	4	There are still losses. However, the edges of the books are reproduced very precisely. You can read the letters on the cover. Solid of pine cones is still drenched and slurred	30	7	The model does not contain cavities. Cards books and shape of cones begin to look realistic. Fragments of objects continue to merge, and the corners are sharp

(continued)

Table 5 (continued)

No. photos	Time (min)	3D quality	No. photos	Time (min)	3D quality
40	9	Books have been recreated with many details, you can distinguish the individual cards. Still, there are distortions in the corners. The contours of the cones are distinct from both the side and from above	50	12	The object looks realistic from every perspective. There are slight distortion in areas with complex geometry

Fig. 20 3D model reconstructed in 123D Catch software using a series of 67 images

Based on the test, it was observed that the quality of an object is directly proportional to the amount of images and time needed to generate the model. It can be considered that the optimal parameter for the reconstruction of solid objects, was a series of 50 images. The difference between the quality of the model based on 50 pictures, is comparable to that obtained with the 67 series of photographs. Generation time is, however, shorter by about 30%. During creation of many 3D models it is a significant difference.

5 Conclusions

The idea behind the study was to draw attention to the promising technology, which is augmented reality and show the complexity of the issues associated with it. The individual elements included in the applications based on augmented reality, were analyzed on the basis of an interactive computer game. The following topics were discussed: methodology of taking photographs, algorithms for three-dimensional reconstruction of the scene, processing of graphic objects, computer modeling, synchronization of the world reconstructed with virtual elements, and user interaction with a reconstructed scenery. The article presents implementation of an innovative approach to the topic of computer games based on stereovision. There is a possibility of finding a realistic scenery and its exploration from different perspective, e.g. chess figures. This approach changes the perception of the player, allowing him to interact with objects on a different scale than it is in reality. Application architecture allows expanding of the game in the future.

The work showed a spectrum of problems associated with the different phases of creating applications based on augmented reality. To reconstruct scenes of complicated construction often is needed several types of methods as stereovision, or laser scanning. To restore objects from above e.g. the use of drones. The concept of augmented reality involves processing of events in real time, therefore processing of such stimuli is a challenging task which requires intensive computational performance hardware and optimized algorithms.

We conclude that mixed reality technology, in particular augmented reality is a rapidly growing field that still leaves room for many discoveries and improvements. The possibilities of applications based on augmented reality exceeds even the most traditional programs and have the chance to become a solution revolutionizing the technology market.

References

1. Bieda, R., Jaskot, K., Jędrasiak, K., & Nawrat, A. (2013). Recognition and location of objects in the visual field of a UAV vision system. In: *Vision based systems for UAV applications* (pp. 27–45). Springer.
2. Babiarz, A., Bieda, R., & Jaskot, K. (2013). Vision system for group of mobile robots. In *Vision based systems for UAV applications* (pp. 139–156). Springer.
3. Kus, Z., & Nawrat, A. (2013). Object tracking in a picture during rapid camera movements. In *Vision based systems for UAV applications* (pp. 77–91). Springer.
4. Azuma, R. (1997). A survey of augmented reality. *Presence: Teleoperators and Virtual Environments, 6*, 355–385.
5. Milgram, P., Takemura, H., Utsumi, A., & Kishino, F. (1994). *Augmented reality: A class of displays on the reality-virtuality continuum* Vol. 2351, pp. 282–292). SPIE.
6. Daniec, K., Jedrasiak, K., Koteras, R., & Nawrat, A. (2013). Embedded micro inertial navigation system. In *Applied mechanics and materials* (Vol. 249, pp. 1234–1246). Trans Tech Publications.
7. Bereska, D., Daniec, K., Fras, S., Jedrasiak, K., Malinowski, M., & Nawrat, A. (2013). System for multi-axial mechanical stabilization of digital camera. In *Vision based systems for UAV applications* (pp. 177–189). Springer.
8. Jedrasiak, K., Nawrat, A., Daniec, K., Koteras, R., Mikulski, M., & Grzejszczak, T. (2012). A prototype device for concealed weapon detection using IR and CMOS cameras fast image fusion. In *International Conference on Computer Vision and Graphics* (pp. 423–432). Berlin: Springer.
9. Jarusirisawad, S., & Hosokawa, T. (2010). Diminished reality using plane-sweep algorithm with weakly-calibrated cameras. *Progress in Informatics,* (7), 11–20.
10. Josinski, H., Switonski, A., Jedrasiak, K., & Kostrzewa, D. (2012). Human identification based on gait motion capture data. In *Proceedings of the 2012 International MultiConference of Engineers and Computer Scientists, IMECS* (Vol. 12).
11. Switonski, A., Josinski, H., Jedrasiak, K., Polanski, A., & Wojciechowski, K. (2010). Classification of poses and movement phases. In *ICCVG 2010. Lecture Notes in Computer Science*. Springer.
12. Azuma, R., Baillot, Y., Behringer, R., Feiner, S., Julier, S., & MacIntyre, B. (2001). Recent advances in augmented reality. *IEEE Computer Graphics and Applications,* 1.
13. Mann, S., & Fung, J. (2001). Video orbits on eye tap devices for deliberately diminished reality or altering the visual perception of rigid planar patches of a real world scene. In *International Symposium on Mixed Reality*, March 14–15 2001.
14. Jeżewski, S., & Jaros, M. (2008). Skanowanie trójwymiarowej przestrzeni pomieszczeń, Katedra Informatyki Stosowanej, Politechnika Łódzka, AUTOMATYKA, Tom 12, Zeszyt 3.

15. Kraus, K. (2000). *Photogrammetry: Geometry from images and laser scans* (Vol. 1). New York: Walter de Gruyter GmbH.
16. Bahadori, S., & Iocchi, L. (2003). A stereo vision system for 3D reconstruction and semi-automatic surveillance of museum areas, 2.
17. Gilson, S., Fitzgibbon, A., & Glennerster, A. (2005). An automated calibration method for non-see-through head mounted displays. *IEEE Transactions on Medical Imaging*, 1492–1499.
18. Jedrasiak, K., Andrzejczak, M., & Nawrat, A. (2014). SETh: The method for long-term object tracking. *Computer Vision and Graphics, 8671*, 302–315. Lecture Notes in Computer Science, 316.
19. Ryt, A., Sobel, D., Kwiatkowski, J., Domzal, M., Jedrasiak, K., & Nawrat, A. (2014). Real-time laser point tracking. In *International Conference on Computer Vision and Graphics*, pp. 542–551. Springer.
20. Nawrat, A., & Jedrasiak, K. (2008). Fast colour recognition algorithm for robotics. *Problemy Eksploatacji*, pp. 69–76.
21. Rolland, J. (2000). Optical versus video see-through head-mounted displays in medical visualization. *Presence,* 287–309.
22. Babiarz, A., Bieda, R., Jedrasiak, K., & Nawrat, A. (2013). Machine vision in autonomous systems of detection and location of objects in digital images. In *Vision based systems for UAV applications* (pp. 3–25). Springer.
23. Daniec, K., Iwaneczko, P., Jedrasiak, K., & Nawrat, A. (2013). Prototyping the autonomous flight algorithms using the prepar3D® simulator. In *Vision based systems for UAV applications*, pp. 219–232. Springer.
24. http://evenamed.com/ (16. 01.2016).
25. https://classic.urbasee.com/apps.php (16. 01.2016).
26. Sobel, D., Jedrasiak, K., Daniec, K., Wrona, J., Jurgas, P., & Nawrat, A. (2014). Camera calibration for tracked vehicles augmented reality applications. In *Innovative control systems for tracked vehicle platforms* (pp. 147–162). Springer.
27. Bieda, R., Grygiel, R., & Galuszka, A. (2015). Naive Kalman filtering for estimation of spatial object orientation. In *Methods and Models in Automation and Robotics (MMAR)*, pp. 955–960.
28. Bieda, R., & Grygiel, R. (2014). Wyznaczanie orientacji obiektu w przestrzeni z wykorzystaniem naiwnego filtru Kalmana. *Przeglad Elektrotechniczny, 90,* 34–41.
29. http://www.nasa.gov/centers/armstrong/features/fused_reality.html (06.01.2016).
30. http://www.space.com/16856-mars-rover-curiosity-cameras-vision.html (07.010216).
31. http://www.123dapp.com (04.01.2016).
32. http://pointcloud.pl/algorithms/integration (04.01.2016).
33. Sankowski, D., & Nowakowski, J. (2014). *Computer vision in robotics and industrial applications*. Singapur: World Scientific Publishing Company Pte Limited.
34. Hartley, R., & Zisserman, A. (2003). *Multiple view geometry in computer vision*. United Kingdom: Cambridge University Press.
35. Rzeszotarski, D., Strumiłło, P., Pełczyński, P., Więcek, B., & Lorenc, A. (2005). System obrazowania stereoskopowego sekwencji scen trójwymiarowych, Elektronika: prace naukowe, 165–184.
36. Woods, A., Docherty, T., & Koch, R. (1993). Image distortions in stereoscopic video system, stereoscopic displays and applications IV. *Proceedings of the SPIE, 1915*. San Jose, CA.
37. Szkodny, T. (2012). *Podstawy Robotyki*. Gliwice: Wydawnictwo Politechniki Śląskiej.
38. Julier, S., Lanzagorta, M., & Baillot, Y. (2000). Information filtering for mobile augmented reality. In *Symposium on Augmented Reality, Munich, Germany*, pp. 3–11.
39. Pardel, P. (2009). Przegląd ważniejszych zagadnień rozszerzonej rzeczywistości, Zeszyty Naukowe Politechniki Śląskiej, Seria Informatyka, Volume 30, Number 1 (82), Silesian University of Technology Press, Gliwice 2009.

Part IV
Experimental Investigation of Dynamic Characteristics

The design of innovative control, tracking and monitoring algorithms most often require prior knowledge of dynamical characteristics of equipment being used in many different national security tasks carried out in extremely difficult field conditions. For instance military actions conducted in the twenty-first century in Iraq and Afghanistan have shown that the tactical and technical solutions applied so far are not sufficiently effective in wars, in which one side of the conflict has a significant advantage over the other. This type of warfare is called asymmetrical warfare the primary aim of combatants is to maintain the status quo and threat potential against the enemy. Adapting to the situational conditions, the fighting of Islamic militants have taken on a guerrilla warfare character, which is characterized by the weaker side of the conflict, with inferior technical equipment adopts a defensive position and conducts attacks from concealment rather than engaging in open field combat. Examples of such tactics include suicide attacks, ambushes in large groups, as well as engaging in various forms of mine warfare. This results in significant losses among military forces of the ISAF coalition (International Security Assistance Force). It is therefore necessary to search for new technical solutions which would serve as a good answer to the challenges of asymmetrical warfare. The proposed set of articles allows trac the whole spectrum of issues related to the discussed topic. Starting from the study of effects of the IED explosion on the health of passengers of an armored vehicle to the study of the effects of temperature on the characteristics of the sensory systems.

The availability of information often decides on the superiority on the battlefield. Therefore systems allowing to determine and track the position of objects are crucial. Unfortunately, satellite navigation systems do not work in buildings or underground therefore it is extremely important to develop new technologies and solutions to solve this problem. As part of the chapter, one of the promising solutions based on personal navigation using inertial measurement sensors integrated with shoes is presented.

Another important issue is to study the impact of IED on vehicles like Shiba special vehicle. The test was performed by Military University of Technology and Military Institute of Engineer Technology. A number of strain gauges and camera markers were placed on the arm to allow recording strains and movements of specific construction points. The arm's motion was recorded using high speed camera. The analysis showed, that maximum stresses in examined construction parts did not exceed yield stress of material. Numerical model of an arm will be developed and validated using data obtained during tests. This will help visualize stress distribution in each arm's part.

Valuable suggestions, conclusions and a number of important challenges in the fields mentioned above are presented in the chapter.

Proving Ground Tests of Selected Energy Absorbing Structure Variants Under a Shock Wave Load

Tadeusz Niezgoda, Grzegorz Sławiński, Paweł Bogusz and Marek Świerczewski

1 Introduction

Military actions conducted in the XXI century in Iraq and Afghanistan have shown that the tactical and technical solutions applied so far are not sufficiently effective in wars, in which one side of the conflict has a significant advantage over the other. This type of warfare is called asymmetrical warfare the primary aim of combatants is to maintain the status quo and threat potential against the enemy. Adapting to the situational conditions, the fighting of Islamic militants have taken on a guerrilla warfare character, which is characterized by the weaker side of the conflict, with inferior technical equipment adopts a defensive position and conducts attacks from concealment rather than engaging in open field combat. Examples of such tactics include suicide attacks, ambushes in large groups, as well as engaging in various forms of mine warfare. This results in significant losses among military forces of the ISAF coalition (International Security Assistance Force). It is therefore necessary to search for new technical solutions which would serve as a good answer to the challenges of asymmetrical warfare.

The most effective weapon employed in the strategy of combat engagements against the ISAF are Improvised Explosive Devices, so called IEDs. They are most frequently placed on roads travelled by allied forces vehicle convoys. According to the statistics [1], they were the cause of death of approximately 40% of all soldiers killed when fighting in Afghanistan and Iraq between 2003 and 2011. The stabilization missions are primarily attended by vehicles with light and medium armour, in the form of transporters and light patrol vehicles. Combat experience has shown, that structural solutions employed currently to ensure appropriate protection to the

T. Niezgoda (✉) · G. Sławiński · P. Bogusz · M. Świerczewski
Faculty of Mechanical Engineering, Department of Mechanics and Applied Computer Science, Military University of Technology, Gen. S. Kaliskiego 2St., 00-908 Warsaw, Poland
e-mail: tadeusz.niezgoda@wat.edu.pll

A. Nawrat et al. (eds.), *Advanced Technologies in Practical Applications for National Security*, Studies in Systems, Decision and Control 106,
https://doi.org/10.1007/978-3-319-64674-9_16

vehicles' crew against mines are insufficient [2–4]. For that reason, the primary course of research is currently intended to limit as much as possible the effect of mine explosions on the passengers of an armoured vehicle.

2 Threats to an Armoured Vehicle's Crew During an IED Explosion

Combat experience in Afghanistan and Iraq has shown high effectiveness of improvised explosive devices, measured in the number of soldiers killed. Figure 1 presents the number of fatalities caused by IED explosions in comparison with other causes of death [5].

In order to increase the safety of a vehicle's crew against the shock wave of a mine and IED explosion, the primary protective structures are enhanced by appropriate structural solutions, such as: deflectors, increased vehicle clearance, modular construction. The listed solutions may not always be applied to an already existing vehicle employed by the military. This arises from the necessity to maintain high mobility and combat capabilities on the battlefield, or the vehicle's structure itself. Due to the above, it is advisable to use energy-absorbing panels. An additional layer in the form of a protective panel will be aimed at decreasing the inertial force affecting the vehicle's passenger during an AT mine or IED explosion, which is the primary cause of injuries sustained by vehicle personnel.

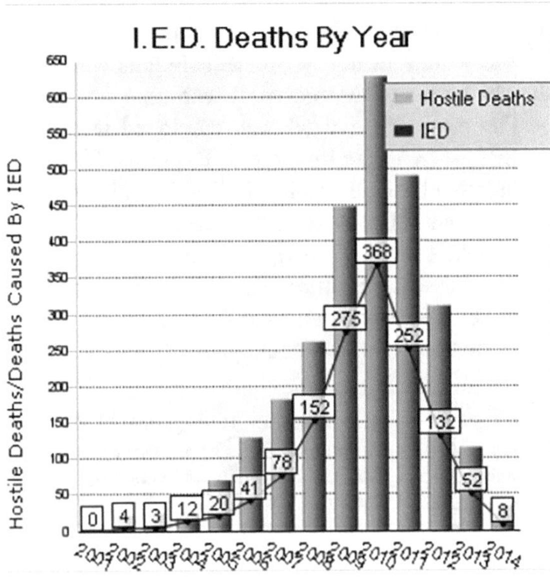

Fig. 1 Percentage share of IEDs compared to other causes of death, as a result of enemy actions, during OEF/ISAF missions in Afghanistan [5]

3 Origin of Using Energy-Absorbing Panels

A significant share of military vehicles taking part in armed conflicts feature a flat shaped floor panels. The event of a mine or IED explosion underneath the vehicle directly affects the floor panel. This results in the generation of high inertial force which causes numerous vehicle personnel injuries. Increasing the safety of passengers in a vehicle with a flat shaped bottom and low clearance may only be realised through the use of energy-absorbing panels. These types of energy-absorbing structures are most often built as hybrid structures which:

- secure the bottom of a vehicle's hull against rupture due to an AT mine or IED explosion [6],
- absorb the shock wave energy of an AT mine or IED explosion through various destruction mechanisms,
- block perforation of the hull's bottom,
- ensure protection of the soldiers' life and health in a vehicle at the required level (STANAG 4569, AEP-55).

Implementation of new solutions requires the performance of a series of experimental studies in order to evaluate the properties of proposed structures.

4 Assumptions, Conception and Aim of Shock Wave Effect Proving Ground Tests

The aim of the designed shielding is the protection of a military vehicle's crew against the shock wave generated by the explosion of AT mines and IEDs. The design assumptions for the shielding are as follows:

- shielding installation without changes to the undercarriage and bodywork of the protected vehicle,
- shielding intended for upgrading existing models of LV and LAV vehicles,
- simple installation and uninstallation of the shielding and rest-plate,
- interchangeability of shielding segments in case of destruction,
- ensure a level of protection of vehicle personnel in case of a shock wave generated by a mine or IED explosion in accordance with STANAG4569,
- inflammability, resistance to weather effects and chemical resistance of the shielding,
- simple manufacturing technology,
- low material and manufacturing costs.

Energy-absorbing structures have been created in the form of 440 × 440 cuboids with a variable thickness, mounted onto a witness plate with dimensions of 550 × 550 × 2. Absorption of the explosion shock wave energy is realised by three primary progressive destruction mechanisms: lateral shear strain, delamination, bending.

The conducted proving ground and experimental tests were intended to investigate the effectiveness of shielding and their resistance to the effects of a shock wave caused by explosive material detonation. The evaluation criteria for shielding effectiveness were:

- acceleration as a function of time,
- surface reaction (force) as a function of time,
- permanent deformation of the witness plate.

All the above valued were recorded for the witness plate integrated with the protective panel, installed on a dedicated station for measuring the surface reaction force and acceleration. Results recorded for individual variants were in the following stage subjected to normalization and compared with a system without the protective layer (witness plate—Panel 1) in order to assess its effectiveness.

5 Description of the Test Rig and Study Methodology

The conducted proving ground studies incorporated an explosive test rig which enables performance of tests, in which the maximum mass of an explosive charge equals 2.0 kg of TNT.

The axonometric view of the proving ground rig for testing the effects of shock waves is presented in Fig. 2. This rig enables performance of explosion load tests using an explosive charge (max. 2 kg of TNT) installed above the tested object, with simultaneous measurement of force in the rig's supports and acceleration of the tested protective panel's centre.

The reaction force values are measured on four cylindrical supports of the lower element—Fig. 2, using the electrical resistance strain gauge measurement method,

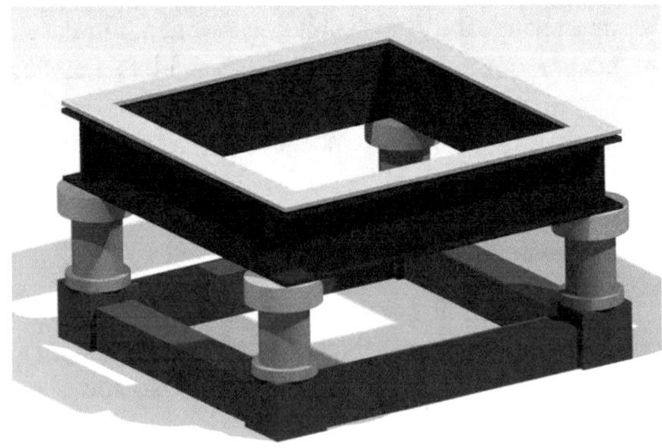

Fig. 2 Axonometric view of the proving ground rig

through tension gauges applied in the axial direction of the supports. In the studies the resultant force has been determined as the sum of forces on all four supports.

Additionally, a piezoelectric acceleration sensor was mounted to the lower plate for each tested protective variant. The sensor was always mounted in the central point of the witness plate, on the opposite side in relation to the placed explosive charge. The scaling of acceleration sensors equalled 0.005 mV/m/s2.

Measurement of the accelerometer sensor signal and surface reaction forces was carried out using high frequency measurement aperture.

The studies used the Semtex 1A explosive material with a mass of 0.75 kg, suspended for each of the selected panel variants at a distance of 430 mm from the witness palate (Fig. 3).

A detailed description of samples for explosion shock wave load studies is presented in Table 1. For each variant the charge distance equalled 430 mm from the upper surface of the tested panel. The charge mass equalled 750 g. During the first trial, only the witness plate was tested and its result served as a reference.

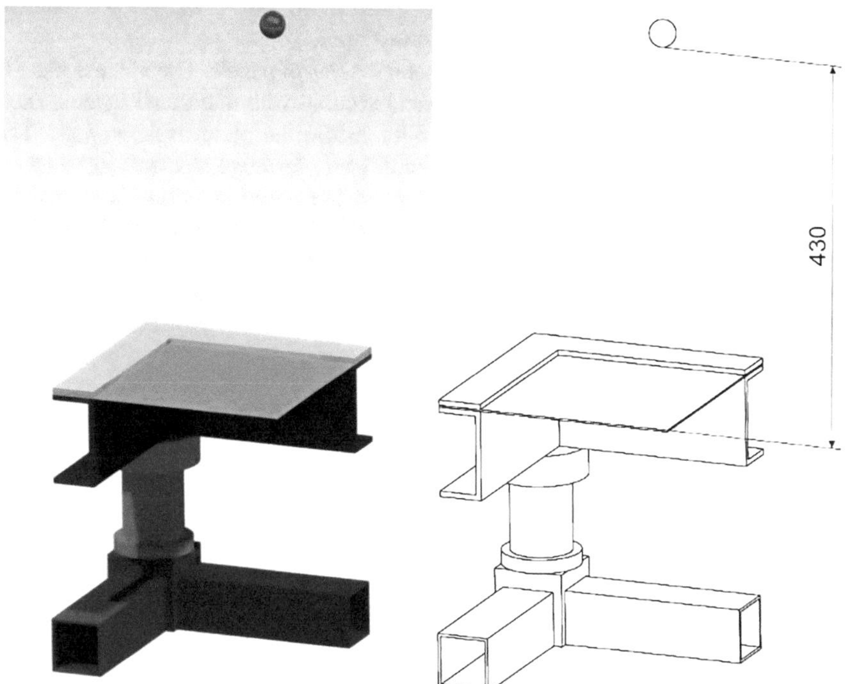

Fig. 3 Axonometric view of ¼ of the proving ground testing rig and charge installation method

Table 1 Description of test sample parameters for explosive shock wave load studies

Sample description	Charge mass [g]	Sampling frequency [Msamples/s]	Height of the charge over the panel plate [mm]
Sample_1–9	750	0.5	430

6 Protective Panel Description

The conducted proving ground studies of explosion shock wave loads involved a total of 9 detonation trials. The characteristics of all panels (witness plate and protective panels) are compiled in Table 2. The values given include panel layers, their materials, thickness and sequence of placement.

Protective panels 2, 3, 4 were covered at the top with a 1 mm thick sheet. Additional sheets have also been used as dividing elements between individual layers in panels 4, 5, 6 and 8.

A very important factor, apart from the panel's absolute effectiveness, is also its mass. For that reason, the study analyses the mass parameters of tested panels, such as: total mass and area density. Based on Table 2 we have compiled comparative bar graphs of these parameters, presented in Fig. 4.

Figure 4a shows a comparison of the total masses of panels. The witness plate is approximately two times lighter than panels secured with additional layers. Their total mass is within the range of 12–16 kg, while the plate itself is 7 kg, The difference of these masses is the mass of additional protective elements installed on the given structure. The lowest weight among the tested solutions is shown by panels 6, 7 and 9. The highest mass panels—5 and 8) contain elastomer layers. The mass of panel 5 reaches nearly 16 kg. Panel 8 contains elastomer and cork layers. Figure 4b presents a comparison of the surface density of panels 2–9. The mass of the lower plate has been subtracted from the total mass of the panel. The result has been divided by the surface of the installed protective element with dimensions of 440 × 440 mm. Due to this, Panel 1 is not included in this compilation. The relative results are analogous to the comparison of panel masses.

7 Study Results

After conducting a series of experimental tests, the resulting changes in force and accelerations have been subjected to scaling. Additionally, the plate centre acceleration and total force test results in the supports have been normalized for all samples, which involved relating those values to the maximum acceleration and total force values per module measured for the witness plate tested during the first trial (marked Panel_1).

The acceleration change graphs were filtered using a low-pass Butterworth filter of the 6th order for the limit frequency value of 2560 Hz.

Table 2 Protective panel layer construction description

Trial no.	Panel component description	Protective system mass [kg]	Area density [kg/m^2]	Explosive charge mass [kg]
Panel_1	**Witness plate (steel sheet) g = 3 mm**	**7.0**	–	0.75 kg TNT
Panel_2	L1-Upper lining (steel sheet) (g = 1 mm) L2-CYMAT 0.36 (g = 2 * 25 mm) L3-Witness plate (g = 3 mm)	12.2	26.86	
Panel_3	L1-Upper lining (steel sheet) (g = 1 mm) L2-CYMAT 0.51 (g = 2 * 25 mm) L3-Witness plate (g = 3 mm)	13.8	35.12	
Panel_4	L1-Upper lining (steel sheet) (g = 1 mm) L1-CYMAT 0.36 (g = 25 mm) L1-Divider (steel sheet) (g = 1 mm) L2-CYMAT 0.51 (g = 25 mm) L3-Witness plate (g = 3 mm)	14.5	38.74	
Panel_5	L1-Elast. ASMA 55ShA (g = 20 mm) L2-Divider (steel sheet) (g = 1 mm) L3-CYMAT 0.51 (g = 25 mm) L3-Witness plate (g = 3 mm)	15.8	45.45	
Panel_6	L1-Corkboard (g = 10 mm) L2-Divider (steel sheet) (g = 1 mm) L3-CYMAT 0.51 (25 mm) L4-Witness plate (g = 3 mm)	12.1	26.34	
Panel_7	L1-Elast. ASMA 55ShA (g = 10 mm) L2-Air cushion (g = 45 mm) L3-Witness plate (g = 3 mm)	12.8	29.96	

(continued)

Table 2 (continued)

Trial no.	Panel component description	Protective system mass [kg]	Area density [kg/m^2]	Explosive charge mass [kg]
	L1-Elast. ASMA 55ShA (g = 5 mm) L2-Divider (steel sheet) (g = 1 mm) L3-Corkboard (g = 10 mm) L4-Divider (steel sheet) (g = 1 mm) L5-CYMAT 0.51 (25 mm) L6-Witness plate (g = 3 mm)	15.3	42.87	
Panel_9	L1-Openwork plate (g = 2 mm) L2-Hemisphere system W3-Witness plate (g = 3 mm)	12.6	28.93	

The graphs present measurement results of normalized acceleration as a function of time, registered using an ICP type (piezoelectric) accelerometer and normalized total force registered using electric resistance type dynamometer.

The figures present test results of Panel 1, which is comprised of the witness plate without additional protective elements. Results of this trial serve as a reference for all other panels. The following criteria have been adopted in the assessment of panels in relation to the witness plate reference panel:

- the force extreme value and number criterion—less numerous and lower extremes equal a better panel,
- the acceleration extreme value and number criterion—less numerous and lower extremes equal a better panel,
- the overall graph change criterion—flatter course of force and acceleration graphs equal better results shown by the panel.

Test results of the load generated in supports have been normalized for all samples in relation to the maximum force value for the witness plate per module. Test results of the plate centre acceleration have been normalized for all samples in relation to the maximum acceleration value for the witness plate per module.

The figures present Panel_1 installed on the force and acceleration measurement rig before the tests and after the tests. The plate sustained significant deformation as a result of load from the explosion shock wave (Fig. 5).

Further figures present the results of normalized acceleration and force measurements. The relative values equal to 1 correspond to the maximum values per module (force and acceleration) prior to normalization, which served to normalize the course of all graphs (Figs. 6 and 7).

Further figures present photographs of Panel_2 (Photos in Fig. 7b).

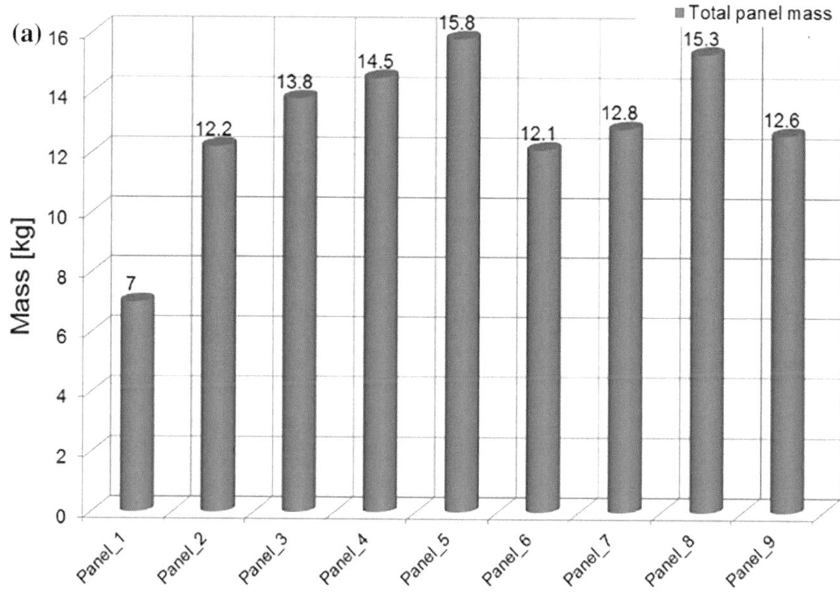

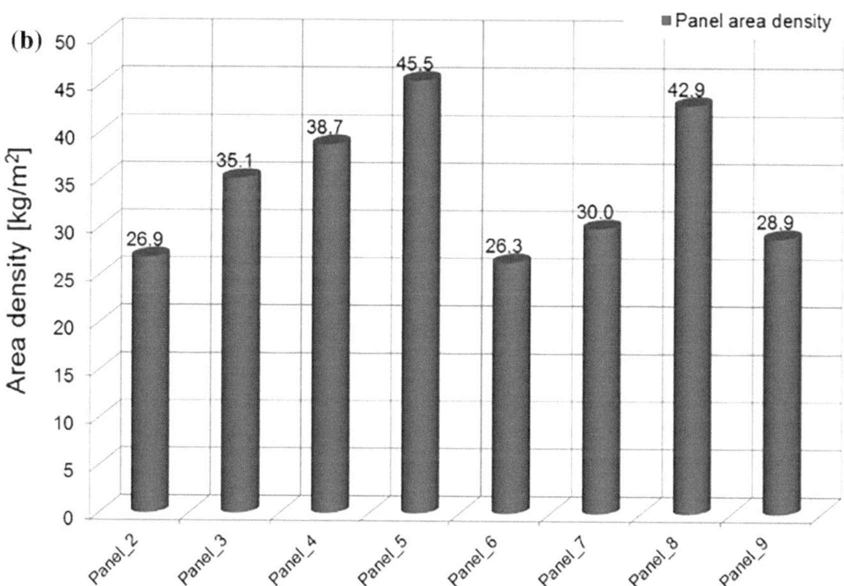

Fig. 4 Mass parameters comparison for all panels: **a** total panel mass; **b** panel area density

Panel_5 installed on the measuring rig after the test is shown in Fig. 7d.

In Panel_6 the witness plate has been reinforced with a layer of cork (NL25) and a layer of type 0.51 aluminum foam installed underneath it; with a thickness of 25 mm each (Table 2).

Fig. 5 Panel_1, witness plate, installed on the shock wave load testing rig: **a** pre-test; **b** post-test

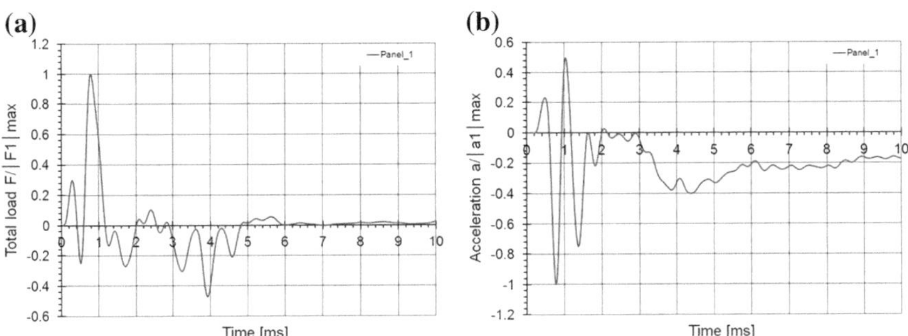

Fig. 6 Comparative graph of **a** normalized force; **b** normalized acceleration, as a function of time for the Panel_1 reference sample

Panel 7 has been made using an air filled metal cushion and reinforced on the top with an elastomer (Table 2). Its installation and condition after the trial is shown in Fig. 7f.

(a)
(b)
(c)
(d)
(e)
(f)
(g)
(h)

Fig. 7 Panels after proving ground tests. **a** Installed on the force and acceleration measurement rig after the tests. In this shielding variant, the witness plate has been reinforced with 2 layers of aluminum foam type 0.36, with a thickness of 2 × 25 mm, manufactured by CYMAT (Table 2). Due to the explosion shock wave, the aluminum layers were subject to indentation (closing of pores) and delamination from one another and the lower plate. **b** Present Panel_3 installed on the force and acceleration measurement rig after the tests. In this shielding variant, the witness plate has been reinforced with layers of type 0.51 aluminum foam manufactured by CYMAT, also with a thickness of 2 * 25 mm (Table 2). Due to the explosion shock wave, similarly to Panel 2, the aluminum layers were subject to indentation (closing of pores) and delamination from one another and the lower plate. **c** Presents Panel_4 installed on the force and acceleration measurement rig after the tests. The witness plate has been reinforced with layers of a type 0.36 and 0.51 aluminum alloy foam with a thickness of 25 mm each (Table 2). Similarly to panels 2 and 3 it was concluded that the layers of aluminum foam were subject to indentation and delamination from each other and the rig due to the explosion. **d**. The witness plate has been reinforced with layers of a type 0.51 aluminum and ASMA 55° ShA elastomer alloy, with a thickness of 25 mm each (Table 2). After the explosion, the elastomer delaminated. **e** Present photographs of the panel after the test. The aluminum foam and cork layers delaminated from the rig. The cork was subjected to defragmentation—it separated into smaller sections. **f** The cushion was subject to deformation. The elastomer was also indented. **g** Shows the panel installed on the measuring rig in a state after the test. The layers did not suffer separation, but were significantly indented by the explosion shock wave. Panels after proving ground tests. **h**—features an openwork metal construction with a layer of steel hemispheres set within it (Table 2). The aim of the structure was to disperse the shock wave caused by the explosive charge and in consequence reduce its negative impact [reference to the article]. As a result of the explosion, the outer surface of the openwork structure became indented. Most hemispheres separated from the surface—Fig. 7h

Panel 8 has been made using three layers identical to the previous trials: an outer elastomer layer, a middle cork layer and a lower aluminum foam layer (Table 2). It was tested in the same way as previous panels.

Panel 9—Further figures show measurement results of normalized force and acceleration, in relation to the reference measurement, for all panels.

The application of foamed materials limited and decreased extreme force and acceleration values.

Further figures show force and acceleration measurement results, normalized in relation to the reference panel. The comparison of graphs with the witness plate is more unfavourable than for panel 2. The course of the force graph shows many extremes during the first 5 ms. A large peak in acceleration has been noted (higher than for the witness plate). The course of acceleration is similar to the one shown for the witness plate.

Figure 8 Comparative graph of (a) normalized force; (b) normalized accelera-tionshow the acceleration and force measurement results for Panel_4, normalized in relation to the reference panel. Panel 4 has significantly improved the force and acceleration readings. For the force as well as the acceleration, the extremes in the initial 2 ms have been significantly decreased, as was their number. The change in force within the entire range and initial acceleration values has been significantly reduced. The extremes have decreased by over 60%.

The measurement results of normalized acceleration and force for Panel_5 are presented in Fig. 8. Panel 5 to a limited extent improved the characteristics of force

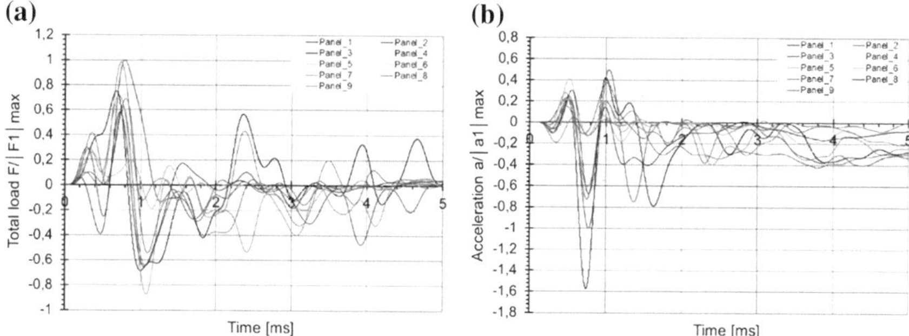

Fig. 8 Comparative graph of **a** normalized force; **b** normalized acceleration

and acceleration graphs. The extremes of force and acceleration in the initial graph phases have also been reduced. A large negative extreme appeared on the force graph.

Panel 6 has significantly suppressed the change of force and acceleration in the entire course of both functions, resulting in reduced extremes of force and acceleration. The maximum force was approximately 30% of the extreme force from the first trial. The acceleration was reduced by as much as 60%.

Panel 7 has not improved the change in force significantly. It has very noticeably suppressed accelerations throughout the entire scope of the process. The maximum acceleration was approximately 20% of the reference acceleration.

Panel 8 has improved the force and acceleration graph characteristics. The positive force extreme was reduced by 40% and the change in the graph during the subsequent phase was suppressed. However, a negative extreme appeared. The initial acceleration extremes are lower. The graph change was significantly suppressed in the phase after 2 ms. The maximum acceleration was 70% of the acceleration in Trial 3.

The application of Panel 9 on the plate has improved the force and acceleration graph characteristics—Fig. 8. The positive force extreme was reduced by 30% and the change in the graph during the middle and end phase was suppressed. The initial acceleration extremes are significantly lowered. The graph change was significantly suppressed in the further phase. The maximum acceleration is approximately 40% of the maximum value in Trial 3.

8 Study Result Analysis and Conclusions

Table 3 presents the maximum and minimum normalized total force and acceleration values. In order to graphically compare these two results, bar graphs were compiled and presented in Figs. 9 and 10. The first of them shows the force extremes, while the second shows acceleration extremes.

Table 3 Maximum and minimum normalized total force and acceleration values

No.	Panel	Max	Max	Min	Min
1	Panel_1	1.00	0.50	−0.47	−1.00
2	Panel_2	0.63	0.25	−0.65	−0.73
3	Panel_3	0.75	0.42	−0.68	−1.58
4	Panel_4	0.47	0.24	−0.14	−0.32
5	Panel_5	0.72	0.28	−0.88	−0.61
6	Panel_6	0.26	0.41	−0.20	−0.20
7	Panel_7	0.98	0.05	−0.32	−0.31
8	Panel_8	0.58	0.30	−0.66	−0.68
9	Panel_9	0.69	0.37	−0.54	−0.38

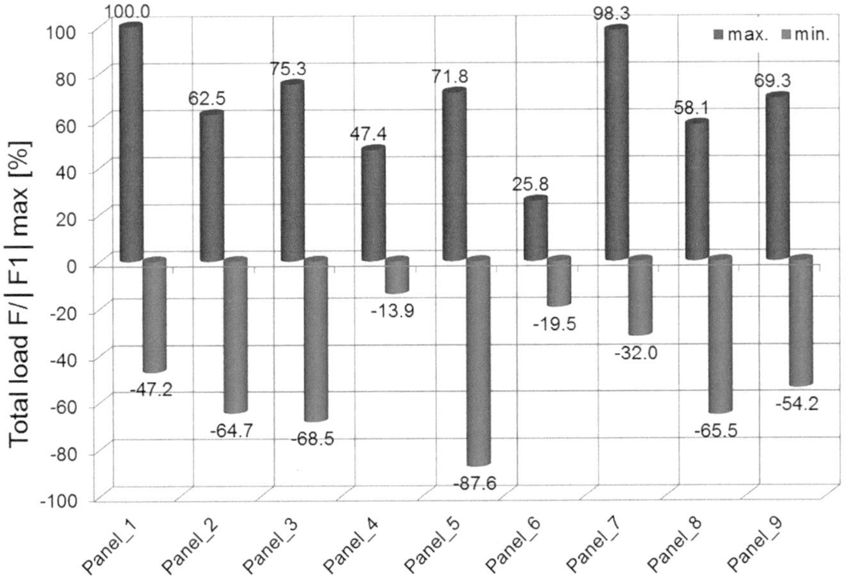

Fig. 9 Normalized total load extremes comparison for all panels

A significant acceleration reduction was achieved for panels 4 and 7—Fig. 10. The extremes were reduced by 70% in relation to Panel_1. In these terms panels: 5, 6, 8 and 9 have also proven to be effective, however to a lesser degree. Panel 3 has shown the weakest protective capabilities, as a large minimum extreme was recorded.

The use of panels resulted in the reaction force in the supports being reduced to a lesser degree compared to acceleration—Fig. 9. One of the best solutions is Panel 4, which has not only shown good properties in terms of acceleration reduction, but has also decreased the total load on supports by over 50%. Panel 6

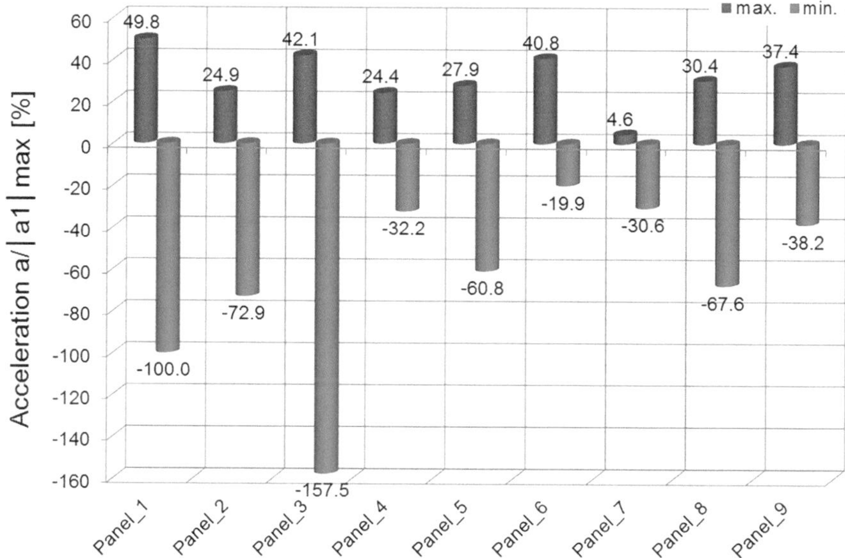

Fig. 10 Normalized acceleration extremes comparison for all panels

decreased the extremes by over 70%. The weakest panels 5 and 7 reduced the force by less than 20%. In the remaining cases (Panels 2, 3, 8 and 9) a reduction of reaction values of approximately 25–35% was achieved.

In the further analysis the acceleration and plate support load values were related to the mass. It is a very important factor, which often determines the possibility or impossibility to use a panel in the structure.

Figure 11 shows the extreme normalized force values (Table 3) divided by the panel's area density (Table 2). Panel 1 was excluded from the compilation. In these compilations Panels 4 and 6 have also proven the most effective. Good relative values were also achieved by Panel 8. The lowest results were shown by Panel 7.

Figure 12 shows the extreme ratio of normalized acceleration (Table 3) to the panel's area density (Table 2). Panel 1 was excluded from the compilation. Here Panel 4 has also shown the best suppressing properties. Good relative values were also achieved by Panels 5 and 7. The worst values relative to mass were achieved by Panels 2 and 3.

Difficulties in selecting the appropriate solution arise from the fact, that there was no possibility to directly observe the Panel's indentation process during the explosion shock wave impact. Therefore, interpretation of the results is very difficult and vague.

When selecting the best panel, it is necessary to consider all the tested parameters. A choice based on only one factor may prove to be a mistake, as not all aspects of the panel's properties are reflected.

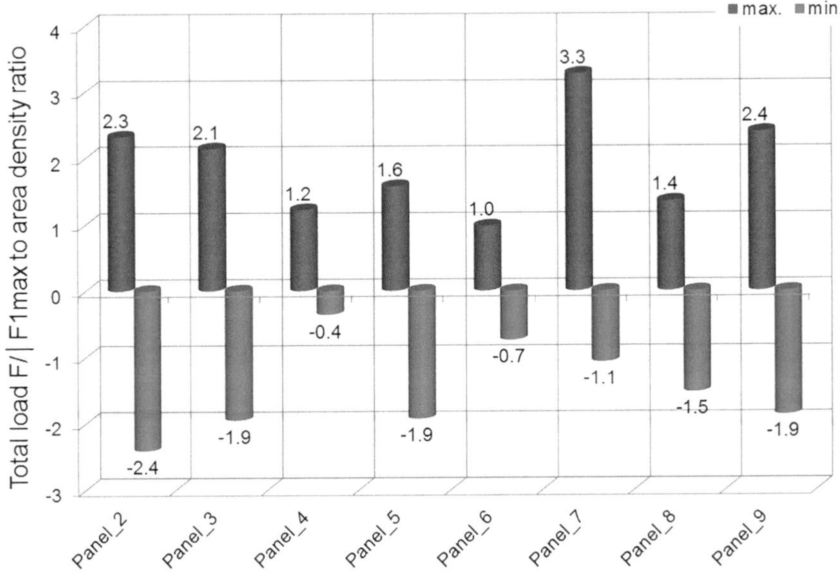

Fig. 11 Comparison of normalized load extremes to area density ratio (panels 2–9)

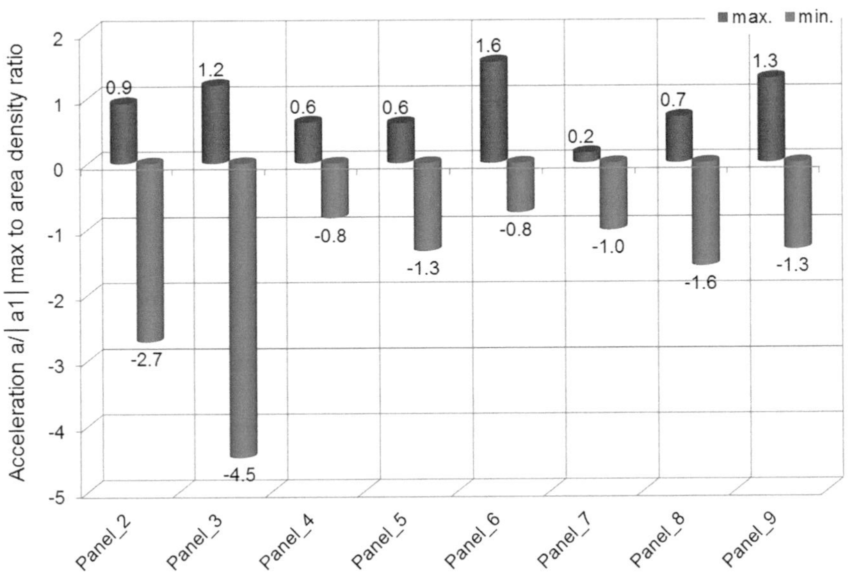

Fig. 12 Comparison of normalized acceleration extremes to area density ratio (panels 2–9)

Conclusions drawn from testing panels loaded with a shock wave caused by the Semtex 750 g charge detonation are as follows:

(1) Panels_2 and 3 show weak protective properties. The force and acceleration extremes are high, although lower than for the witness plate. The changes in graphs are similar to those shown for the individual witness plate. For Panel 3 the changes in graphs are more unfavourable than for Panel 2. the normalized force and acceleration values to mass ratio are also unfavourable compared to other panels;
(2) Panel 4 has significantly improved the force and acceleration readings. For the force as well as the acceleration, the extremes in the initial 2 ms have been significantly decreased, as was their number. The change in force within the entire range and initial acceleration values has been significantly reduced. Also, when relating the measured values to the area density, this panel has shown very good parameters;
(3) Panel 5 has improved the force and acceleration graph characteristics to a limited degree. The extremes of force and acceleration in the initial graph phases have been reduced. A large negative extreme appeared on the force graph;
(4) Panel 6 has significantly suppressed the change of force and acceleration in the entire course of both functions, resulting in reduced extremes of force and acceleration. It showed good results when relating its properties to the area density (Fig. 11);
(5) Panel 7 has not improved the change in force. It has very noticeably suppressed accelerations throughout the entire scope of the process;
(6) Panel 8 has improved the force and acceleration graph characteristics. The positive force extreme was reduced and the change in the graph during the subsequent phase was suppressed. However, a negative extreme appeared. The initial acceleration extremes are lower. The graph change was significantly suppressed in the phase after 2 ms. It showed positive results in comparisons of mass.
(7) Panel 9 has improved the force and acceleration graph characteristics. The positive force extreme was reduced and the change in the graph during the middle and end phase was suppressed. The initial acceleration extremes are significantly lowered. The graph change was significantly suppressed in the later phase. It shows inferior results in relation to other panels when comparing the measured parameters (particularly load) in relation to its mass.

Acknowledgements The study has been supported by the National Centre for Research and Development, Poland, as a part of a research—project No. DOBR-BIO4/022/13149/2013. This support is gratefully acknowledged.

References

1. Materiały informacyjne Foreign Policy at Brookings Tracks Security and Reconstruction in Afghanistan, Iraq and, Pakistan. Retrieved November 30, 2011 from http://www.brookings.edu/~/media/centers/saban/iraq%20index/index20111130.pdf.
2. Krzystała, E., Mężyk, A., & Kciuk, S. (2011). Analiza zagrożenia załogi w wyniku wybuchu ładunku pod kołowym pojazdem opancerzonym. Zeszyty naukowe WSOWL. Nr 1 (159).
3. Kania, E. (2009). Development tendency of landmine protection devices. *Modelling and optimization of physical systems* (pp. 67–72) Gliwice.
4. Modrzewski, J. (2010). Development of new generation of M-ATV class mine-proof vehicles resistant against warheads IED and EFP and PG-7 missiles. *Problemy Techniki Uzbrojenia* R. 39, z. 116, s. 67–79.
5. Brzozowski, R., Guła, P., & Sanak, T. (2014). Obrażenia powybuchowe tkanek miękkich. W: Bezpieczeństwo wojsk w aspekcie zagrożeń wynikających z użycia improwizowanych urządzeń wybuchowych (IED), Wydaw. Akademii Obrony Narodowej. s. 117–124. ISBN: 978-83-7523-376-6.
6. Klasztorny, M., & Swierczewski, M. (2013). Use of ALFC shield for passive protection of occupants of light armoured vehicles. *Computer Methods in Mechanics*.

Temperature Correction of Measurements Results of 3-Axis Accelerometers in IMU Modules

Witold Ilewicz, Damian Bereska, Marcin Pacholczyk and Aleksander Nawrat

1 Introduction

One of the most important factors affecting the accuracy of measurements obtained using acceleration sensors made in MEMS technology (Micro-Electro-Mechanical System) is the ambient temperature. Hence, the production of IMU modules (Inertial Measurement Unit) equipped with this type of sensor, requires studies on the effects of temperature changes on the measurements and the temperature calibration procedure to be included in the IMU firmware [1–4]. This study used data of temperature parameters of 3-axis acceleration sensors used in the IMU modules developed at the Institute of Automatic Control, Silesian University of Technology. Studies were conducted using the climate chamber (Feutron, type 3007), in which we placed six IMU modules containing different sensors. The tests included changes of temperature in range from −10 to +60 °C with 10 °C step. We monitored the effect of temperature change on the value of zeros ($\Delta Z_{acc}(T)$—change of the position of the sensor zero related to the zero at the reference temperature) and gains ($\delta W_{acc}(T)$—relative change of the sensor gain, related to the gain in the reference temperature) of tested sensors and collected data for correcting the influence of temperature. A detailed description of the method for determining the temperature characteristics of zeros and gains of tested sensors in presented

W. Ilewicz (✉) · D. Bereska · M. Pacholczyk · A. Nawrat
Institute of Automatic Control, Silesian University of Technology,
Akademicka 16, 44-100 Gliwice, Poland
e-mail: witold.ilewicz@polsl.pl

A. Nawrat et al. (eds.), *Advanced Technologies in Practical Applications for National Security*, Studies in Systems, Decision and Control 106,
https://doi.org/10.1007/978-3-319-64674-9_17

Table 1 Parameters of the sensors used in IMU modules

Module name	Accelerometer type, A/D resolution	Housing
IMU1	ADXL203, 16 bits	
IMU2	ADXL203, 16 bits	
IMU3	ADXL203, 16 bits	
IMU4	ADXL203, 16 bits	
IMU5	LSM303DLM, 12 bits	
IMU6	LSM303DLM, 12 bits	

elsewhere [5–7]. Table 1 shows the detailed description of the sensors used in tested IMU modules.

Table 2 shows how the change of temperature influences the gains of tested accelerometers.

In Table 3 the measured zeros of tested sensors are presented as a function of temperature.

Table 2 Thermal characteristics of gain of accelerometers $\delta W_{acc}(T)$, ppm

Module	Axis	Temperature (°C)							
		−10	0	10	20	30	40	50	60
IMU1	X	−498	−268	−115	0	153	306	402	383
	Y	−692	−442	−231	0	231	346	442	577
	Z	−575	−268	−115	0	192	192	422	441
IMU2	X	−379	−227	−152	0	76	76	−38	−379
	Y	762	209	−57	0	990	1143	−1486	−1676
	Z	−615	−288	−154	0	−77	−346	−653	−615
IMU3	X	−382	−229	−38	0	76	191	191	38
	Y	−687	−439	−229	0	76	−248	−439	−573
	Z	−496	−267	−153	0	−38	−38	−248	−343
IMU4	X	−1153	−769	−423	0	192	192	269	538
	Y	−723	−418	−190	0	0	38	−114	−304
	Z	−459	−172	−96	0	96	440	650	1243
IMU5	X	−606	−273	61	0	667	667	−637	−2001
	Y	7655	5217	2284	0	−2346	−4352	−7655	−10495
	Z	−1379	−1379	−1199	0	−600	−450	−1289	−3238
IMU6	X	−2166	−879	−534	0	502	502	−1193	−2417
	Y	11490	7516	3758	0	−3265	−6500	−9088	−12384
	Z	1084	361	301	0	−903	−3132	−6354	−9727

Reference temperature $T_0 = 20$ °C

Table 3 Thermal characteristics of zeros of accelerometers $\Delta Z_{acc}(T)$ [$^{o}/_{oo}$]

Module	Axis	Temperature (°C)							
		−10	0	10	20	30	40	50	60
IMU1	X	−5.5	−3.6	−1.7	0.0	1.5	2.9	4.3	5.7
	Y	0.4	0.4	0.2	0.0	−0.2	−0.1	0.1	0.1
	Z	4.4	2.9	1.7	0.0	−1.3	−1.3	−4.4	−5.8
IMU2	X	−16.5	−9.4	−2.8	0.0	1.5	1.5	4.4	6.1
	Y	3.6	2.9	1.3	0.0	0.8	−0.4	−7.5	−9.0
	Z	6.1	3.7	1.8	0.0	−1.7	−3.3	−5.4	−7.2
IMU3	X	−6.6	−3.8	−1.9	0.0	1.5	2.9	4.5	6.2
	Y	−1.8	−1.1	−0.5	0.0	0.6	1.4	2.0	3.1
	Z	−1.2	−0.6	−0.3	0.0	0.2	0.2	1.0	1.6
IMU4	X	2.2	1.7	1.0	0.0	−0.8	−0.8	−2.0	−3.5
	Y	−2.0	−1.1	−0.5	0.0	0.4	0.6	1.3	1.9
	Z	−4.4	−2.5	−0.9	0.0	1.1	2.6	3.9	5.1
IMU5	X	−17.6	−12.6	−5.5	0.0	7.7	7.7	20.3	22.0
	Y	67.2	42.3	20.8	0.0	−16.8	−30.3	−48.8	−61.1
	Z	20.7	12.8	5.8	0.0	−2.6	−4.8	−5.0	0.6
IMU6	X	134.5	83.1	38.9	0.0	−25.7	−25.7	−76.0	−79.1
	Y	207.5	128.6	63.0	0.0	−41.5	−72.9	−89.9	−101.2
	Z	−45.2	−30.4	−12.1	0.0	12.8	24.5	35.0	41.2

Reference temperature $T_0 = 20$ °C

2 Temperature Correction of Acceleration Measurement

After calibration in the reference temperature (T_0 = 20 °C), characteristic of linear acceleration sensor is given by:

$$a = b_0 + b_1 \cdot ACC(T_0) \tag{1}$$

where a—is the measured acceleration; b_0, b_1—are coefficients of the linear model of the sensor identified at the reference temperature; ACC—raw reading of the sensor at the reference temperature (digital output from the A/D converter). Readings of sensor calibrated using the above procedure are correct (up to the accuracy of calibration errors) at the reference temperature. Due to temperature change the raw reading of the sensor changes according to the relationship:

$$ACC(T) = (1 + \delta W_{acc}(T)) \cdot ACC(T_0) + \Delta Z_{acc}(T) \tag{2}$$

Knowing the characteristics $\delta W_{acc}(T)$ and $\Delta Z_{acc}(T)$ one can determine correctly the acceleration value based on raw sensor reading $ACC(T)$ at the temperature T that differs from the T_0 according to the formula:

$$a = b_0 + b_1 \cdot \frac{ACC(T) - \Delta Z_{acc}(T)}{1 + \delta W_{acc}(T)} \tag{3}$$

The procedure does not require the coefficients of the linear model of the sensor b_0 and b_1 recalibration—it is sufficient to calibrate the sensor once, at the reference temperature T_0.

Experimental characteristics of $\delta W_{acc}(T)$ and $\Delta Z_{acc}(T)$ are known at discrete points in the domain of temperature T (from −10 to 60 °C with 10 °C step), as shown in Tables 2 and 3. In order to apply the procedure defined by the formula (3) it is necessary to restore the values of these characteristics for the points in the domain of temperature located between nodes with known values. Most often an approximation of the temperature characteristic by polynomial of appropriate degree is used in such way that the error does not exceed a predefined value.

3 Approximation of the Temperature Characteristics by Polynomial Functions

Shapes of characteristics for the acceleration sensors based on the data from Tables 2 and 3 are shown in Figs. 1 and 2 (connected dotted line). The numbers 1–3 represent acceleration sensors in the IMU1 module in the X, Y and Z axis respectively, numbers 4–6 represent acceleration sensors in the IMU2 module, etc. Presented temperature characteristics were approximated by 1st, 2nd and 3rd degree

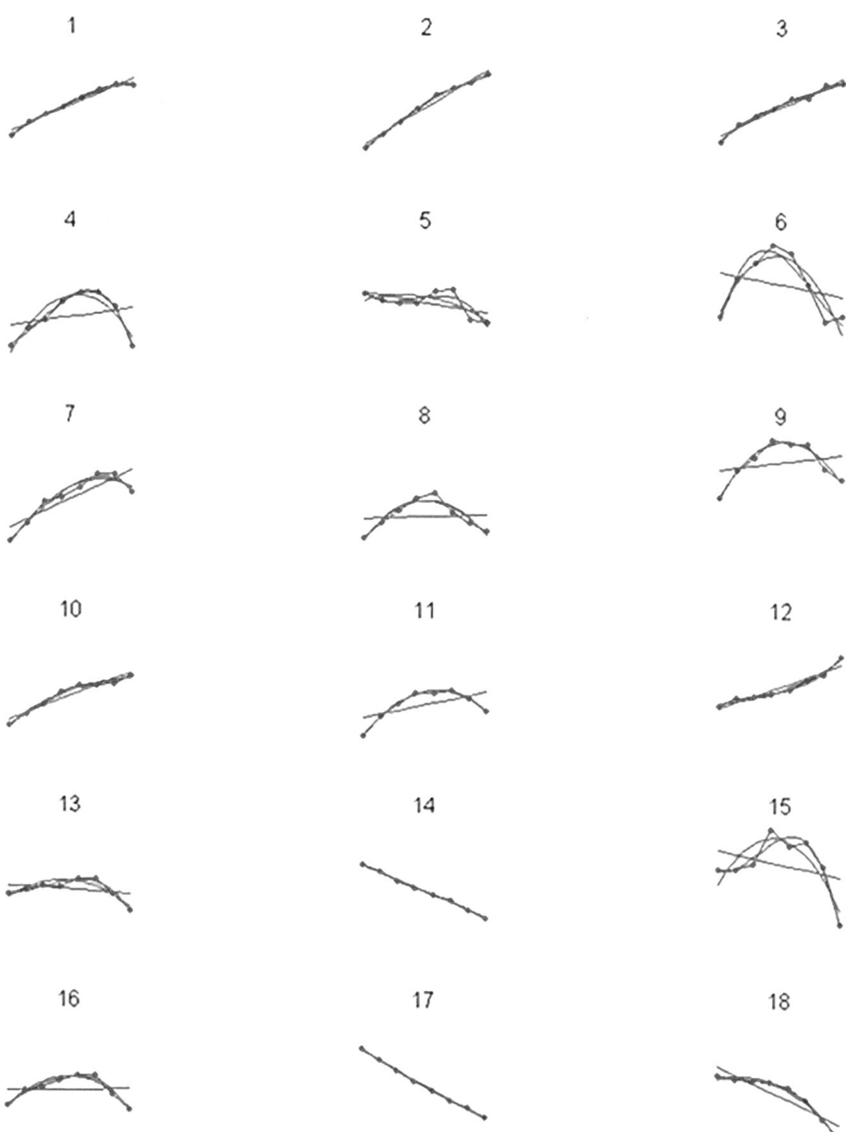

Fig. 1 Shapes of temperature characteristics of gain $\delta W_{acc}(T)$ and their approximations by 1st, 2nd and 3rd degree polynomials. Data in Table 4

polynomials. In order to visually assess the quality of approximation the shapes of approximating polynomials has been added (red, solid line) to the charts in Figs. 1 and 2. The values of the maximum error of approximation expressed in % of the range of variation of given characteristic are given in Tables 4 and 5.

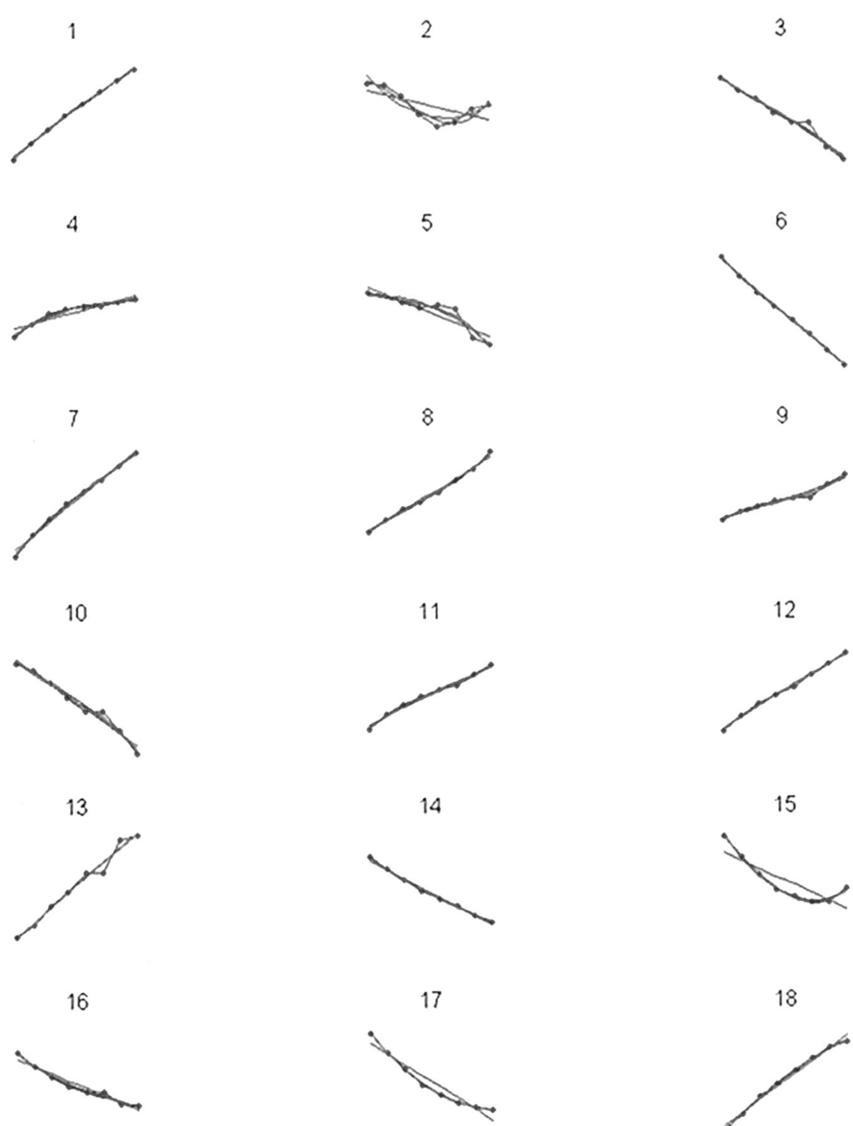

Fig. 2 Shapes of temperature characteristics of zeros $\Delta Z_{acc}(T)$ and their approximations by 1st, 2nd and 3rd degree polynomials. Data in Table 5

Table 4 Maximal approximation error of temperature characteristics $\delta W_{acc}(T)$ of tested sensors for 1st, 2nd and 3rd degree polynomials in % of $\delta W_{acc}(T)$ range

Module	Axis	Number	1st order polynomial	2nd order polymomial	3rd order polynomial
IMU1	X	1	7.3	4.8	3.9
	Y	2	8.7	3.3	3.1
	Z	3	8.5	5.4	6.7
IMU2	X	4	42.6	16.1	8.3
	Y	5	54.3	44.3	25.5
	Z	6	50.3	23.2	14.1
IMU3	X	7	18.6	9.9	9.1
	Y	8	50.9	17.2	18.3
	Z	9	41.7	7.9	8.2
IMU4	X	10	15.2	6.3	5.4
	Y	11	32.1	4.0	4.0
	Z	12	17.1	6.8	5.3
IMU5	X	13	41.0	24.5	10.4
	Y	14	3.7	3.3	1.9
	Z	15	34.8	15.0	15.9
IMU6	X	16	43.6	21.1	13.0
	Y	17	3.0	1.4	1.2
	Z	18	19.5	5.5	3.2

Table 5 Maximal approximation error of temperature characteristics of zeros $\Delta Z_{acc}(T)$ of tested sensors for 1st, 2nd and 3rd degree polynomials in % of $\Delta Z_{acc}(T)$ range

Module	Axis	Number	1st order plynomial	2nd order polymomial	3rd order polynomial
IMU1	X	1	3.0	0.8	0.6
	Y	2	36.6	21.4	19.8
	Z	3	12.3	11.3	8.8
IMU2	X	4	15.1	8.6	2.9
	Y	5	25.3	18.4	12.8
	Z	6	3.0	2.1	0.6
IMU3	X	7	4.1	2.1	0.6
	Y	8	6.1	3.1	1.7
	Z	9	9.4	5.8	4.8
IMU4	X	10	11.1	9.2	6.5
	Y	11	5.1	3.8	2.4
	Z	12	4.3	3.6	1.8
IMU5	X	13	7.5	7.6	7.5
	Y	14	5.5	2.3	1.5
	Z	15	31.8	4.7	3.2
IMU6	X	16	11.1	10.1	8.0
	Y	17	13.4	1.1	1.0
	Z	18	4.0	1.5	2.0

4 Results and Discussion

The results in Figs. 1 and 2 show that the temperature error characteristics of zeros and gains are non-linear. One can distinguish 3 types of such characteristics. First—some of the characteristics are almost linear (e.g. characteristics 14 and 17 shown in Fig. 1, or the characteristics 1 and 6 in Fig. 2). The maximum error of approximation of these characteristics by 1st degree polynomial is about 3–4% of the range of their value and the use of 2nd and 3rd degree polynomials only slightly improves the accuracy of approximation. Second—weakly non-linear characteristics for which the error of approximation by 1st degree polynomial is in the range of 7–20% (e.g. characteristics 7 and 10 in Fig. 1). For such characteristics the approximation by 2nd and 3rd degree polynomial leads to a significant decrease in the approximation error as compared to the 1st degree polynomial. Third—some of the characteristics are strongly non-linear and error of approximation by 1st degree polynomial exceeds 30% of the variation range. It is apparent that the strong non-linearity is much more frequently observed in the case of temperature characteristics of gains (8 cases) as compared to the characteristics of zeros (2 cases).

Temperature characteristics shown in Figs. 1 and 2 can also differ in terms of shape regularity. Some of the characteristics can be irregular, which makes them difficult to approximate correctly. This applies particularly to the characteristics 5, 6 and 15 in Fig. 1 and 2, 3, 5 and 13 in Fig. 2. The reason for some of the irregularities may be measurement errors during the temperature tests (e.g. 3 and 13 in Fig. 2)—in this case outliers should be removed from the characteristics or a robust method to estimate the parameters of the polynomial should be used. The characteristics of type 5 shown in Fig. 1 and 2 in Fig. 2 require polynomials of degree higher than 3 or different type of approximating function (e.g. non-linear with respect to parameters) in order to improve the accuracy of approximation.

The above considerations relate to temperature characteristics for which ranges of variation are large, because if small ranges are considered, even a strong non-linearity has no significant effect on the results and the correction may not be necessary.

5 Summary and Conclusions

Currently, the global market is dominated by the sensors manufactured with semiconductor technologies, correction of temperature influence is an essential aspect of handling this type of sensors. As a function for approximating temperature characteristics (based on which the correction is made), most often a polynomials are used in practice, due to their simplicity. An interesting alternative can be approximating functions non-linear with respect to the parameters, which would reduce the number of model parameters. There has also been a strong tendency to include some sort of software correction procedure directly into the logic circuit of temperature sensor, as opposed to hardware solutions with passive or active correction circuits.

Acknowledgements This work has been supported by the Polish National Centre for Research and Development as a project DOB-BIO/7/13/05/2015 "WIMA—a virtual mast as a platform for observation sensors adopted for the needs of the National Border Guards".

References

1. Gang, D., Mei, L., Xiaoping, H., Lianming, D., Beibei, S., & Wei, S. (2011). Thermal drift analysis using a multiphysics model of bulk silicon MEMS capacitive accelerometer. *Sensors and Actuators, A: Physical, 172*(2), 369–378.
2. Jedrasiak, K., Daniec, K., & Nawrat, A. (2013). The low cost micro inertial measurement unit. In *8th IEEE Conference on Industrial Electronics and Applications (ICIEA)* (pp. 403–408).
3. Nawrat, A., Jedrasiak, K., Daniec, K., & Koteras, R. (2012). Inertial navigation systems and its practical applications. In *New approach of indoor and outdoor localization systems*. InTech.
4. Shiau, J. K., Huang, C. X., & Chang, M. Y. (2012). Noise characteristics of MEMS Gyro's null drift and temperature compensation. *Journal of Applied Science and Engineering, 15*(3), 239–246.
5. Bereska, D., Ilewicz, W., Daniec, K., Koteras, R., Fraś, S., Jędrasiak, K., et al. (2014) Badanie wpływu temperatury na wskazania 3-osiowych czujników przyspieszenia w modułach IMU. Automatyzacja procesów dyskretnych. Teoria i zastosowania. Pod red. Andrzeja Świerniaka i Jolanty Krystek. Gliwice: Wydaw. Pracowni Komputerowej Jacka Skalmierskiego (s. 33–40) ISBN 978-83-62652-68-6.
6. Daniec, K., Ilewicz, W., Bereska, D., Koteras, R., Fraś, S., Jędrasiak, K., & Nawrat, A. (2014) Badanie wpływu temperatury na wskazania czujników prędkości kątowej modułach IMU. Automatyzacja procesów dyskretnych. Teoria i zastosowania. Pod red. Andrzeja Świerniaka i Jolanty Krystek. Gliwice: Wydaw. Pracowni Komputerowej Jacka Skalmierskiego, (s. 75–82) ISBN 978-83-62652-68-6.
7. Bereska, D., Daniec, K., Ilewicz, W., Jędrasiak, K., Koteras, R., Nawrat, A., & Pacholczy, K. M. (2016). Influence of temperature on measurements of 3-axial accelerometers and gyroscopes embedded into inertial measurement unit. In *Signals and Electronic Systems (ICSES), 2016 International Conference on Krakow* (pp. 200–205).

Experimental Investigation of IED Interrogation Arm During Normal Operation and Mine Flail Structure Subjected to Blast Loading

Wiesław Barnat, Andrzej Kiczko, Paweł Gotowicki, Paweł Dybcio, Marcin Szczepaniak and Wiesław Jasiński

1 Introduction

This article is dedicated to the issue of protection of lightly armored vehicles against the adverse effects of improvised explosive devices. Light armored vehicles are widely used due to many advantages such as high mobility. Unfortunately due to the light armor vehicles of this type are often the victims of attacks using IEDs, whereby the vehicle is destroyed (Fig. 1). The current technical solutions to protect the lightly armored vehicles can be divided into the following categories:

- Ballistic protection,
- Protection against mines,
- Protection against IEDs.

It is recognized that the design of the outer casing of special vehicles in a manner that protects the crew and the vehicle [1] from the destructive effects of different types of IEDs, is a difficult problem. It is an active research area.

An important research issue is the search for new types of solutions for solving a given problem of neutralizing IEDs. One potential solution to the problem is to use techniques of passive protection [2–5]. Alternative solutions can be developed through simulation studies, numerical and field experiments.

W. Barnat (✉) · A. Kiczko · P. Gotowicki · P. Dybcio
Department of Mechanics and Applied Computer Science,
Military University of Technology, Gen. S. Kaliskiego 2 Street,
00-908 Warsaw, Poland
e-mail: wbarnat@wat.edu.pll

M. Szczepaniak · W. Jasiński
Military Institute of Engineer Technology, Obornicka 136 Street,
50-961 Wrocław, Poland
e-mail: szczepaniak@witi.wroc.pl

A. Nawrat et al. (eds.), *Advanced Technologies in Practical Applications for National Security*, Studies in Systems, Decision and Control 106,
https://doi.org/10.1007/978-3-319-64674-9_18

Fig. 1 The results of IED explosion

One of the possible ways to disable the deadly capabilities of the IEDs is to early detect and neutralize them. There are various vehicles designed for the purpose (Fig. 2).

One of goals of the research project was to design an interrogation arm capable of neutralizing the IEDs. The concept and analytical work was documented (Fig. 3).

In order to measure the displacement of the selected points located on the interrogations arm computer model was developed. Using the computer model and numerical calculations the deformation under the IED explosion was simulated. Further field tests were required to verify the acquired numerical results from the simulation.

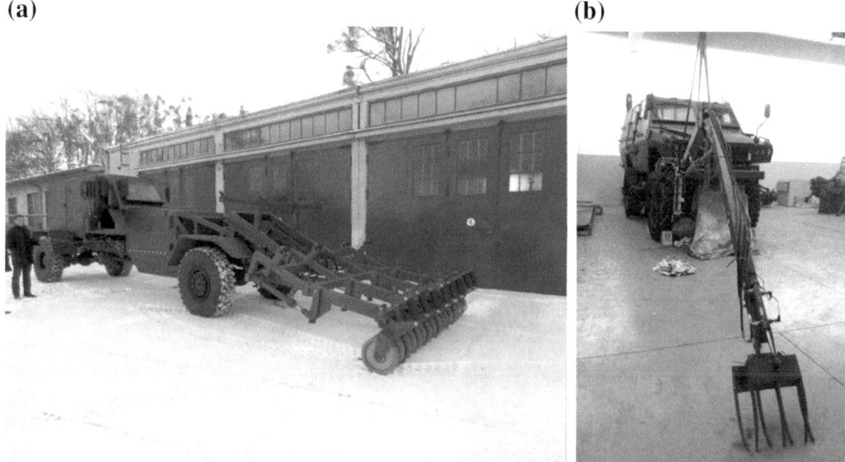

Fig. 2 **a** SHIBA anti-mine vehicle, **b** ŻUBR anti-mine vehicle

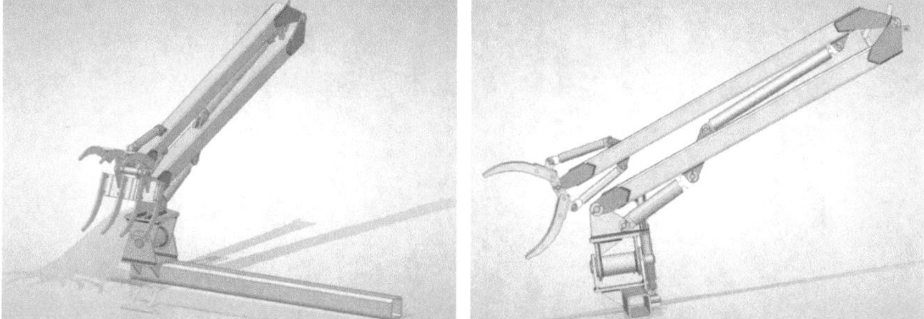

Fig. 3 The design of the anti-mine arm. It's goal is to nullify the IEDs

2 Goal of the Experiments

The main goal of the performed tests were to measure the strain of the Shiba's interrogation arm during standard operation. For the purpose the electroresistance based sensors were selected. The Vishay EA 06 060LZ 120 sensor together with ESAM Traveller CF was used in order to measure the strain (Figs. 4 and 5).

The sensors for measuring linear strain were installed on the interrogation arm. The placements of sensors (Fig. 6) guaranteed measuring strains normal to the

Fig. 4 The devices installed within the special purpose vehicle

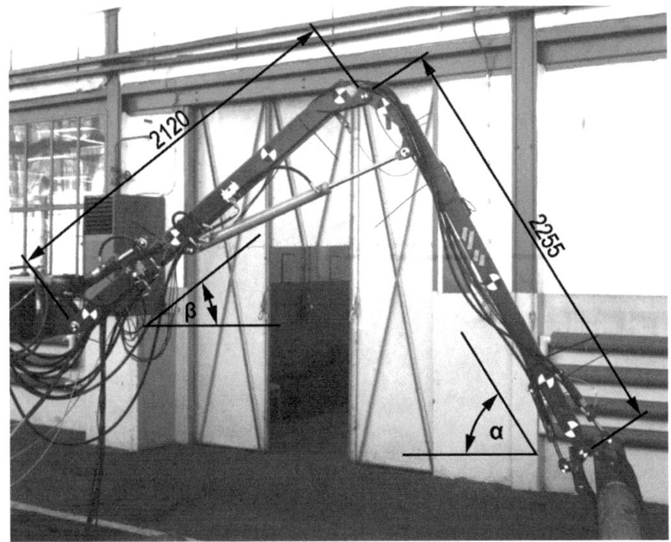

Fig. 5 The dimensions of the device overlaid on the photography of the interrogation

profile of the axis of the arm. The frequency of sensors was selected to 50 Hz. The data acquired during tests is presented in Fig. 5. It can be observed that the measurements were performed with 1 mm precision.

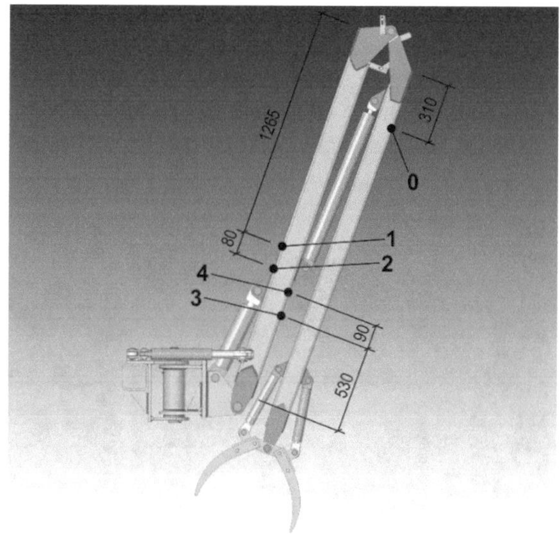

Fig. 6 The placement of sensors for strain measurement

3 Experimental Tests

Data specifying construction points of the vehicle's arm were set in the software. The device was tested by moving up and down a 2.7 m pipe with diameter 0.22 m and thickness of the wall 0.01 m. Middle of the pipe was selected as the grab point.

During another test the arm was installed on mobile station for testing. Parker hydraulic control system was used to control the arm. The pump was installed inside the special vehicle (Fig. 7).

The equipment used to measure the strains was installed on the ground. Results of the tests with microstrains visible (10–6 m/m) are presented in charts. During test

Fig. 7 The photography of hydraulic pump. The equipment was installed inside

preparation special tags were placed on the construction (α—arm 2, β—arm 1). The tracking software "TEMA" was used to track the tags. Photography's from the tests are presented in Fig. 8.

The results of the strain test for the selected 9 points is presented in Fig. 9. Figure 10. The angle located between horizontal level and arms.

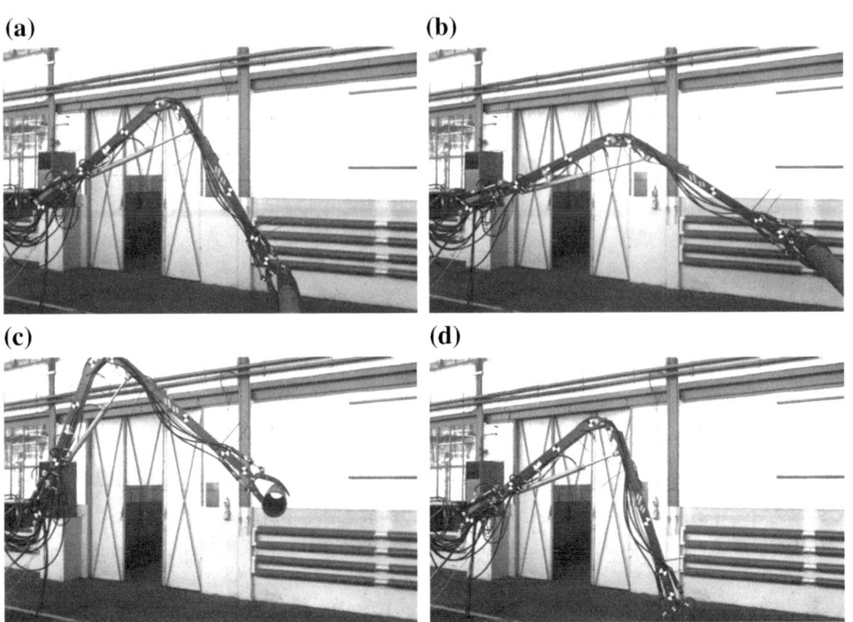

Fig. 8 **a** 31 s, **b** 42 s, **c** 56 s, **d** 104 s

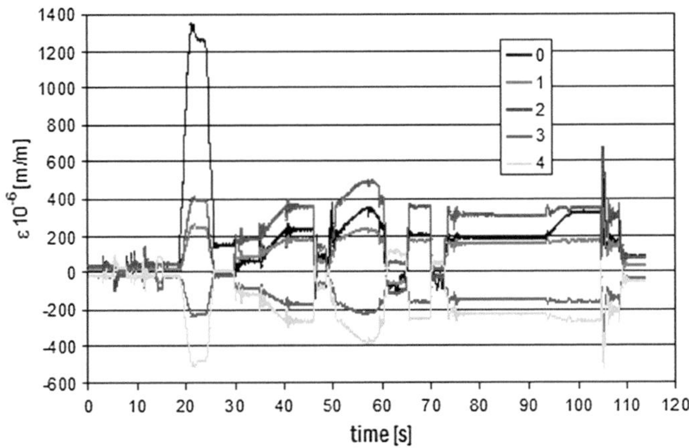

Fig. 9 The diagram of ε-t. Discrete points 0÷4 were selected

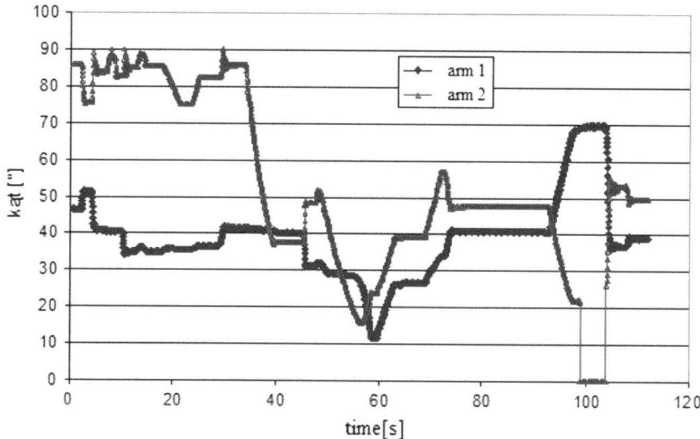

Fig. 10 Vertical angles of arms (1 and 2) in relation to time

4 Anti-mine Vehicle Experimental Research

The aim of research was experimental measurement of deformation of selected structural elements of anti-mine vehicle Shiba in terms of loads during operation. The work was carried out in cooperation of WAT—WITI for military training. Figure 11 is a diagram of the test subject—trawl and distribution of measuring points with marked characteristic distances. During the study deformation of the structure in measuring points was measured.

The study used trawl method for electro resistive strain measurement using strain gauges Vishay EA 06 060LZ 120 and the bridge ESAM Traveller CF.

Strain gauges were glued in pairs on both sides of the plane of symmetry of the trawl segment. The measurement points positioned at the centre of the wall sections of the upper side.

Strain line were provided in such a way as to use them to measure the deformation of the normal to the axis of the profiles. The research was prepared according to the time-deformation (ε-t), as shown in Figs. 16, 17, 18, 19, 20, 21 and 22 was used sampling rate of 100 kHz.

Figure 12 show the main views: a side view and the top view of the anti-mine vehicle with installed trawl.

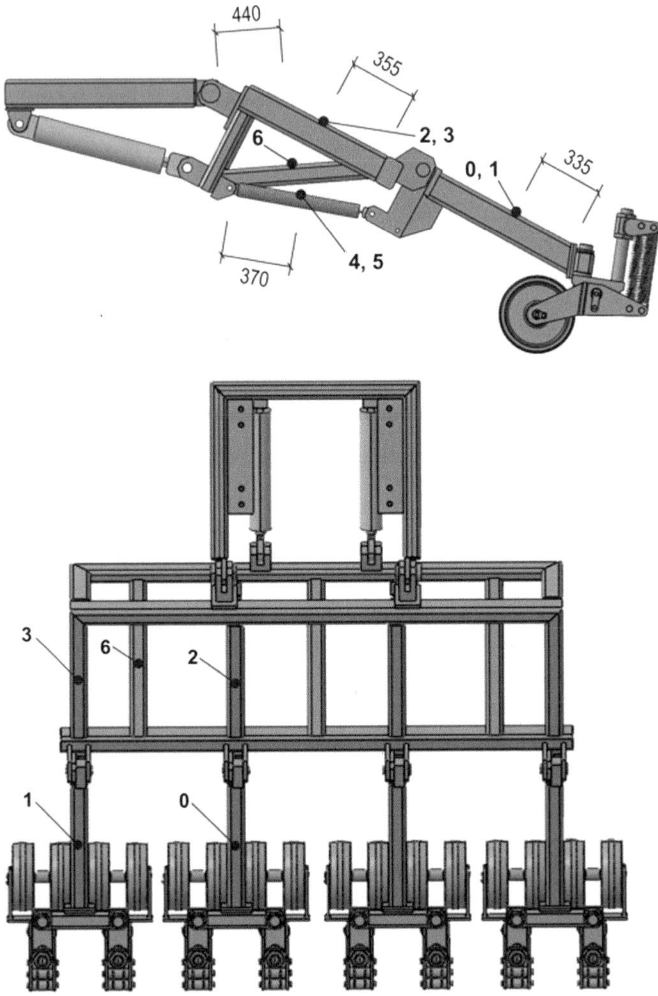

Fig. 11 Schematic layout of measuring points

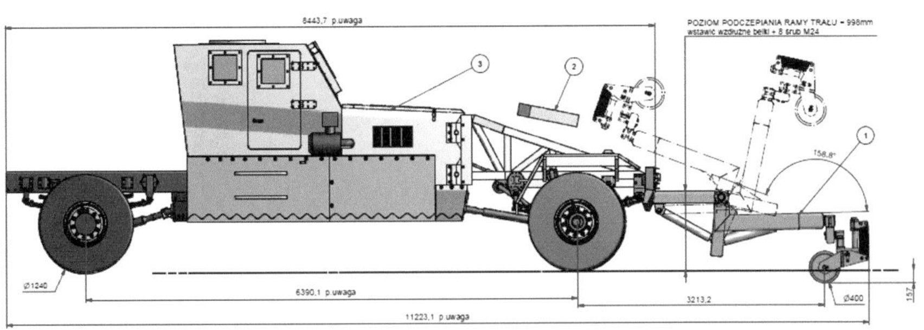

Fig. 12 Shiba vehicle with mounted the pressing trawl

5 Experimental Results

The research program included measurements of deformation of selected parts of the structure of anti-mine Shiba during the shock (Fig. 13).

The research position consisted of trawl attached to the subframe of the loaded weight of 26.7 kN. Figures 14 and 15 show the sequence of selected frames of recorded images while trying polygonal dynamic load caused by the detonation 8 kg of TNT cast in the form of 20 blocks of 400 g., Installed directly below the wheel of anti-mine vehicle (Fig. 13).

Images were recorded using high-speed cameras. The following charts show the recorded deformation values for the individual measuring points, expressed in units of μm/m (microstrain). The recorded waveforms were processed with averaging filter (Figs. 16, 17, 18, 19, 20, 21 and 22).

Fig. 13 8 kg load installed under the wheel of anti-mine vehicle

Fig. 14 Images recorded during the experiment by the fast camera no. 2

Fig. 15 Images recorded during the experiment by the fast camera no. 1

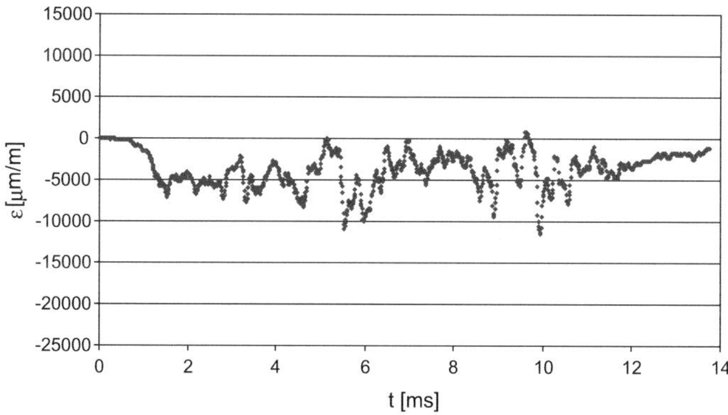

Fig. 16 The dependence of ε-t for measuring point 0

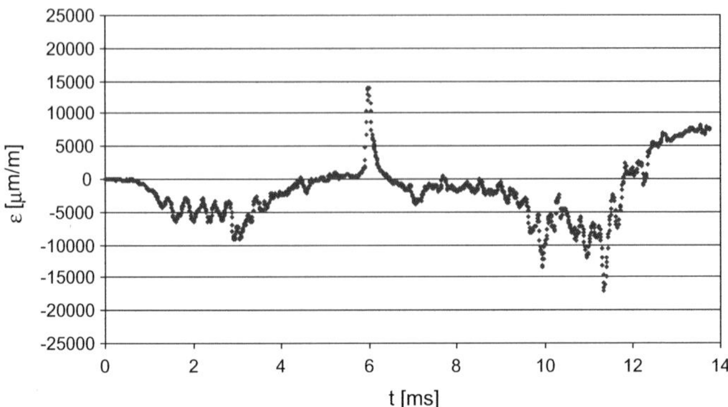

Fig. 17 The dependence of ε-t for measuring point 1

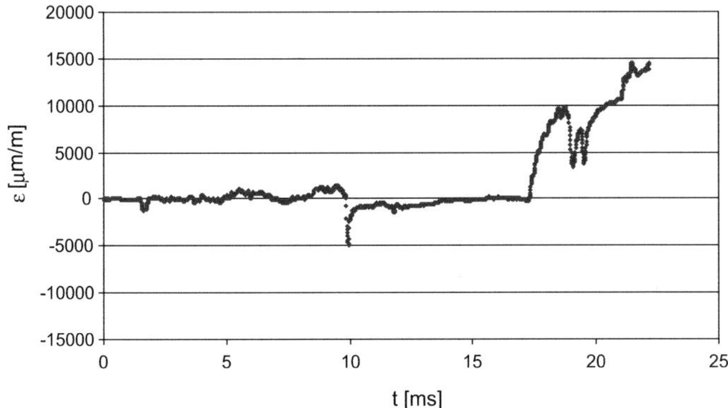

Fig. 18 The dependence of ε-t for measuring point 2

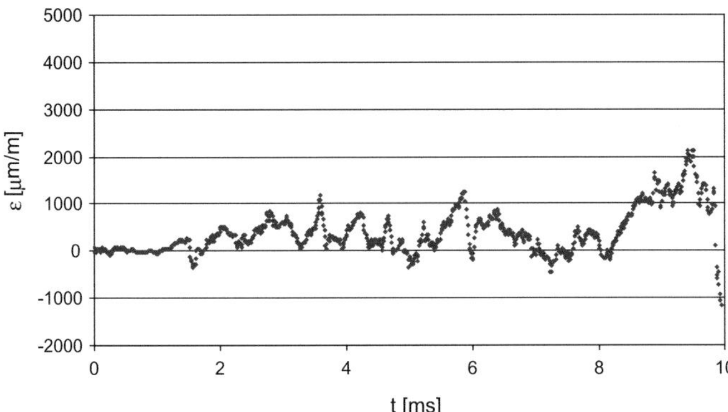

Fig. 19 The dependence of ε-t for measuring point 3

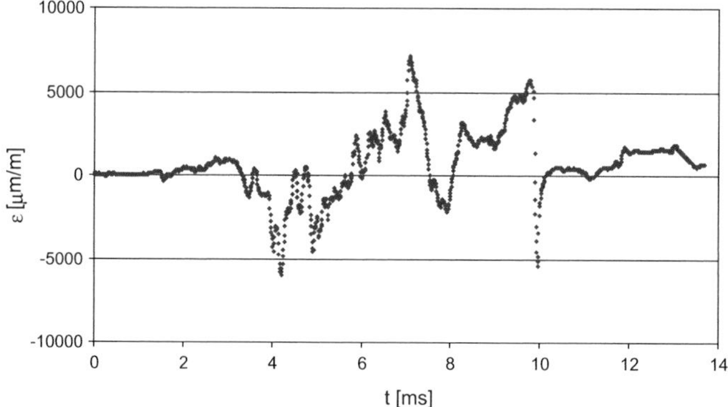

Fig. 20 The dependence of ε-t for measuring point 4

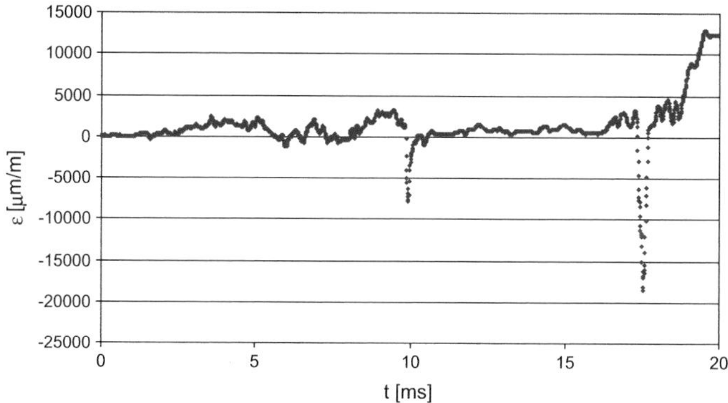

Fig. 21 The dependence of ε-t for measuring point 5

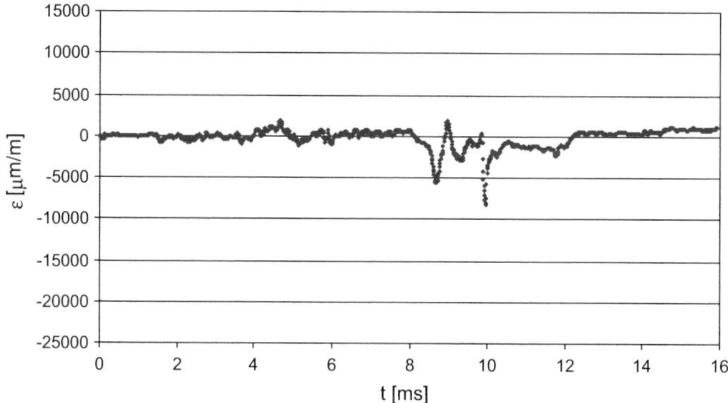

Fig. 22 The dependence of ε-t for measuring point 6

Fig. 23 Anti-mine vehicle after destructive testing

6 Conclusions

Authors in the chapter presented numerical models that successfully verified results of the field tests of the designed equipment.

The main problem in the design of pieces of equipment that works on pulse pressure or force is adequate modelling of physics phenomena.

There is a possibility that strain gauge method could measure the deformation of the structure dynamically loaded. The advantage of this method is the relatively low price of sensors—that when you try to be destroyed (Fig. 23). In addition, the test results are very important because they make the experimental results that in numeric be authenticated.

During the study of explosive recorded sequences of frames with high-speed cameras can be used only in the initial period. In further period gases and pollution coming from the ground reduces visibility.

Experimental studies of this type of construction are expensive and require a lot of time to prepare. Debated issue is very important in terms of safety of special constructions and troops (users).

References

1. Prace minerskie i niszczenia, Sztab Generalny, szefostwo Wojsk inżynieryjnych Warszawa 1995.
2. Barnat, W. (2011). Wybrane zagadnienia ochrony życia i zdrowia załóg pojazdów przed wybuchem, Wydawnictwo MilitaryRok.
3. Barnat, W. (2009). Numeryczno doświadczalna analiza złożonych warstw ochronnych obciążonych falą uderzeniową wybuchu, Bell Studio.

4. Kania, E. (2010). Projektowanie środków ochrony przeciwminowej pojazdów specjalnych, Górnictwo Odkrywkowe 4/2010.
5. Mężyk, A. (2010). Nowoczesne technologie w projektowaniu pojazdów specjalnych, 66 Inauguracja Roku Akademickiego w Politechnice Śląskiej, Gliwice.

Indoor Navigation with Micro Inertial Navigation Technology

Paweł Iwaneczko, Karol Jędrasiak and Aleksander Nawrat

1 Introduction

Navigation is a field of study that focuses on the process of monitoring and controlling the movement of a craft, vehicle or human from one place to another. Navigation is divided into: satellite navigation, astronomy navigation, pilot and radio navigation, coastal navigation, radar navigation and inertial navigation [1]. Micro Inertial Navigation Technology (MINT) is an indoor navigation system, which is using basically inertial measurement units (IMU) and which can be used inside of the buildings (closed constructions), where there is no access to the reference systems. The best example of the reference system, that can not be used inside of the building is the Global Positioning System (GPS). IMU sensor is an electronic device that integrates 3-axis accelerometers, 3-axis gyroscopes and 3-axis magnetometers, each created in MEMS technology. Sensors data is acquired by build-in processor, where are implemented complementary algorithms used to estimate the orientation relative to a plane tangent to the surface of the earth. This paper presents prototype of the device used to perform a reckoning navigation algorithms test in building of Silesian University of Technology. Device is using two inertial navigation units, that are mounted in sole of the polish army shoe, two integrated micro-switches and can-to-bluetooth adapter which is responsible for data transmission between device and PC computer. Reckoning navigation algorithms, used in this article are authorial solution and are called as odometry algorithms. These algorithms applies only for the pedestrians (person

P. Iwaneczko (✉) · K. Jędrasiak · A. Nawrat
Institute of Automatic Control, Silesian University of Technology, Akademicka 16, 44-100 Gliwice, Poland
e-mail: pawel.iwaneczko@polsl.pl

K. Jędrasiak
e-mail: karol.jedrasiak@polsl.pl

A. Nawrat
e-mail: aleksander.nawrat@polsl.pl

A. Nawrat et al. (eds.), *Advanced Technologies in Practical Applications for National Security*, Studies in Systems, Decision and Control 106,
https://doi.org/10.1007/978-3-319-64674-9_19

moving on foot) in a stationary frame of reference. Reckoning algorithms require start location, which for the purposes of the results preparation was made by hand from the map. Developing new algorithms [2–5] and computer simulation [6] are classical means to achieve progress of the technology.

2 Literature Review

Existing solutions of the inertial navigation can be divided to Micro Inertial Navigation Technology (MINT) and Inertial Magnetic Motion Capture (IMMCAP) systems. MINT systems are using IMU sensors, that are generally mounted only inside or outside of human shoes, however IMMCAP systems are composed of multiple IMU sensors that are placed on each of the human limbs, but for the odometry algorithms purpose there are used just this sensors, which are placed in the bottom part of human body. It has to be emphasize, that aspects related with inertial navigation are developed for many years in the University of Cambrige [7], Gdansk University of Technology [8], and in other institution around the world, what can be read in the following papers: [9–15]. Most of these articles describe problems with control of the unmanned aerial vehicles (UAV) and mobile robots, positioning of manned aircrafts and ships, or even control of near and long range rockets. To determine the correct course of vessels and ships in marine navigation, commonly are used devices named gyrocompass [1]. In commercial airplanes, jets and in the army UAV planes, nowadays are used laser and optic gyroscopes, which give a very accurate measurement of the angular velocity, or analog accelerometers, which give much more accurate measurements than MEMS sensors. Unfortunately because of the size of this type of sensors, they are generally not used in inertial navigation. From this reason the main and widely analyzed problem is to design adequate complementary filter which is necessary to estimate orientation of the sensor, especially to the determination of the yaw angle (ψ), otherwise azimuth, which is an angle between direction of movement and earth north pole.

Complementary filtration was raised in many of research centers and in the literature can be found for example Sebastian's Madgwick authorial algorithm of complementary filter: [16–19]. Another example can be found in the following sources: [20–23], which describe solutions proposed by Robert Mahony. However, the most common algorithms of inertial sensors filtration are based on Kalman filters [24–31] and Extended Kalman Filters [32–38]. Pedestrians navigation is applicable both in civil and military industries. In the literature can be found many scientific publications in which the inertial sensors and other necessary electronic devices such as force sensors and microprocessors are placed inside and outside of the shoe: [19, 32, 39–41]. Good results of inertial navigation can be found in article that was presented at the International Conference on Automation and Robotics in 2007 by Xiaoping'a Yun [39], where are described surprising results of square path (Fig. 1a) and path traveled up to the stairs (Fig. 1b). Next example that can be cited in this article is

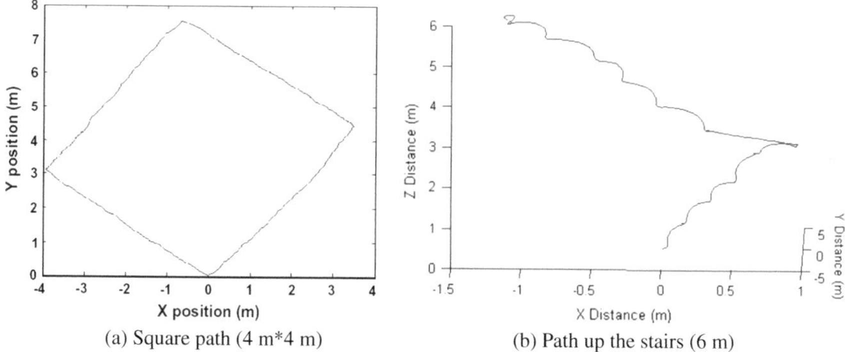

(a) Square path (4 m*4 m)

(b) Path up the stairs (6 m)

Fig. 1 Inertial navigation in Xiaoping Yun's paper [39]

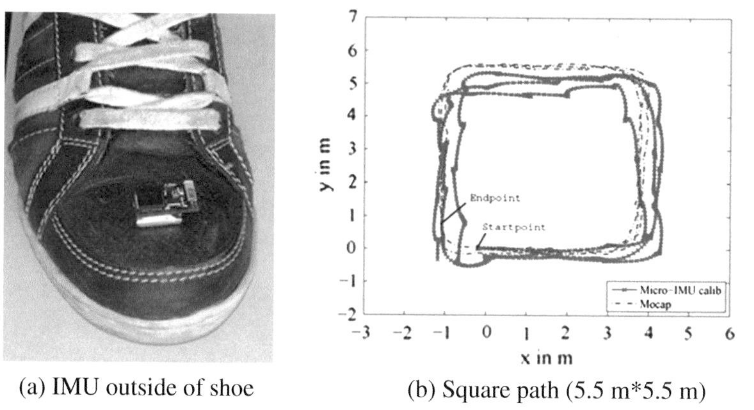

(a) IMU outside of shoe

(b) Square path (5.5 m*5.5 m)

Fig. 2 Inertial navigation using wireless IMU sensor [41]

paper of Fabian Höflinger [41], who uses wireless inertial measurement unit located on human foot (Fig. 2a) to estimate the square path, that is compared to MOCAP system (Fig. 2b).

3 Indoor Navigation Algorithm

For the purpose of design and implementation of odometry algorithms and performing tests of these algorithms, was used an inertial measurement unit, that was developed and created at the author university (Fig. 3) [42]. Sensor consists 3-axis MEMS: accelerometers, gyroscopes and magnetometers and has build in microprocessor, which main task is to estimate orientation data from extended Kalman filter. Sensor returns orientation data in the Euler angles representation (ϕ-roll, θ-pitch, ψ-yaw, Fig. 3a) and in the quaternion representation (q_r, q_i, q_j, q_k), which is used in inertial navigation algorithms.

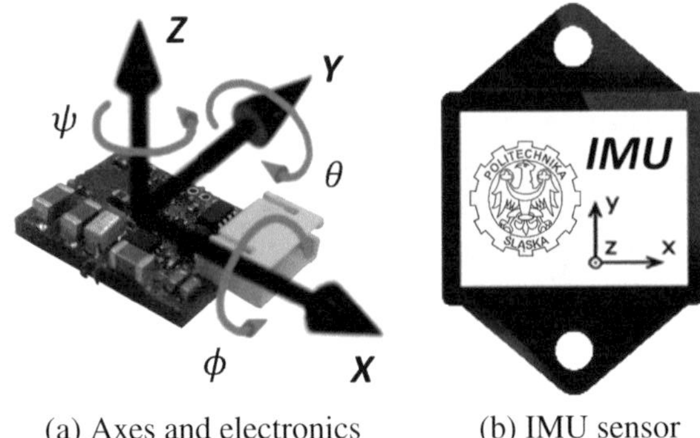

(a) Axes and electronics (b) IMU sensor

Fig. 3 IMU sensor used for testing algorithms

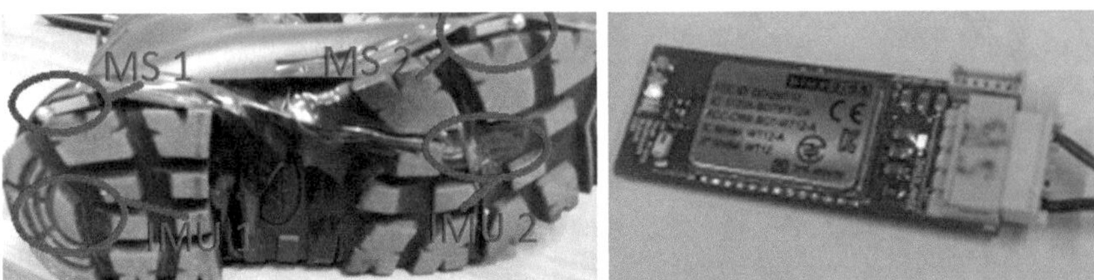

(a) Micro-switches and IMU location (b) Can to Bluetooth adapter

Fig. 4 Prototype of the pedestrian navigation system

Like it was mentioned in the introduction section in the prototype of pedestrians navigation system (Fig. 4) are mounted two IMU sensors in the sole of the army shoe, which are integrated with two micro-switches (Fig. 4a) and with can-to-bluetooth device (Fig. 4b) used to communication between computer or mobile device, where the sensors data can be recorder. The micro-switches are needed to determine the moment of contact the shoe sole with the floor and to specify the moment of the start and end time of foot movement. These time points are a border of an double integral calculation of the acceleration data. Micro-switch number one determines time moments for the first IMU sensor, while analogously second micro-switch specify the start and end time point for the second inertial measurement unit. Two inertial sensors are used separately to calculate two different paths. These paths are compared to each other to choose better result and also are used to average data fusion. Another reason, that sensors were placed in the back and front parts of the shoe was to perform analyze and choose the proper location of the sensor in this sole.

Micro switches are connected to microprocessor electronic device, that transmits information about switch state to CAN 2.0A bus. IMU sensors are connected to the same CAN bus wire, which contains also two power supply conductors routed from the batteries with 5 V power source. CAN wire is connected to CAN-to-Bluetooth adapter, which transmit sample time and data received from the IMU sensors and from the micro-switches synchronously every 10 ms.

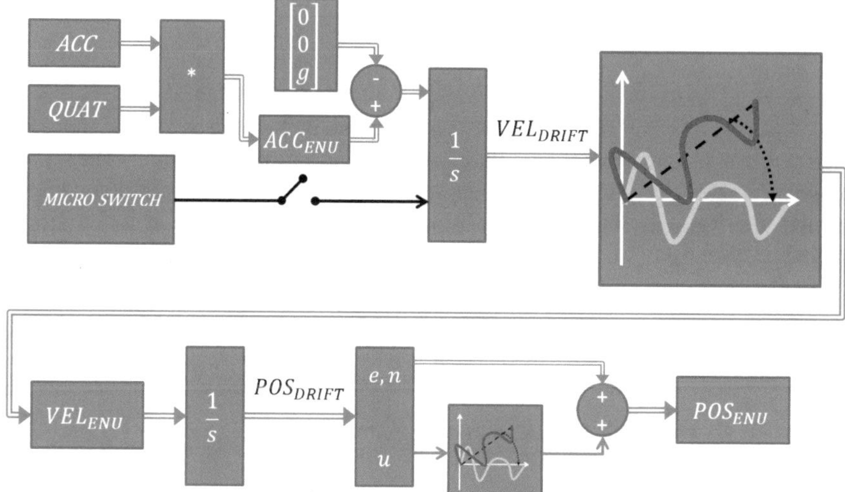

Fig. 5 MINT algorithm schematic

After receiving data from wireless adapter or after loading registered data from the file, measurements can be exploited into the navigation odometry algorithm, which is presented in the Fig. 5.

Firstly, the acceleration vector has to be rotated to the plane tangent of the earth surface (Eq. 1), so that the result acceleration could be interpreted as East-North-Up acceleration vector (acc_{enu}). When the orientation change does not influence on the acceleration (acc_{enu}) measurements, non-zero value of up acceleration part acc_u has to be reduced by the gravity value g (Eq. 2).

$$acc_{enu} = acc_{xyz} * q \tag{1}$$

$$acc_u = acc_u - g \tag{2}$$

Afterwards, when up part of the acceleration vector acc_u only contains the information about acceleration change and it is independent of gravity non-zero value, the first integral part can be computed (Eq. 3):

$$vel_d = \int_{t_1}^{t_2} acc_{enu}(t)dt \tag{3}$$

where: t_1 is the step start time, t_2 is the step end time and vel_d is velocity with drift. Velocity from the first integral results, at the start point of the step is equal to zero, but at the end point of foot step has non-zero value (bias), which is the accumulated error caused by MEMS accelerometers sensors. This error is called drift, and can be removed by using linear drift function described in Eq. 4.

$$vel_{enu}(t) = vel_d(t) - vel_d(t_2) * \frac{t}{t_2 - t_1} \tag{4}$$

Here is the place to enable the second integral—the velocity integral, which gives the calculation of position relative to the start location, and it is described in the Eq. 5. This position is also encumbered with errors and the results, especially the up part of the calculated position, has non-zero value, even in the case of walking on the flat ground. Analogously, the next step is to remove the drift movement in the horizontal axis of position (pos_u) (Eq. 6)

$$pos_d = \int_{t_1}^{t_2} vel_{enu}(t)dt \tag{5}$$

$$pos_u(t) = pos_{du}(t) - (pos_{du}(t_2) - pos_{du}(t_1)) * \frac{t}{t_2 - t_1} \tag{6}$$

where: pos_{du} is an up part of drift position acquired from the second integral. Finally, the position relative to the start location is calculated and can be used as inertial navigation position inside and outside of the buildings, highlighting, that there is no used reference positioning systems, such as GPS system.

4 Results and Conclusions

In order to compare algorithm quality in this article, first performed and presented tests is square path, that was registered during the walking around the square pavement outside of the building of the author university. As it can be seen in the Fig. 6, obtained position in the graphical environment (Fig. 6a) is adequate to its real representation (Fig. 6b). Test was performed several times in the same place, and results are very similar, emphasizing that the maximum distance difference between them is up to 10 cm. Surprising is, that in each of the tests results, euclidean distance error between the path calculated from the odometry algorithms and the actual route measured by hand are maximum up to the half of a meter. Data visible on the 3D graphical environment (Fig. 6a) is calculated from the back sensor mounted in the shoe sole by proposed and described algorithm. The first research and tests was performed outside of the building, only because of the magnetic disturbances, which are typically much more higher in reinforced concrete buildings, where the walls are also filled with a large number of wires and cables, than at the outside on the pavement.

Next experiments was performed inside of the building of Silesian University of Technology, at the Institute of Automatic Control. Graphical environment presented in the Fig. 7 was written in C++ language using the OpenGL libraries and is mainly used to generate results associated exactly with the inertial navigation. As can be seen in the Fig. 7a, position computed from the algorithm does not goes beyond the

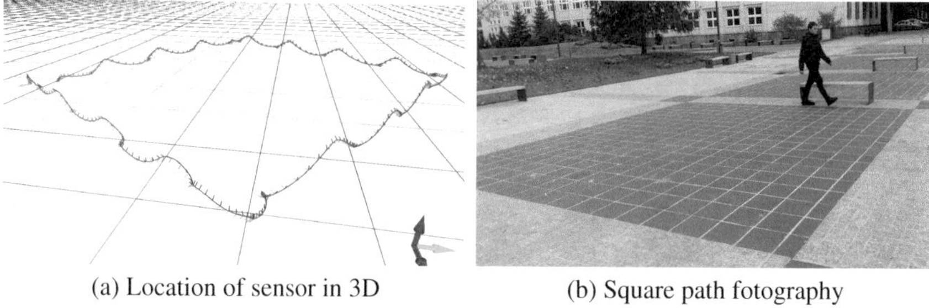

(a) Location of sensor in 3D (b) Square path fotography

Fig. 6 Square route (5.5m × 5.5m) outside of author university building

(a) Traveled path visible from the top (b) Path with drift seen from the side

Fig. 7 Route inside of the Silesian university of technology building

areas of the building and is approximate with the path traveled in reality, but just in the case of red path color. Red path is calculated from the data, that was received from the back sensor, whereas the magenta color belongs to the sensor mounted in the front part of the shoe sole. When the part, that is responsible for position drift removing, will be disabled in the navigation algorithm, what is presented in the Fig. 7b, experiment shows, that results from the front sensor, accumulates very big error of vertical position, which equal about 15 m, unlike from the back sensor, where the vertical error equals up to half of the meter.

As follows from performed tests and analyses, the navigation algorithm determines the relative position with an accuracy of up to 1 m error (among other mentioned examples, in particular results from the front shoe sensor). Algorithms are still testes and are improved as much as possible. In the future, it is planned to improve the detection of the start and end step moment by using the force sensing resistors

(FSR) sole, which is often used to determine the force of human foot [43] applied to the floor. Another possibility of algorithms improvement is to use other complementary filters, which are cited in the section of literature review, because in all tests up to this time, it was used only Extended Kalman Filter developed at the author university and implemented on the IMU build-in microprocessor.

References

1. Gucma, M., & Montewka, J. (2006). *Podstawy morskiej nawigacji inercyjnej*. Akademia Morska.
2. Switonski, A., Josinski, H., Jedrasiak, K., Polanski, A., & Wojciechowski, K. (2010). Classification of poses and movement phases. *Lecture notes in computer science*.
3. Ryt, A., Sobel, D., Kwiatkowski, J., Domzal, M., Jedrasiak, K., & Nawrat, A. (2015). Real-time laser point tracking. In *International Conference on Computer Vision and Graphics* (pp. 542–551).
4. Sobel, D., Jedrasiak, K., Daniec, K., Wrona, J., Jurgas, P., & Nawrat, A. (2014) Camera calibration for tracked vehicles augmented reality applications. In *Innovative Control Systems for Tracked Vehicle Platforms* (pp. 147–162).
5. Nawrat, A., & Jedrasiak, K. (2008) Fast colour recognition algorithm for robotics. *Problemy Eksploatacji*, 69–76.
6. Daniec, K., Iwaneczko, P., Jedrasiak, K.,& Nawrat, A. (2013) Prototyping the autonomous flight algorithms using the prepar3d simulator. *Vision based systems for UAV applications* (pp. 219–232).
7. Woodman, O. J. (2007) *An introduction to inertial navigation*. University of Cambridge, Computer Laboratory, Technical report UCAMCL-TR-696 (Vol. 14, p. 15).
8. Bonisławski, A., Juchniewicz, M., & Piotrowski, R. (2014) Projekt techniczny i budowa platformy latającej typu quadrocopter. *Pomiary Automatyka Robotyka* (Vol. 18).
9. Parvin, R. H. (1962). Inertial navigation systems: Prelaunch alignment. *IRE Transactions on Aerospace and Navigational Electronics*, *3*, 141–145.
10. Stieler, B., & Winter, H. (1982) Agard flight test instrumentation series. volume 15. *Gyroscopic instruments and their application to flight testing*. Technica Report DTIC Document.
11. King, A. (1998). Inertial navigation-forty years of evolution. *GEC Review*, *13*(3), 140–149.
12. Britting, K. R. (2010) *Inertial navigation systems analysis*. Artech House.
13. Titterton, D., & Weston, J. L. (2004). *Strapdown inertial navigation technology* (Vol. 17). IET.
14. Weston, J., & Titterton, D. (2000). Modern inertial navigation technology and its application. *Electronics & Communication Engineering Journal*, *12*(2), 49–64.
15. Barshan, B., & Durrant-Whyte, H. F. (1995). Inertial navigation systems for mobile robots. *IEEE Transactions on Robotics and Automation*, *11*(3), 328–342.
16. Madgwick, S. O., Harrison, A. J., & Vaidyanathan, R. (2011). Estimation of imu and marg orientation using a gradient descent algorithm. In *2011 IEEE International Conference on Rehabilitation Robotics* (pp. 1–7). IEEE.
17. Madgwick, S. (2010). *An efficient orientation filter for inertial and inertial/magnetic sensor arrays*. Report x-io and University of Bristol (UK).
18. Carberry, J., Hinchly, G., Buckerfield, J., Tayler, E., Burton, T., Madgwick, S., & Vaidyanathan, R. (2011). Parametric design of an active ankle foot orthosis with passive compliance. In *2011 24th International Symposium on Computer-Based Medical Systems (CBMS)* (pp. 1–6). IEEE.
19. Madgwick, S. Gait tracking with x-imu. Zasoby sieciowe, dost p 24.07.2016.
20. Euston, M., Coote, P., Mahony, R., Kim, J., & Hamel, T. (2008) A complementary filter for attitude estimation of a fixed-wing uav. In *2008 IEEE/RSJ International Conference on Intelligent Robots and Systems* (pp. 340–345). IEEE.

21. Mahony, R., Hamel, T., & Pflimlin, J.-M. (2008). Nonlinear complementary filters on the special orthogonal group. *IEEE Transactions on Automatic Control*, *53*(5), 1203–1218.
22. Mahony, R., Hamel, T., & Pflimlin, J.-M. (2005). Complementary filter design on the special orthogonal group so (3). In *Proceedings of the 44th IEEE Conference on Decision and Control* (pp. 1477–1484) IEEE.
23. Baldwin, G., Mahony, R., Trumpf, J., Hamel, T., & Cheviron, T. (2007) Complementary filter design on the special euclidean group se (3). In *Control Conference (ECC), 2007 European* (pp. 3763–3770). IEEE.
24. Kędzierski, J., & Konar, K. N. R. (2008) Filtr kalmana-zastosowania w prostych układach sensorycznych," *Artykuł koła naukowego KoNaR Politechnika Wrocławska.*
25. Wnuk, M. (2014) Filtracja komplementarna w inercyjnych czujnikach orientacji. *Granth S30080* (Vol. SPR 3).
26. Grygiel, R., Bieda, R., & Wojciechowski, K. (2014). Metody wyznaczania kątów z żyroskopów dla filtru komplementarnego na potrzeby określenia orientacji imu. *Przegląd Elektrotechniczny*, *9*, 217–224.
27. Bieda, R., & Grygiel, R. (2014). Wyznaczanie orientacji obiektu w przestrzeni z wykorzystaniem naiwnego filtru kalmana. *Przeglad Elektrotechniczny*, *90*, 34–41.
28. Won, S.-H. P., Melek, W. W., & Golnaraghi, F. (2010). A kalman/particle filter-based position and orientation estimation method using a position sensor/inertial measurement unit hybrid system. *IEEE Transactions on Industrial Electronics*, *57*(5), 1787–1798.
29. Won, S.-H. P., Golnaraghi, F., & Melek, W. W. (2009). A fastening tool tracking system using an imu and a position sensor with kalman filters and a fuzzy expert system. *IEEE Transactions on Industrial Electronics*, *56*(5), 1782–1792.
30. Kolecki, J. (2012). Wykorzystanie jednostki imu typu mems do określenia przybliżonych elementów orientacji zdjęć naziemnych. *Archiwum Fotogrametrii, Kartografii i Teledetekcji* (Vol. 24).
31. Kolecki, J. (2013) *Wyznaczanie elementów orientacji zewnetrznej zdjęć naziemnych z wykorzystaniem obserwacji fotogrametrycznych i inercyjnych oraz satelitarnego systemu pozycjonowania*. Wydawnictwa AGH.
32. Jiménez, A. R., Seco, F., Prieto, J. C., & Guevara, J. (2010) Indoor pedestrian navigation using an ins/ekf framework for yaw drift reduction and a foot-mounted imu. In *2010 7th Workshop on Positioning Navigation and Communication (WPNC)* (pp. 135–143). IEEE.
33. Mirzaei, F. M., & Roumeliotis, S. I. (2008). A kalman filter-based algorithm for imu-camera calibration: Observability analysis and performance evaluation. *IEEE tRansactions on Robotics*, *24*(5), 1143–1156.
34. Ruiz, A. R. J., Granja, F. S., Honorato, J. C. P., & Rosas, J. I. G. (2012). Accurate pedestrian indoor navigation by tightly coupling foot-mounted imu and rfid measurements. *IEEE Transactions on Instrumentation and Measurement*, *61*(1), 178–189.
35. Mirzaei, F. M. & Roumeliotis, S. I. (2007). 1| a kalman filter-based algorithm for imu-camera calibration. In: *2007 IEEE/RSJ International Conference on Intelligent Robots and Systems* (pp. 2427–2434). IEEE.
36. Filter, E. E. K. (2007) Vision-aided navigation for small uavs in gps-challenged environments. In *AIAA Infotech.*
37. Hellmers, H., Norrdine, A., Blankenbach, J. & Eichhorn, A. (2013) An imu/magnetometer-based indoor positioning system using kalman filtering," in *2013 International Conference on Indoor Positioning and Indoor Navigation (IPIN)* (pp. 1–9). IEEE.
38. Nützi, G., Weiss, S., Scaramuzza, D., & Siegwart, R. (2011). Fusion of imu and vision for absolute scale estimation in monocular slam. *Journal of Intelligent & Robotic Systems*, *61*(1–4), 287–299.
39. Yun, X., Bachmann, E. R., Moore, H., & Calusdian, J. (2007). Self-contained position tracking of human movement using small inertial/magnetic sensor modules. In *Proceedings 2007 IEEE International Conference on Robotics and Automation* (pp. 2526–2533). IEEE.
40. Kelly, A. (2011). Personal navigation system based on dual shoe-mounted imus and intershoe ranging. In *Proceedings of the Precision Personnel Locator Workshop.*

41. Höflinger, F., Müller, J., Zhang, R., Reindl, L. M., & Burgard, W. (2013). A wireless micro inertial measurement unit (imu). *IEEE Transactions on Instrumentation and Measurement, 62*(9), 2583–2595.
42. Jędrasiak, K., Daniec, K., & Nawrat, A. (2013). The low cost micro inertial measurement unit. In *2013 IEEE 8th Conference on Industrial Electronics and Applications (ICIEA)* (pp. 403–408) IEEE.
43. Josinski, H., Switonski, A., Jedrasiak, K., & Kostrzewa, D. (2012) Human identification based on gait motion capture data. *Proceedings of the 2012 International MultiConference of Engineers and Computer Scientists*.

The Concept of RFID-Based Positioning System for Operational Use

Mateusz Opuchlik, Karol Jędrasiak, Jarosław Cymerski, Damian Bereska and Aleksander Nawrat

1 Introduction

The main aim of this work is to present the concept of a complete set of the radio identification of the location of objects and detailing its components. As part of the article we will discuss the proposed technological solution with emphasis on the choice of components. The primary objective of the planned system was implemented to monitor and collect information from the stitched unique ID RFID numbers to the figures. RFID is a radio frequency identification technology (Radio Frequency IDentification) [1–3]. The technology is based on the use of radio waves of different frequencies to read coded information from the RFID tag to identify the object. RFID is used in many areas of heavy industry [4–6] which is mining, the use of civilian-type contactless payment or documents with stitched information [7, 8].

The object of any RFID system is to store a certain amount of data in convenient device transceivers. Then read this data in an automated manner at the right time and place in order to obtain the desired result for a given application. Information included in the tag may describe e.g. different parts of the production line, the goods during transport, the location of objects [9–11], identify vehicles, animals or people [12–15]. By attaching to tag additional information it is possible to enrich applications [16] with capabilities to support its operation, taking into account specific information about the object to which the tag belongs (Fig. 1).

M. Opuchlik · K. Jędrasiak (✉) · D. Bereska · A. Nawrat
Institute of Automatic Control, Silesian University of Technology,
Akademicka 16, 44-100 Gliwice, Poland
e-mail: karol.jedrasiak@polsl.pl

J. Cymerski
Government Protection Bureau, Warsaw, Poland
e-mail: j.cymerski@bor.gov.pl

A. Nawrat et al. (eds.), *Advanced Technologies in Practical Applications for National Security*, Studies in Systems, Decision and Control 106,
https://doi.org/10.1007/978-3-319-64674-9_20

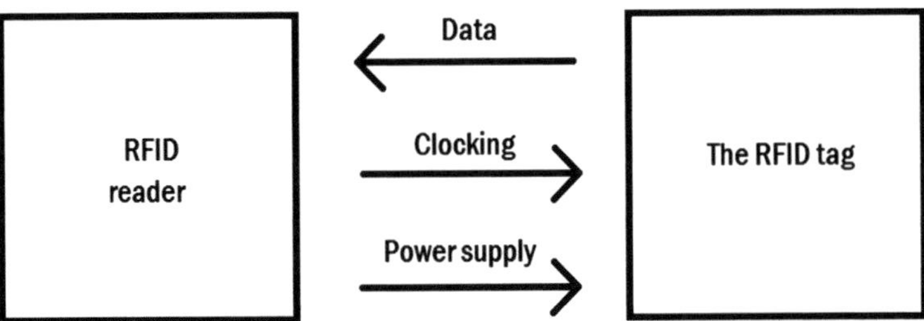

Fig. 1 The principle of operation of the RFID

RFID systems always consists of at least two components

- tag placed on an object identified;
- reader, whose role is to read the data on the tag.

The reader device comprises

- reader comprising a transmitter, receiver and decoder,
- transmitting and receiving antenna or two separate antennas: transmit and receive.

The tag consists of:

- an electronic circuit—integrated circuit without a housing. These systems may be sized at 0.4 × 0.4 mm. Systems usually do not have their own power supply —they are passive,
- antenna.

The reader comprises a radio transmission module (this is both a transmitter and receiver), the control element and coupling with the tag. In addition, many readers includes an interface connecting to a PC, allowing the transfer to and from the PC all kinds of application data and system. The RFID reader is a device that can read and send data to other compatible RFID tags. For readers are also classified as RFID printers enable printing smart labels.
The RFID reader is usually made from the following parts:

- Transmitter,
- Receiver,
- Microprocessor,
- Memory,
- Channels input/output for optional sensors, actuators,
- Controller (may also be external),
- Interface,
- Power supply.

An RFID tag (Fig. 2) comprises a microchip and a connecting element of the reader. The tag is a data bearer RFID system. The tag is generally a passive device

Fig. 2 RFID tags

when it is outside the area of influence of the reader. Activation of the tag occurs when you place it in the zone of influence of the reader. The energy required to power the tag is transmitted wirelessly to the reader through the connector, the antenna. In the same way, a clock signal is transmitted—and timing data.

Depending on the type of label may be of more or less complex structure, which consists of the following elements:

- Microchip,
- Antenna,
- The power supply (not included in passive tag),
- Additional electronics (not present in the passive tag).

Both tags and RFID readers are equipped with the antennas. In both types of device antennas are used to receive and send information transmitted by radio waves, however, differ among themselves structure. These common tags are usually a few centimeters and are directly connected to the microchip tag. However, in the case of the most common reader is separately connected device. There are also the reader with a built-in antenna. The length of the connecting cable is reduced to range starting from approx. 2 to approx. 8 m. A single reader can support up to four antennas. Generally, RFID reader antennas are square or rectangular (Fig. 3). The antenna systems are divided according to the criterion of polarization. There are two types of antennas:

- Linear polarized—Linear polarization excels in applications where the orientation of the RFID tag is known. Setting the antenna polarization can be precisely adjusted to the labels and reduce the power of the device. Polarization can

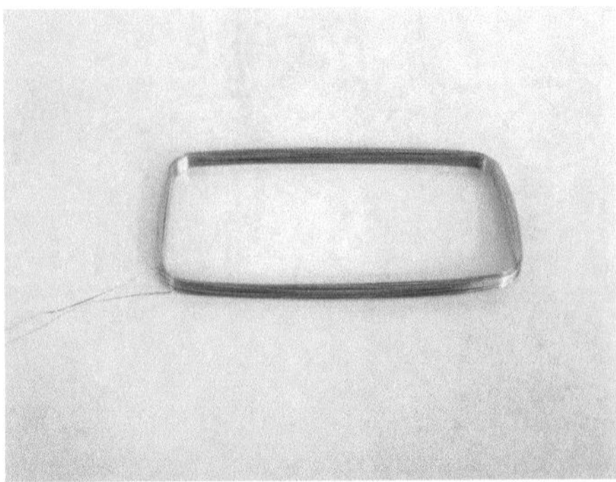

Fig. 3 The rectangular external antenna

also be used to separate labels that are close to each other, but with a different orientation. This increases reading speed and reduces noise generated by the processes carried out in the vicinity.

- Circularly polarized—In applications where the orientation of the label is not known or is variable, best suited for circular polarization, because it allows reliable detection of the label regardless of their orientation. An example of such an application might be to identify the unsorted goods or processes, in which the spatial position of the label may change.

The power signal emitted by the antenna is measured in units of ERP Effective Radiated Power, used in Europe, while in the United States in units EIRP Effective Isotropic Radiated Power. The maximum power of the antenna is governed by the relevant legislation.

RFID systems are systems that generate and-radiating electromagnetic waves and therefore are legally classified as radio systems. Therefore, the measure cannot under any circumstances affect or interfere with other radio systems. This requires strict adherence to only the permitted frequency ranges. Usually it comes down to the use of RFID systems in frequency bands standardized throughout the world under the name of ISM (Industrial-Scientific-Medical), which ranges for applications in industry, science (research), and medicine. In addition, the frequency range below 135 kHz is also available.

Available for RFID systems frequency ranges are thus: 0–135 kHz ISM frequency within 6.78, 13.56, 27.125, 40.68, 433.92, 869.0, 915.0 MHz (outside Europe), 2.45, 5.8 and 24.125 GHz.

Because of the technical implementation of RFID (encoding type, the size of the tag memory, speed of transmission, distinctness of multiple tags within the reader, etc.), There are many different standards:

- Tiris—one of the earliest systems, based on the FM transmission
  - Application: trade.
- Unique—the simplest and most widely used at present RFID system. Passive tags, unique code originally written during production, now appear duplicated cards. The frequency of 125 kHz, transmission speed 2 kb/s.

- Application of access control, time and attendance.

- Q5—a system using programmable tags reacting e.g. Specific password.
- Hitag—standard for industrial applications, lets you read and write messages in tags; passive tags. The frequency of 125 kHz, the baud rate 4 kb/s. Anti-collision algorithm, the ability to encode data.
  - Application: charging systems (e.g. ski lifts), systems for marking products, marking animals.
- Mifare—standard includes the ability to use both simple tag memory, as well as very complex—containing processors that support encryption. The frequency of 13.56 MHz, the transmission speed 106 kb/s. A standard developed by Philips.
  - Application: bank cards (smart-cards); identification cards; tickets.
- Icode—standard with very flat tags; tags allow you to read and write (512b capacity). The frequency of 13.56 MHz. The ability to support up to 30 tags per second.
  - Application: retail, libraries, control the flow of shipments, records of equipment.

2 System Architecture

The idea behind the study was to draw attention to the promising technology, which is augmented reality [17–20].

Planned to implement the system should consist of the following components (Fig. 4):

- RFID tag,
- low-level software,

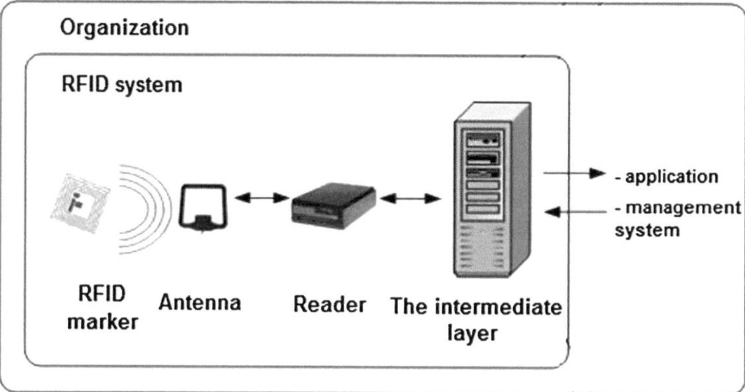

Fig. 4 The concept of system the architecture

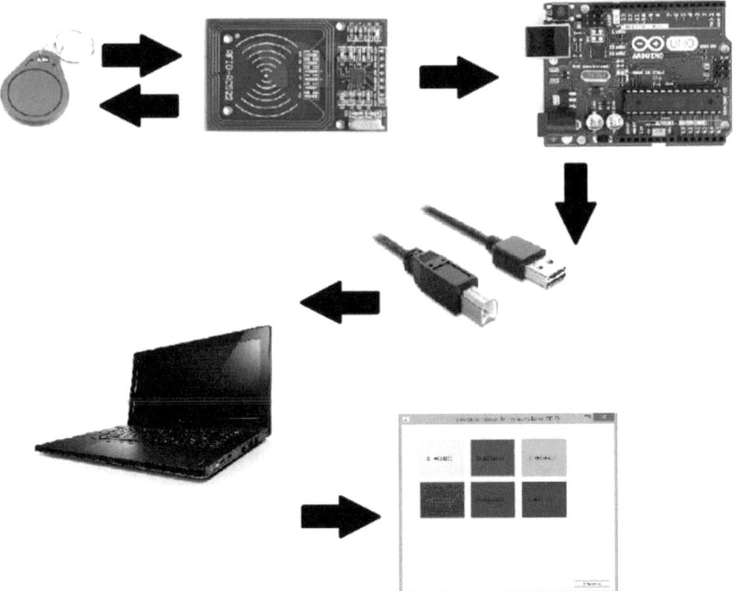

Fig. 5 The general principle of operation of the RFID system

- Antenna,
- RFID reader,
- Middleware,
- Visualization software.

It is assumed that the system is as follows: a reader via an antenna of the transmitter generates an electromagnetic wave, the same or the second antenna receives the electromagnetic waves, which are then filtered and decoded, so as to read and respond tags (Fig. 5).

Passive tags do not have their own power when they find themselves in the electromagnetic field of the resonant frequency of the receiving accumulate received energy in a capacitor included in the structure of the tag. After receiving sufficient energy response is sent containing the code tag. In most applications, broadcasting wave of the charging system and informing about finding in the field of the reader is interrupted, and the transponders respond in moments of interruption of transmission. Tag does not respond immediately, but after some time, and if you answered and remains in the field of electromagnetic wave, it remains inactive for a specified time, which allows you to read many tags that are also in the field of reading.

It is assumed that in the first step of the transmitter/receiver expects to find in its field of action the transponder. Upon detection of a transponder and reading therefrom, a unique 5-bit ID code (it is also possible to overwrite the ID code, for example, to verify, on which the last station was tag). Read the ID code and number of the receiver/transmitter that it has been detected by the reader is then saved in a prepared 6-bit register in the temporary memory of the microcontroller via USB serial port is sent to a computer that acts as an intermediate layer between the microcontroller and application. The last step is to read up by the application and the corresponding matching ID tag to the corresponding item in the database, and then present it by lighting the corresponding color in the application at the appropriate place responsible for the ID reader that the signal received.

It is assumed that the hardware layer will consist of a set of six receivers RFID and one breadboard (Fig. 6). How to connect a single sensor RFID rc522 plate prototype is shown in Fig. 7. Schematic six receivers for proper operation assumes that all beyond SDA receivers are shared on the same pin in the plate prototype, in

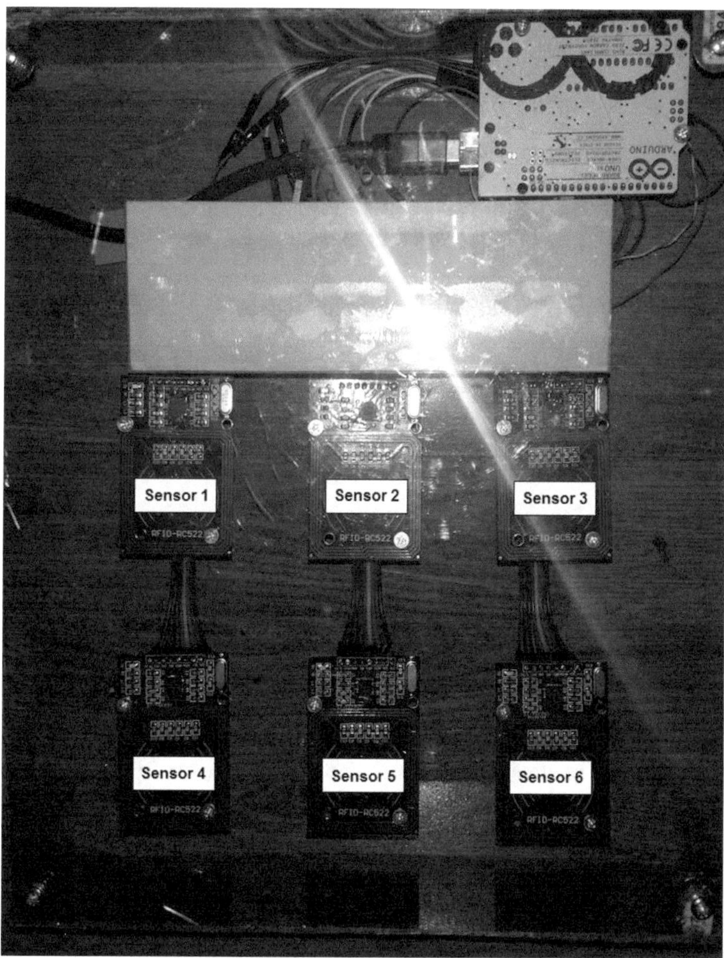

Fig. 6 The station for hardware tests

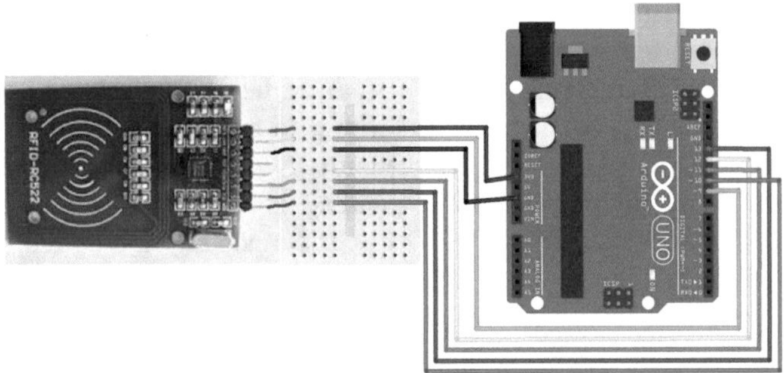

Fig. 7 Connection diagram RFID rc522

	RFID1	RFID2	RFID3	RFID4	RFID5	RFID6
RST	0	0	0	0	0	0
SDA	10	9	8	7	6	5
MOSI	11	11	11	11	11	11
MISO	12	12	12	12	12	12
SCK	13	13	13	13	13	13
GND	GND	GND	GND	GND	GND	GND
VCC	3.3 V	3.3 V	3.3 V	3.3 V	3.3 V	3.3 V

Fig. 8 Wiring diagram of RFID sensors

addition to said pin SDA, after which the information is received on the applied tag (Fig. 8).

Middleware, mediates between the reader and other information systems. The functions of middleware include:

- Management of readers and devices,
- Data management,
- Application Integration.

Management of readers and devices—intermediate layer RFID allows users to configure, monitor, and start sending orders directly to readers through the common interface. Data Management—RFID intermediate layer takes data from the readers. It can intelligently filter and direct them to the appropriate destinations. Application Integration—RFID middleware solutions provide opportunities for messaging, routing and communication required to integrate RFID data with existing systems, supply chain management, warehouse management, and customer relationship management. The application layer is divided into two parts:

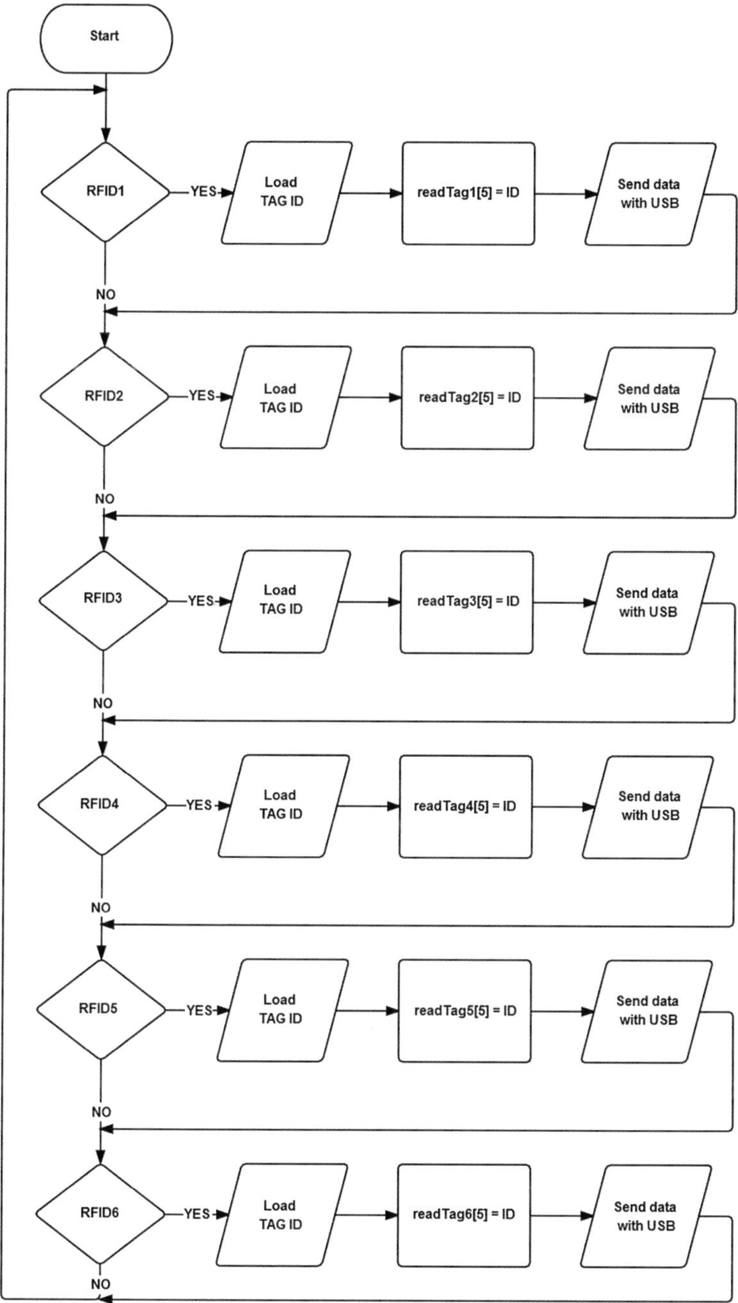

Fig. 9 The block diagram of the software executed by the microcontroller

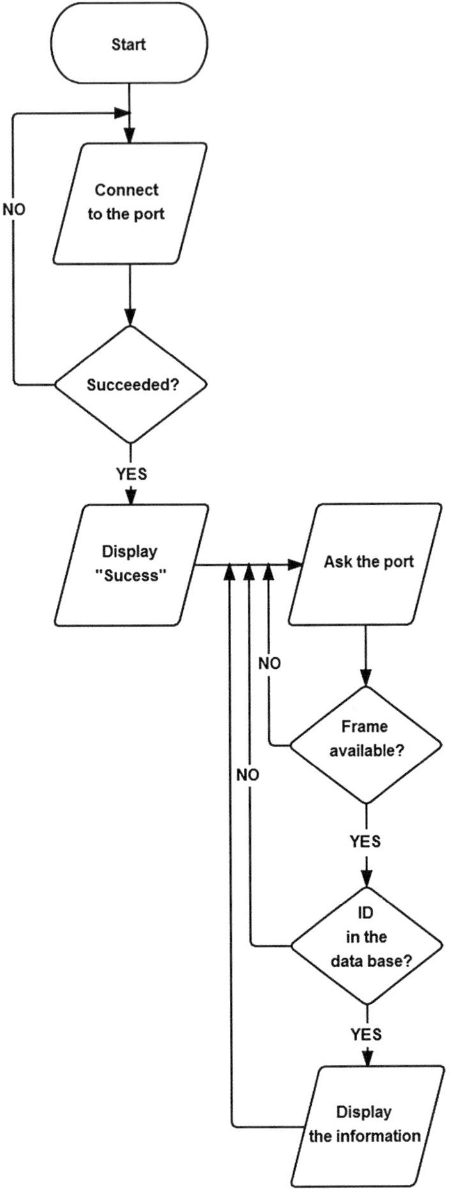

Fig. 10 Block diagram of the software executed by the computer

- low-level software implemented in the device reader,
- high-level software for use on your computer.

Low-level software is executed on the microcontroller in the following manner (Fig. 9):

1. Microcontroller sequentially polls each further sensor in order of 1, 2, 3, 4, 5, 6, 1, 6 … and so on.
2. If the sensor detects within the action applied TAG, the microcontroller goes to the function responsible for taking the ID tag and save it to a prepared register
3. Information from the register together with the number of the receiver is sent to the serial port to an intermediate layer,
4. The program proceeds to check the next sensor.

The software on the computer have been implemented in the Java programming language. The software operates in the following manner (Fig. 10):

1. obtain access to the port which is connected to the microcontroller,
2. then waits for the availability of a data frame at a time when the output of the microcontroller appears in the data frame, it shall be received and compared with information entered in the database, the first bit indicates the frame number detector, which has transmitted the frame, the next 5 bits contains a unique ID Tag,
3. upon receipt of information on the screen are represented data on the tag is applied to the appropriate sensor.

It is assumed that in order to visualize the position of the RFID tags in the space there should be a graphical user interface for visualizing the position of the tags detected.

It was assumed division of space into 6 squares. Each of the squares represents another of the sensors, while the colors correspond to the tags applied to the sensor. Each RFID tag is assigned an individual color for itself represented in the program. To facilitate the verification of the correctness of the software RFID tags could have the same color as their representation in the program (Fig. 11).

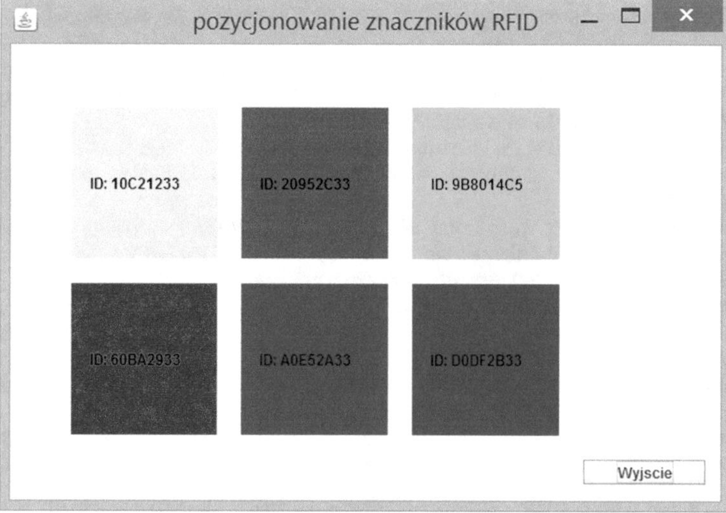

Fig. 11 Hardware representation in software GUI

3 Conclusions

In this paper the concept of the system to locate objects in real time based on RFID tags was presented. We present the concepts of: system, software and the use of sensors. Certain assumptions have been experimentally verified in a lab environment. The obtained promising results encourage further development of technology. As the development direction it was determined to the use the RFID system to monitor objects moving in the building. For this purpose the principle of operation could be reversed and the transponders could create a matrix. The RFID reader could be used as a mobile scanner (a kind of compass) determining its position in space created by a matrix of RFID tags.

References

1. Szczurkowski Rozprawa Doktorska, M. (2010). Wytwarzanie i sterowanie polami magnetycznymi dla radiowej identyfikacji obiektów RFID oraz indukcyjnego przekazu energii, AGH Kraków.
2. Portal RFID—Radio Frequency Identification. Retrieved June 15, 2016, from www.portalrfid.pl.
3. RFID. Retrieved January 17, 2015, from http://pl.wikipedia.org/wiki/RFID.
4. Daniec, K., Jedrasiak, K., Koteras, R., & Nawrat, A. (2013). Embedded micro inertial navigation system. In *Applied mechanics and materials* (Vol. 249, pp. 1234–1246). Trans Tech Publications.
5. Bereska, D., Daniec, K., Fras, S., Jedrasiak, K., Malinowski, M., & Nawrat, A. (2013). System for multi-axial mechanical stabilization of digital camera. In *Vision based systems for UAV applications* (pp. 177–189). Springer International Publishing.
6. Jedrasiak, K., Nawrat, A., Daniec, K., Koteras, R., Mikulski, M., & Grzejszczak, T. (2012, September). A prototype device for concealed weapon detection using IR and CMOS cameras fast image fusion. In *International Conference on Computer Vision and Graphics* (pp. 423–432). Berlin, Heidelberg: Springer.
7. Josinski, H., Switonski, A., Jedrasiak, K., & Kostrzewa, D. (2012). Human identification based on gait motion capture data. In *Proceedings of the 2012 International Multi Conference of Engineers and Computer Scientists, IMECS* (Vol. 12).
8. Switonski, A., Josinski, H., Jedrasiak, K., Polanski, A., & Wojciechowski, K. (2010). Classification of poses and movement phases, ICCVG 2010. In Lecture notes in computer science. Springer.
9. Daniec, K., Iwaneczko, P., Jedrasiak, K., & Nawrat, A. (2013). Prototyping the autonomous flight algorithms using the Prepar3D® simulator. In *Vision based systems for UAV applications* (pp. 219–232). Springer International Publishing.
10. Bieda, R., Grygiel, R., & Galuszka, A. (2015). Naive Kalman filtering for estimation of spatial object orientation. In *Methods and models in automation and robotics (MMAR)* (pp. 955–960).
11. Bieda, R., & Grygiel, R. (2014). Wyznaczanie orientacji obiektu w przestrzeni z wykorzystaniem naiwnego filtru Kalmana. *Przeglad Elektrotechniczny, 90,* 34–41.
12. Bieda, R., Jaskot, K., Jędrasiak, K., & Nawrat, A. (2013). Recognition and location of objects in the visual field of a UAV vision system. In *Vision based systems for UAV applications* (pp. 27–45). Springer International Publishing.
13. Babiarz, A., Bieda, R., & Jaskot, K. (2013). Vision system for group of mobile robots. In *Vision based systems for UAV applications* (pp. 139–156). Springer International Publishing.

14. Babiarz, A., & Jaskot, K. (2013). The concept of collision-free path planning of UAV objects. In *Advanced technologies for intelligent systems of national border security* (pp. 81–94). Berlin, Heidelberg: Springer.
15. Kus, Z., & Nawrat, A. (2013). Object tracking in a picture during rapid camera movements. In *Vision based systems for UAV applications* (pp. 77–91). Springer International Publishing.
16. Sobel, D., Jedrasiak, K., Daniec, K., Wrona, J., Jurgas, P., & Nawrat, A. (2014). Camera calibration for tracked vehicles augmented reality applications. In *Innovative control systems for tracked vehicle platforms* (pp. 147–162). Springer International Publishing.
17. Babiarz, A., Bieda, R., Jedrasiak, K., & Nawrat, A. (2013). Machine vision in autonomous systems of detection and location of objects in digital images. In *Vision based systems for UAV applications* (pp. 3–25). Springer International Publishing.
18. Jedrasiak, K., Andrzejczak, M., & Nawrat, A. (2014). SETh: The method for long-term object tracking. In *Computer vision and graphics.* Lecture notes in computer science (Vol. 8671, pp. 302–315).
19. Ryt, A., Sobel, D., Kwiatkowski, J., Domzal, M., Jedrasiak, K., & Nawrat, A. (2014). Real-time laser point tracking. In *International Conference on Computer Vision and Graphics* (pp. 542–551). Springer International Publishing.
20. Nawrat, A., & Jedrasiak, K. (2008). Fast colour recognition algorithm for robotics. *Problemy Eksploatacji*, 69–76.

About the Book

The reader of this book will be presented with advanced technologies used in practice to enable early recognition and tracking of various threats for National Security. Undeniably fast advances in development of sophisticated sensory devices, significant increase of computing power available to embedded designs and development of airborne and ground unmanned vehicles give almost unlimited possibilities to fight various types of pathologies affecting our societies. This book presents practical applications, examples and recent challenges in these mentioned application fields. Scientists, researchers, engineers, officers and graduate students involved in computer vision, image processing, data fusion, control algorithms, mechanics, data mining, navigation and IC can find many valuable, useful and practical suggestions and solutions.

A. Nawrat et al. (eds.), *Advanced Technologies in Practical Applications for National Security*, Studies in Systems, Decision and Control 106,
https://doi.org/10.1007/978-3-319-64674-9

Author Index

A. Nawrat et al. (eds.), *Advanced Technologies in Practical Applications for National Security*, Studies in Systems, Decision and Control 106,
https://doi.org/10.1007/978-3-319-64674-9

Printed in the United States
By Bookmasters